The Quantum Theory of Magnetism

Second Edition

The Quantum Theory of Magnetism

Second Edition

Norberto Majlis

McGill University, Canada

World Scientific

NEW JERSEY · LONDON · SINGAPORE · BEIJING · SHANGHAI · HONG KONG · TAIPEI · CHENNAI

Published by

World Scientific Publishing Co. Pte. Ltd.

5 Toh Tuck Link, Singapore 596224

USA office: 27 Warren Street, Suite 401-402, Hackensack, NJ 07601

UK office: 57 Shelton Street, Covent Garden, London WC2H 9HE

British Library Cataloguing-in-Publication Data
A catalogue record for this book is available from the British Library.

First published 2007 (Hardcover)
Reprinted 2016 (in paperback edition)
ISBN 978-981-3203-25-9

THE QUANTUM THEORY OF MAGNETISM (2nd Edition)

ISBN-13 978-981-256-792-5
ISBN-10 981-256-792-5

Printed in Singapore

Preface to the Second Edition

For the second edition I have added three new chapters, namely on magnetic anisotropy, on coherent magnon states and on local moments.

Besides, I have enlarged the chapter on itinerant magnetism, by including a section on paramagnons.

I have corrected several errors which appeared in the first edition. Some of these were indicated to me by friends and colleagues, for which I thank all of them.

Norberto Majlis

e-mail: majlisn@physics.mcgill.ca

Centre for the Physics of Materials, McGill University

Montréal, April 2007.

Preface to the First Edition

This is an advanced level textbook which grew out of lecture notes for several graduate courses I taught in different places over several years. It assumes that the reader has a background of Quantum Mechanics, Statistical Mechanics and Condensed Matter Physics. The methods of Green's functions, which are standard by now, are used fairly extensively in the book, and a mathematical introduction is included for those not very familiar with them.

The selection of subjects aims to present a description of the behaviour of systems which show ordered magnetic phases. This, plus the necessary limitation of the extension within reasonable limits, imposed the exclusion of many important subjects, among them diamagnetism, the Kondo effect, magnetic resonance, disordered systems, etc.

In turn, the reader will find a detailed presentation of the mean-field approximation, which is the central paradigm for the phenomenological description of phase transitions, a discussion of the properties of low-dimensional magnetic systems, a somewhat detailed presentation of the RKKY and related models of indirect exchange and a chapter on surface magnetism, among other characteristics which make it different from other texts on the subject.

This book can be used as a text for a graduate course in physics, chemistry, chemical engineering, materials science and electrical engineering and as a reference text for researchers in condensed matter physics.

Many exercises are included in the text, and the reader is encouraged to take an active part by trying to solve them.

I hope readers who find errors in the book or want to suggest improvements get in touch with me.

It is a great pleasure to acknowledge the moral support I enjoyed from my wife and my children, the generous and extremely competent help with the software during the preparation of the manuscript from Luis Alberto Giribaldo and my daughter Flavia, and the careful reading of several chapters by June Gonçalves. I want also to express deep recognition to the Centre for the Physics of Materials of McGill University for their support, particularly to Martin J. Zuckermann, Martin Grant and Juan Gallego.

Norberto Majlis

Centre for the Physics of Materials, McGill University

Montréal, March 7th, 2000.

Contents

Chapter 1

Paramagnetism

1.1   Introduction

Some examples of atomic systems with a permanent magnetic moment in the ground state are:

- atoms, molecules, ions or free radicals with an odd number of electrons, like $H, NO, C(C_6H_5)_3, Na^+$, etc.;

- a few molecules with an even number of electrons, like O_2 and some organic compounds;

- atoms or ions with an unfilled electronic shell. This case includes:

  - transition elements ($3d$ shell incomplete);

  - the rare earths (series of the lanthanides) ($4f$ shell incomplete);

  - the series of the actinides ($5f$ shell incomplete).

We shall consider in the rest of this chapter that the atomic entities carrying angular momenta occupy sites on a perfect crystalline insulator, that they are very well localized on their respective sites, and that their mutual interactions are negligible. This implies that we can neglect the unavoidable dipole-dipole interactions, which we assume are so weak that they could only affect the behaviour of the system at extremely low temperatures. If such a system is placed in an external uniform magnetic field $\mathbf{B}$ the Zeeman energy term is:

$$V = -\mathbf{B} \cdot \sum_{i=1}^{N} \mu_i \tag{1.1}$$

where μ_i is the magnetic moment at site i. If the magnetic moments are replaced by classical vectors, as in the semi-classic, large J limit, the corresponding partition function for this system in the canonical ensemble is:

$$\mathcal{Z} = \int d\Omega_1 d\Omega_2 \cdots d\Omega_N \, \exp\left(\beta\mu_0 B \sum_{i=1}^{N} cos\theta_i\right) = (z(a))^N \qquad (1.2)$$

where we have defined

$$z(a) = \int \exp\left(a \, cos\theta\right) d\Omega = \frac{\pi}{a} \sinh(a) \qquad (1.3)$$

with $a = \beta\mu_0 B$, $\mu_0 =$ magnetic moment of each atom, $\beta = 1/k_B T$, $d\Omega_i$ is the differential element of solid angle for the i-th dipole and θ is the angle between that dipole and the applied magnetic field.

The Gibbs free energy per particle is:

$$f = -k_B T \log z \qquad (1.4)$$

and the average magnetic moment $\bar{\mu}$ per atom along the applied field direction is, in units of μ_0:

$$\frac{\bar{\mu}}{\mu_0} = -\frac{1}{\mu_0}\frac{\partial f}{\partial B} = \coth(a) - 1/a \equiv L(a) \qquad (1.5)$$

here $L(a)$ is the Langevin function [1].

Exercise 1.1
Prove that for $a \ll 1$, that is for $B \to 0$ or $T \to \infty$ or both, the magnetization approaches

$$m_z \approx (N/V)\mu_0^2 B/3k_B T \qquad (1.6)$$

(Curie's law) [2].

Equation 1.5 describes *Langevin paramagnetism*. The typical value of μ_0 is a few Bohr magnetons $\mu_B \approx 10^{-20}$ erg gauss $^{-1}$, so that only at very low temperatures or very high fields, like those produced with superconducting and/or pulsed refrigerated magnets, can saturation effects be observed in the $m_z(T)$ curve. We show in Fig. 1.1 a comparison of Langevin theory with experimental measurements of the magnetization of Cr potassium alum. These measurements were performed in fields of up to $50,000$ gauss and at temperatures down to 1.29 K, which allowed for a large saturation degree. It is evident from Fig. 1.1 that Langevin's theory, which assumes continuity of the observable values of the magnetic dipolar moment of an ion or atom, does not fit the experiments, except at high temperatures and/or low fields.

In order to reach agreement with experiment one must incorporate those changes which are due to the quantum nature of the ions. The main effects of angular momentum quantization are twofold:

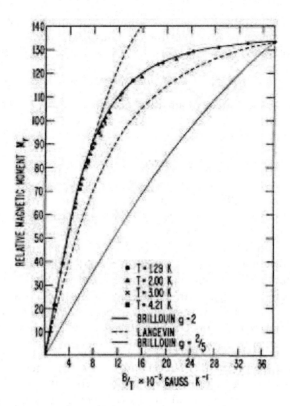

Figure 1.1: *Plot of $\overline{\mu}/\mu_0$, in arbitrary units, vs. B/T for potassium chromium alum. The heavy solid line is a Brillouin curve for $g = 2$ (complete quenching of L) and $J = S = 3/2$, fitted to experimental data at the highest value of B/T. The thin solid line is a Brillouin curve for $g = 2/5$, $J = 3/2$ and $L = 3$ (no quenching). The broken lines are Langevin curves fitted at the highest value of B/T (lower curve) and at the origin (slope fitting) (upper curve). From Ref.* [5].

- the discrete character of the eigenvalue spectrum of the vector components of angular momentum operators, or *space quantization*, leads to a statistical distibution for the magnetization different from that obtained by Langevin. The consequence is the substitution of Langevin's function $L(a)$ by Brillouin's function $B_J(a)$ (Sect. 1.3).

- the paramagnetic substances we are considering in this chapter are ionic crystals which contain some ions with non-zero permanent magnetic moment in the ground-state. In the solid, they have of course an electronic spectrum different from that of the free ion. The main effect of the crystalline environment that will concern us here is known as *quenching* of the orbital angular momentum under certain circumstances. This effect is observed for instance in magnetization measurements. We define the effective magneton number p as the modulus of the ionic magnetic moment in units of Bohr magnetons:

$$p = g\sqrt{J(J+1)}$$

where J = total angular momentum of ion in units of $\hbar$ and g = gyromagnetic ratio (see Sect. 1.3). One verifies that agreement with experiments in the estimate of p for the Cr^{3+} ion is only achieved if the ground state expectation value of $\mathbf{L}$ is assumed to vanish in the crystal, although for the free ion $L = 3$. This *quenching* of $\mathbf{L}$ is the result of the local symmetry of the electrostatic potential at the Cr^{3+} ion site in the solid. This potential generates the so called *crystal-field* , which, having a symmetry lower than spherical, in general mixes degenerate atomic orbitals with different M_L values, and lifts the degeneracy of the orbital manifold of states. The new orbital ground state will be the lowest energy one among those arising from the original ground state multiplet under the effect of the crystal-field potential. When the symmetry of the crystal-field admits a singlet orbital ground state one can prove that time reversal invariance leads to $< \vec{L} >= 0$. This theorem is proven in Section 1.5 and Appendix B. We shall see that a non-degenerate ground state requires an even number of electrons in the outer shell of the ion.

In the experiments we refer to in this chapter the Zeeman term in the Hamiltonian lifts the $2J + 1$ degeneracy of the ground state of the ions. For low fields the level separation is proportional to $g\mu_B B$ which is typically of the order of $1K$, much smaller than the level separation with the excited states, so that to a good approximation we can neglect all excited states and consider, as in Sect. 1.3, a problem very similar to Langevin paramagnetism.

Before we discuss the quantum theory of paramagnetism, we shall make a brief review of the quantum mechanics of atoms.

1.2   Quantum mechanics of atoms

1.2.1   L-S (Russel-Saunders) coupling

Let us write the Hamiltonian for an atom with Z valence electrons (that is, Z electrons in shells exterior to a filled atomic core of total charge $+Z$) as:

$$H = H_0 + V_1 + V_2 \tag{1.7}$$

where

$$H_0 = \sum_{i=1}^{Z} p_i^2/2m + V_c(r_i) \tag{1.8}$$

and $V_c(r)$ is a central effective potential, which is usually calculated in the *Hartree-Fock* approximation. The next two terms in Eq. (1.7) are the corrections to this approximate one-electron potential [3]. The second one contains all the Coulomb interaction corrections to the effective self-consistent potential $V_c(r_i)$:

$$V_1 = \sum_{i<j}^{Z} \frac{e^2}{\mid \mathbf{r_i} - \mathbf{r_j} \mid} - \sum_{i=1}^{Z} \frac{Ze^2}{r_i} - V_c(r_i) \qquad (1.9)$$

Finally, V_2 contains the magnetic interactions resulting from relativistic effects, of which the dominant ones are the spin-orbit coupling terms.

The relative quantitative importance of V_1 and V_2 varies along the periodic table. V_1 reflects the fluctuations of the exact electrostatic potential relative to the *Hartree–Fock* potential. The average fluctuations should be negligible if the effective one-electron potential had been well chosen. Their root mean square value is roughly proportional to $\sqrt{Z}$ for large Z [3].

The contribution of the spin-orbit coupling can be estimated through a simple calculation based on the *Thomas-Fermi* approximation for the many-electron atom. This yields [3] $V_2 \propto Z^2$.

As an immediate consequence, one expects that the spin-orbit contributions to the energy become comparable to -or even greater than- the Coulomb corrections given by Eq. (1.9), only for the heavier atoms. In the case $V_1 \gg V_2$, which applies in the transition elements,

$$H \approx H_0 + V_1 \qquad (1.10)$$

which, being spin independent, commutes with $\mathbf{S}$ and with S_z. Besides, rigid coordinate rotations around the nucleus leave the Hamiltonian (1.10) invariant, so it commutes with the total angular momentum $\mathbf{L}$ and with L_z. One can take advantage of the fact that $\mathbf{L}$ and $\mathbf{S}$ are independently conserved by separately adding $\mathbf{L} = \sum_i^Z \mathbf{l_i}$ and $\mathbf{S} = \sum_i^Z \mathbf{s}_i$, and afterwards combining both to obtain the total angular momentum $\mathbf{J} = \mathbf{L} + \mathbf{S}$, which commutes with the total Hamiltonian. This is the *L-S* or *Russel-Saunders* coupling. The Hamiltonian

$$H^{LS} = H_0 + V_1 \qquad (1.11)$$

can then be diagonalized in the many-electron basis of states

$$\{ \mid L\,S\,M_L\,M_S > \}$$

where M_L and M_S are the z components of $\mathbf{L}$ and $\mathbf{S}$ respectively.

The whole subspace spanned by the basis above, with L and S fixed and all possible values of M_L and M_S is called a spectroscopic *term*. This manifold, of dimension $(2L + 1)(2S + 1)$ is customarily symbolized by

$$^{2S+1}(Spectroscopic\ symbol\ for\ L)$$

where the symbols for L are the same used for the one electron states, in capital letters. For instance, ^1P $(S = 0, L = 1)$, ^3D $(S = 1, L = 2)$, etc.

Since V_1 commutes with $\mathbf{L}$ and $\mathbf{S}$, it is diagonal in the subspace of a given term and it does not mix different terms [3]:

$$\langle\, L\,S\,M_L\,M_S \mid V_1 \mid L'\,S'\,M_L'\,M_S'\,\rangle = \delta_{LL'}\delta_{SS'}\delta_{M_L M_L'}\delta_{M_S M_S'}\,\mathcal{V}_1^{LS} \qquad (1.12)$$

When all one electron occupied atomic shell states are specified one generates a manifold which in general contains several different states. This manifold is called *configuration*. Let us look at a simple example and see how we can decompose a given configuration into several terms. For *Si* ($Z = 14$) the configuration is $1s^2 2s^2 2p^6 3s^2 3p^2$. The only shell partially occupied is $3p$. Let us now construct the terms arising from the configuration $3p^2$, disregarding the closed shells. Two $l = 1$ electrons can be combined to form states with $L = 2, 1$ or 0. On the other hand, the spins of both electrons can be combined into a singlet ($S = 0$) or a triplet ($S = 1$). According to Pauli's principle the total wave function must be antisymmetric under electron exchange. Since within the LS coupling scheme we consider a product of an orbital wave function times a spin wave function, they must accordingly have opposite parity under electron exchange.

We need to know the parity of the orbital states. We can construct an orbital wave function with given L and M for the two-electron system by making an appropriate linear combination of products of one-electron orbital states. The linear coefficients are the *Clebsch-Gordan coefficients* (C-G in the following) [3]:

$$\mid l_1\, l_2\, L\, M\rangle = \sum_{m_1\, m_2} \mid l_1\, l_2\, m_1\, m_2\rangle\langle l_1\, l_2\, m_1\, m_2 \mid L\, M\rangle \qquad (1.13)$$

where $\mid l_1\, l_2\, m_1\, m_2\rangle = \mid l_1\, m_1\rangle \mid l_2\, m_2\rangle$ is a direct product of one-electron kets. Under electron exchange the C-G coefficients have the parity of L [3], so that the same is true of the resulting wave function of Eq. (1.13). In the case of *Si* we are only left with the terms ^1S, ^3P and ^1D, in order to satisfy the total antisymmetry required by the Pauli principle. These terms are L-S subspaces, each of degeneracy $(2S+1)(2L+1)$ respectively, spanned by the eigenfunctions of the Hamiltonian H^{LS}. In Si, the 15 degenerate states available for the $3p^2$ configuration split into three degenerate subspaces of smaller dimension, corresponding to the following terms: a singlet ^1S, the term ^3P of dimension 9 and the term ^1D of dimension 5. The effect of the Coulomb interaction V_1 is generally to lift the degeneracy of the configuration, since the expectation values of V_1, which are degenerate within each term, can in principle be different for different terms. The term degeneracy can further be lifted by the spin-orbit interaction V_2.

1.2.2   Hund's rules

As to the ordering of the different eigenvalues $\mathcal{V}_1^{LS}$ corresponding to the various terms (Eq. (1.12)), we can determine which is the ground state of the configuration, that is, the lowest energy term, by applying Hund's rules [3]. Let us briefly state these rules here. They dictate that, in order to find the ground state of ion or atom with a given configuration, one must:

1. choose the maximum value of S consistent with the Pauli principle;

2. choose the maximum L consistent with the Pauli principle and rule 1;

3. to choose J:

    (a) if the shell is less than half full, choose

    $$J = J_{min} = \mid L - S \mid \quad ;$$

    (b) if the shell is more than half full, choose

    $$J = J_{max} = L + S \quad .$$

Rule 1 allows to minimize the intra-atomic Coulomb repulsion among electrons in the same configuration. The second rule is derived as an empirical conclusion from numerical calculations. The magnetic (spin-orbit) interactions are taken into account in rule 3.

Exercise 1.2
Show that according to these rules, the ground state of Si should be $^3P_0$, where the subindex on the right of the term denomination is by convention the value of J.

1.2.3 Spin-orbit splitting

Dirac's relativistic theory of the atom leads to the following spin-orbit correction to the energy [3]:

$$V_2 \simeq \frac{\hbar^2}{2m^2c^2} \sum_{i=1}^{Z} \mathbf{l_i} \cdot \mathbf{s_i} \frac{1}{r_i} \frac{dV(r_i)}{dr_i} \tag{1.14}$$

The contributions involving different electrons, proportional, that is, to the operators $\mathbf{l_i} \cdot \mathbf{s_j}$ are usually neglected [4]. More concisely:

$$V_2 = \sum_{i=1}^{Z} g(r_i)\mathbf{l_i} \cdot \mathbf{s_i} \tag{1.15}$$

We consider now the matrix elements of V_2 within the subspace of a given LS *term*. All terms in the sum (1.15) must have the same matrix elements, since the wave functions are antisymmetric by exchange of any pair of electrons. Then,

$$< L\,S\,M_L\,M_S \mid V_2 \mid L\,S\,M_L'\,M_S' > =$$
$$Z < L\,S\,M_L\,M_S \mid g(r_1)\mathbf{l_1} \cdot \mathbf{s_1} \mid L\,S\,M_L'\,M_S' > \tag{1.16}$$

Let us look now at a product like $l_\mu s_\nu$. This operator transforms under the symmetry group of H, namely the full rotation group in three dimensions, in the same way as the product $L_\mu S_\nu$. According to Wigner-Eckart theorem (Appendix A), corresponding matrix elements of both operators within an invariant subspace of fixed L and S are proportional. Besides, the proportionality constant is the same within the whole manifold, and it only depends on the values of L and S. In particular, for the degenerate manifold $\{\alpha, L, S\}$ of all eigenstates of Hamiltonian H^{LS} with eigenvalue α, we have:

$$< \alpha L\, S\, M_L\, M_S \mid V_2 \mid \alpha L\, S\, M'_L\, M'_S > =$$
$$A(LS\alpha) < \alpha L\, S\, M_L\, M_S \mid \mathbf{L} \cdot \mathbf{S} \mid \alpha L\, S\, M'_L\, M'_S > \qquad (1.17)$$

where $A(LS\alpha)$ is a constant for the whole manifold.   The eigenvalues of the operator $\mathbf{L} \cdot \mathbf{S}$ can be obtained at once, leading to the eigenvalues of V_2:

$$\lambda_{J\alpha} = 1/2\, A_\alpha \hbar^2 \left[J(J+1) - L(L+1) - S(S+1) \right] \qquad (1.18)$$

The degenerate subspace $\{\alpha, L, S\}$ splits into the eigenstates of different values of J. Returning to our Si example, we see that for the term ^1S, $J = 0$. The term ^1D has $J = L = 2$. Finally the term ^3P can yield $J = 2, 1$ or 0, so that it splits, according to (1.18) into three different levels ^3P_J. Hund's rules prescribe that the lowest in energy is $^3P_0$.

Exercise 1.3
Obtain the spin-orbit level splitting for the various terms of $Si\, 3p^2$ configuration, and prove that the center of gravity of the levels remains unchanged.

1.3   The quantum theory of paramagnetism

Let us now consider the eigenstates of the atomic Hamiltonian H in Eq. (1.7), which we shall denote by $\{\mid E_n\, J\, M\rangle\}$. Upon application of a uniform static external magnetic field $\mathbf{B}$ the perturbing term to be added to the Hamiltonian, neglecting a diamagnetic correction of second order in the field, is the Zeeman interaction energy

$$W_Z = -\mu_B \mathbf{B} \cdot (\mathbf{L} + 2\mathbf{S}) \qquad (1.19)$$

Here the angular momentum operators are expressed in terms of $\hbar$, so that they are dimensionless, and the Bohr magneton $\mu_B = e\hbar/2m_e$ has dimensions of magnetic moment. The factor 2 for spin arises from Dirac's relativistic theory of the electron. Each level E_n has a degeneracy $2J + 1$, which will be lifted by W_Z. To obtain the displacements of the energy levels within each multiplet of unperturbed energy E_n it is convenient first of all to choose the quantization axis parallel to the direction of the external field. Then, W_Z will commute with J_z. We can again make use of the Wigner-Eckart theorem, since any vector operator, and in particular $\mathbf{L} + 2\mathbf{S}$, transforms like $\mathbf{J}$. Then, their matrix elements are proportional within the level multiplet (see Appendix A):

$$\langle E_0 J M_J \mid (\mathbf{L} + 2\mathbf{S}) \mid E_0 J M'_J \rangle = g \langle E_0 J M_J \mid \mathbf{J} \mid E_0 J M'_J \rangle \qquad (1.20)$$

where g is the *spectroscopic splitting* factor, or Landé g factor. We get for the z component:

$$< E_0 J M_J \mid (L_z + 2S_z) \mid E_0 J M'_J >= g\, M_J \hbar \delta_{M_J M'_J} \qquad (1.21)$$

Then

$$< E_0 J M_J \mid W_Z \mid E_0 J M'_J >= -g\mu_B B M_J \delta_{M_J M'_J} \qquad (1.22)$$

so that W_Z completely lifts the degeneracy of the atomic levels. When we are within the limits of validity of the LS coupling description of the ionic levels (Eq. (1.10)), that is $V_1 \gg V_2$, we obtain for g the expression:

$$g = 1 + \frac{J(J+1) + S(S+1) - L(L+1)}{2J(J+1)} \qquad (1.23)$$

Exercise 1.4
Prove Eq. (1.23).

Let us now consider the lowest lying Zeeman multiplet in Eq. (1.22), and assume that the spin-orbit splitting is much greater than the Zeeman level separation:

$$A\hbar^2 \gg g\mu_B B \qquad (1.24)$$

We are now in a situation very similar to that described by the Langevin theory of paramagnetism, in which the different possible orientations of the spin of an ion determine its energy levels in the presence of an external magnetic field. The main change is that the quantum theory of the electronic level structure has led to space quantization of the angular momentum components. For a canonical ensemble of identical ions, each with total angular momentum J, we can now calculate the partition function resulting from the discrete energy levels of Eq. (1.22), and then follow the same process as in the classical case in Sect. 1.1, to obtain the z component of the magnetization, as:

$$m_z(a) = (N/V)gJ\mu_B B_J(a) \qquad (1.25)$$

where $a = gJ\mu_B \beta B$, N/V is the volume concentration of magnetic ions and the Brillouin function $B_J(a)$ is defined as:

$$B_J(a) = \frac{2J+1}{2J} \coth\left(\frac{(2J+1)\,a}{2J}\right) - \frac{1}{2J}\coth\left(\frac{a}{2J}\right) \qquad (1.26)$$

Exercise 1.5
Prove Eq. (1.26).

For $a \ll 1$ we obtain the zero field static longitudinal susceptibility:

$$\chi = \frac{\partial m_z}{\partial B} \Big|_{B=0} = (N/V)J(J+1)g^2\mu_B^2\beta/3 \qquad (1.27)$$

which is analogous to the susceptibility obtained from Eq. (1.6). By comparison of both equations, we obtain the expression

$$\mu^2 = J(J+1)(g\mu_B)^2 \qquad (1.28)$$

for the square of the magnetic moment of the ion in terms of its quantized total angular momentum. The Curie law is still valid in the quantum case, as long as the assumptions we made to arrive at the above expression for χ are justified.

In Fig. 1.2, as in Fig. 1.1, the values attributed to J are determined by Hund's rules and g is given by Eq. (1.23). The compounds studied in these experiments contain the trivalent ions Cr^{+++}, Fe^{+++} or Gd^{+++}, which according to Hund's rules have the ground state terms $^4F_{3/2}, {}^6S_{5/2}$ and $^8S_{7/2}$ respectively.

Exercise 1.6
Verify the term assignments for the ground state of Cr^{+++}, Fe^{+++} and Gd^{+++} mentioned above.

For the rare earths, Brillouin formula gives very good agreement with experiment. Measured moments of the rare earths agree very closely with the free ion picture of paramagnetism as presented above, with the exception of Eu^{3+} and Sm^{3+}. We can understand the anomalies exhibited by these two ions if we consider that the theory sketched above assumes that the excited states are sufficiently separated from the ground state by the spin orbit splittings that they can be neglected. This is the same as assuming that $k_BT \ll A\hbar^2$, which implies that the probability that the excited states are occupied at the temperatures of interest is negligible. If this were not the case, we must be prepared to find discrepancies, which can result in an effective magnetic moment different from the one expected for the ground state multiplet. In Eu^{3+} and Sm^{3+}, $A\hbar^2$ is comparable to k_BT at room temperature.

In the case of the iron group the situation is completely different. The agreement with Eq. (1.27) can only be reached if one assumes that $L = 0$, and accordingly $J = S$. This is an example of the phenomenon called *quenching* of the orbital angular momentum. We shall see in the next section that this is a result of the symmetry lowering of the ionic one-electron effective potential due to the charge distribution around the magnetic ion. In the rare earths, the effect of the crystal-field is much weaker, because the electronic orbitals of the f shell, which are the unpaired magnetic electron-states, are much more tightly localized in space as compared with those of the d shell, and they are screened by the s, p and d orbitals.

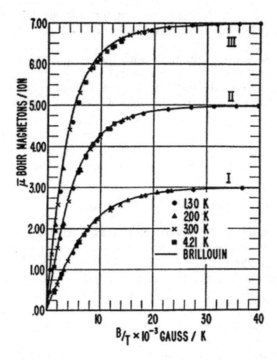

Figure 1.2: *Plot of $g\mu_B\langle J_z\rangle$, in Bohr magnetons per ion vs. B/T for (1) potassium chromium alum ($J = S = 3/2$) , iron ammonium alum ($J = S = 5/2$) and (3) gadolinium sulfate octahydrate ($J = S = 7/2$). In all cases $g = 2$. The normalizing point is at the highest value of B/T. From Ref. [5].*

1.4 Crystal-field corrections

We shall now consider the effect of the electrostatic interactions between the electrons of the paramagnetic ion and the electric charge distribution of the nearby non-magnetic ions surrounding it, which are called *"ligands"*. The latter produce a resultant effective potential V_{cryst} that must be added to the atomic Hamiltonian of Eq. (1.7).

The concept of crystal-field was advanced by Becquerel [7] and developed originally by Bethe [8], Kramers [9], Van Vleck [10] and others. This subject has deserved many reviews [11].

The crystal-field potential is generally weaker than the intra-atomic Coulomb interactions, but it may be comparable with the spin-orbit forces. We distinguish accordingly three cases, classifying the crystal-field as:

- weak, if it is smaller than the spin-orbit interaction;

- intermediate, if it is stronger than the former, but still weaker than the intra-atomic Coulomb electron-electron interactions;

- strong, if it becomes comparable with the intra-atomic interaction.

We do not expect to find the last case to prevail in predominantly ionic compounds, since that would imply the presence of covalent bonding of the magnetic ions with the surrounding ligands. The crystal-field is weak in the rare earths and actinides, since the $4f$ and $5f$ shells are fairly localized near the core and screened from the ligands by the outer shell electrons, as was mentioned at the end of last section. It is intermediate in the iron-group elements.

1.4.1   Effects of crystal-field symmetry

The details of how the energy levels of the paramagnetic ions will be split by the crystal-field, depend on the symmetry of the local environment of the ion. As an example, consider the case of a single $3d$ electron: $S = 1/2$, $L = 2$. Then, according to Hund's rules, $J = 3/2$ for the ground state multiplet, which is the $^2D_{3/2}$ term. Therefore, this term has a $5-$fold orbital degeneracy. Let us remind ourselves of the form of these five wave-functions. In cartesian coordinates they can be written as the product of a radial function $f_2(r)$ times the polynomials: $X = yz$, $Y = zx$, $Z = xy$, $\Phi_1 = x^2 - y^2$, and $\Phi_2 = x^2 + y^2 - 2z^2$, which are linearly independent, as can be verified by noting that they are proportional to linear combinations of the spherical harmonics $\{Y_2^m\}$ [15].

Exercise 1.7
Prove that the previous five functions are linearly independent.

Let us now consider the symmetry operations of a cubic array of ions. We can always place one magnetic ion at the origin of the cartesian coordinate system, and choose the cartesian axes coincident with the cubic axes of the atomic array. One can easily see that a reflection on one of the specular symmetry planes perpendicular to the respective cubic axis, namely one of the transformations $x \to -x$, $y \to -y$, $z \to -z$, leaves the last two functions invariant, while the first three change sign under two of these reflections. On the other hand, by performing a $2\pi/3$ rotation around a body diagonal axis of the cube, also a symmetry operation, we get cyclic permutations of the coordinates, like $(xyz) \to (yzx)$. A permutation is also thereby obtained among the first triplet X, Y, Z, while under the particular permutation above,

$$\Phi_1 \to \Psi_1 = y^2 - z^2$$

$$\Phi_2 \to \Psi_2 = y^2 + z^2 - 2x^2$$

This correspondence can be written as a linear transformation:

$$\Phi_1 = -1/2(\Psi_1 + \Psi_2)$$

$$\Phi_2 = 3/2\Psi_1$$

with the inverse

$$\Psi_1 = 1/2(\Phi_1 - \Phi_1)$$

$$\Psi_2 = 1/2(3\Phi_1 + \Phi_2)$$

The analysis of the effect of all the 48 symmetry operations of the cube (or, which amounts to the same, of the regular octahedron) upon our five dimensional set of d functions, would generalize what is already suggested by the symmetry operations considered above, namely that under this group of point transformations the first triplet of functions spans a subspace which transforms onto itself. The same is true for the doublet. Both subspaces are therefore *closed* under this group. This conclusion can be arrived at in a very straightforward and elegant way by means of the application of the theory of groups. In the language of this theory, the original 5-dimensional subspace spans a **reducible** representation of the cubic point group which decomposes into, or reduces to, two irreducible representations: a three dimensional one, called T_2 (or Γ_5, depending on the notation chosen) and a two dimensional one, called E (or Γ_2) (see Appendix A) [15]. Under each of the continuous rotations compatible with the spherical symmetry of the ionic potencial the original 5 functions obey well defined transformation laws, which carry each of them into a linear combination of them all. The cube, however, has only a discrete, finite set of symmetry operations, which are of course contained within the infinite group of rotations which leave a sphere invariant. Therefore if a cubic perturbation $V_{cryst}(\mathbf{r})$ is added to the spherically symmetric one-electron ion potential for the example considered above of a $2d$ state, the five $L = 2$ states must split into a triplet T_2 and a doublet E which as we have seen do not mix any more under the symmetry operations of the cube. Since both subspaces are now disjoint under the smaller symmetry group, degeneracy of the eigenvalues of both multiplets (the doublet and the triplet) is not to be expected. Whenever degeneracy occurs this situation is called *accidental degeneracy*.

Within the electrostatic approximation, in which the ligands are substituted by point charges, a simple calculation indicates which of the split levels has the lowest energy. For the present example of a single d state in a $6 - fold$ simple cubic (or octahedral) coordination the triplet is the lowest level, while in an $8 - fold$ b.c.c. coordination or in the tetrahedral case the opposite is true. For the square planar symmetry, which is still lower than the cubic one, the triplet splits into a doublet plus the singlet Z state. As regards the relative ordering of levels, determining the position of the Φ_1 state level requires some calculation. A practical rule can be stated: the orbitals of the central ion directed along the axes containing ligand ions will have higher energy than those directed away from the ligands [18].

Bethe [8] and Runciman [12] solved originally the problem of calculating the splitting of the ionic energy levels in crystal-fields of various symmetries, and for different values of J, in the case that the spin-orbit splitting is much larger than the crystal-field effect.

A word of caution is pertinent at this point. One finds in the literature, as already mentioned, different nomenclatures for the various symmetry types. They are denoted Γ_i by Bethe [8]. Mulliken [19] uses the following nomenclature instead: the one dimensional representations with even parity under rotations around the principal axis of symmetry are called A, and the odd ones B; the letters E, T and G are used for two, three and four dimensional representations respectively; a numerical subscript distinguishes different irreducible representations with the same dimension; subscripts g and u denote even (*gerade*) and odd (*ungerade*) representations.

Group theory as applied to Quantum Mechanics establishes that if a Hamiltonian is invariant under a given symmetry group $\mathcal{G}$, each degenerate multiplet of eigenstates belonging to a given eigenvalue is a basis for an irreducible representation of $\mathcal{G}$. As shown in the example above, the superposition of a crystal-field onto the central (spherically symmetric) one electron effective Hamiltonian, will, as a rule, reduce the dimension of the degenerate multiplet subspace, since the symmetry point group of a crystal, being a (discrete) subgroup of the full rotation group, in general contains subspaces of the multiplet which are invariant under the smaller group, but not so under the full group. By *point group* is denoted one that contains rotations and reflection planes (eventually multiplied by the inversion) around a fixed point (Appendix A).

From all the foregoing considerations, one concludes that in general the crystal-field will partially lift the orbital degeneracy of the central ion eigenstates. The resulting multiplets must be characterized by the symmetry types of the crystal point group. We shall denote a state γ of the Γ representation of the point group by $|\Gamma\gamma>$. For instance, $|T_2, X>$, $|E_2, \Phi_1>$, etc.

Whenever covalent effects can be disregarded, a very useful simplification for the discussion of the effects of the crystal-field is its substitution by an electrostatic field with the same symmetry. This implies in general the replacement of the ligand ionic charge distribution around each paramagnetic ion by a set of point charges. The corresponding electrostatic potential satisfies Laplace differential equation, and it admits accordingly an expansion in spherical harmonics, in the region around the central ion. This expansion can be used, in conjunction with the Wigner-Eckart theorem, to obtain a widely used representation of the crystal-field in terms of angular momentum operators, the *Stevens operator equivalents* that we shall discuss in next section.

1.4.2   Stevens operator equivalents

Let us consider the important cases of a ligand field with: a) tetragonal symmetry ($4 - fold$ coordination); b) octahedral symmetry ($6 - fold$ coordination) and c) $8 - fold$, or *bcc* coordination. The set of N ligand ions of charges q_j located at points $\mathbf{R}_j$ produces at the point $\mathbf{r}$ the potential

$$V_{cryst}(\mathbf{r}) = \sum_{j=1}^{N} \frac{q_j}{|\mathbf{R_j} - \mathbf{r}|} \qquad (1.29)$$

We shall now expand this potential in powers of the components of $\mathbf{r}$, assuming that $|\mathbf{r}| \ll |\mathbf{R}_j|$. For most purposes, it is sufficient to expand up to terms of the sixth degree. Let us consider in detail the case of octahedral symmetry, which corresponds to a paramagnetic ion with a simple cubic environment. Assume the coordinates of the surrounding ions are $(\pm a, 0, 0)$, $(0, \pm a, 0)$, $(0, 0, \pm a)$ and all the charges are $q_j = q$.

Exercise 1.8
Expand Eq. (1.29) up to sixth order, and obtain [13]:

$$
\begin{aligned}
V_{cryst}(xyz) =\ & \frac{6q}{a} + \frac{35q}{4a^5}\left(x^4 + y^4 + z^4 - \frac{3}{5}r^4\right) \\
& - \left(\frac{21q}{2a^7}\right)\left[x^6 + y^6 + z^6 + \frac{15}{4}\left(x^2 y^4 + x^2 z^4 + \cdots\right) - \frac{15}{14}r^6\right]
\end{aligned}
$$

$$(1.30)$$

If we adopt spherical coordinates, we must first remind ourselves of the spherical harmonics expansion of the unit point charge potential [20]:

$$
\frac{1}{|\mathbf{R}_j - \mathbf{r}|} = \sum_{l=0}^{\infty} \frac{r^l}{R_j^{l+1}} \sum_{m=-l}^{l} \frac{4\pi}{2l+1} Y_{lm}^*(\theta_r \phi_r) Y_{lm}(\theta_{R_j} \phi_{R_j})
$$

$$(1.31)$$

It is convenient to change to *tesseral*, or real, harmonics, defined as:

$$
Z_{l0} = Y_l^0
$$

$$
Z_{lm}^c = \frac{1}{\sqrt{2}}(Y_l^{-m} + Y_l^m)
$$

$$
Z_{lm}^s = \frac{i}{\sqrt{2}}(Y_l^{-m} - Y_l^m)
$$

with $m > 0$. The expansion (1.30) in tesseral harmonics is then [13]:

$$
V_{cryst}(r, \theta, \phi) = \sum_{l=0}^{\infty} r^l \sum_{\alpha} \gamma_{l\alpha} Z_{l\alpha}(\theta, \phi)
$$

$$(1.32)$$

where

$$
\gamma_{l\alpha} = \frac{4\pi}{2l+1} \sum_{j=1}^{N} \frac{q_j Z_{l\alpha}(\theta_j \phi_j)}{R_j^{l+1}}
$$

$$(1.33)$$

In Eq. (1.32) the index α contains the azimuthal quantum number m and the index c or s defined above.

The above expansion in tesseral harmonics is a very convenient one, since by expressing the $Z_{l\alpha}$ in cartesian coordinates, one can obtain an immediate correspondence with the Stevens operator equivalents, which are extremely useful for evaluating matrix elements of the crystal-field potential. Tables of the most usual tesseral harmonics can be found in the review article by Hutchings already cited [13]. To illustrate the use of Eq. (1.32), let us mention that for the octahedral environment the only non vanishing γ 's are $\gamma_{00}, \gamma_{40}, \gamma_{60}, \gamma_{44}^c$ and γ_{64}^c, so that the potential is:

$$V_{cryst}(r, \theta, \phi) =$$
$$r^4 \left(\gamma_{40} Z_{40} + \gamma_{44}^c Z_{44}^c \right) + r^6 \left(\gamma_{60} Z_{60} + \gamma_{64}^c Z_{64}^c \right) \tag{1.34}$$

or in the notation of Bleaney and Stevens [14]:

$$V_{cryst}(r, \theta, \phi) = D_4^\dagger [Y_4^0 + \sqrt{\frac{5}{14}} (Y_4^{-4} + Y_4^4)] +$$

$$D_6^\dagger [Y_6^0 - \sqrt{\frac{7}{2}} (Y_6^{-4} + Y_6^4)] \tag{1.35}$$

where the coefficients are:

$$D_4^\dagger = \frac{7\sqrt{\pi}}{3} \frac{q}{a^3} r^4 \ , \quad D_6^\dagger = \frac{3}{2} \frac{\sqrt{\pi}}{13} \frac{q}{a^7} r^6 \ .$$

Other expressions are obtained if the quantization axis is not the z axis [13].

The next task is to calculate the matrix elements of the crystal-field potential operator in the basis of the free paramagnetic ion orbital states belonging to the various spectroscopic terms as described in previous sections. In the case of the $3d$ transition elements in ionic compounds, the potential energy of the valence electrons associated with the crystal- field, namely the operator

$$W_c = - \mid e \mid \sum_{i=1}^{N} V_{cryst}(x_i, y_i, z_i) \tag{1.36}$$

is larger than the spin-orbit coupling, so that the appropriate free-ion states are $\mid L S M_L M_S >$. In the $4f$ rare-earth group, they are the states $\{\mid L S J M_J >\}$, as the spin-orbit coupling is larger than W_c. In either case, the crystal-field potential energy operator must be diagonalized within the adequate basis of orbital states.

Let us now briefly describe the fundamentals of Stevens method of operator equivalents. It consists essentially in the application of the Wigner-Eckart theorem to the calculation of matrix elements of the operator which results from the expansion of W_c in cartesian coordinates. One starts from the homogeneous polynomials with the proper symmetry, as in Eq. (1.30) above, and then substitutes the one-electron coordinate operators $\{ x_i, y_i, z_i \}$ by the corresponding components of $\mathbf{J}$, care taken of the non-commuting properties of the latter.

Whenever it applies, one symmetrizes the corresponding product of different components of **J**. Some simple examples are:

$$\sum_i (3z_i^2 - r_i^2) \rightarrow \alpha_J < r^2 > [3J_z^2 - J(J+1)] \equiv \alpha_J < r^2 > \mathcal{O}_2^0$$

$$\sum_i (x_i^2 - y_i^2) \rightarrow \alpha_J < r^2 > [J_x^2 - J_y^2] \equiv \alpha_J < r^2 > \mathcal{O}_2^2$$

$$\sum_i (x_i^4 - 6x_i^2 y_i^2 + y_i^4) \rightarrow$$

$$\beta_J < r^4 > \frac{1}{2}(J_+^4 - J_-^4) \equiv \beta_J < r^4 > \mathcal{O}_4^4$$

A few common low-order operators are:

$$\mathcal{O}_2^0 = 3J_z^2 - J(J+1)$$

$$\mathcal{O}_2^2 = \frac{1}{2}(J_+^2 - J_-^2)$$

$$\mathcal{O}_4^0 = 35J_z^4 - 30J(J+1)J_z^2 + 25J_z^2$$
$$-6J(J+1) + 3J^2(J+1)^2$$

$$\mathcal{O}_4^4 = \frac{1}{2}(J_+^4 + J_-^4)$$

$$\mathcal{O}_6^6 = \frac{1}{2}(J_+^6 + J_-^6)$$

where z is a crystal axis. The constants α_J for second order operators, β_J for $4 - $ th order, γ_J for $6 - $ th order, etc., depend on Z, L, S and J, but not on M_J. The actual calculation of the matrix elements can be done, if necessary, for just one of them in each multiplet, and it involves the radial integral

$$< r^n > = \int r^{n+2} (f(r))^2 dr$$

where $f(r)$ is the electron radial wave function [15]. Since the latter is usually not accurately known, these averages are taken as parameters and determined by fitting experimental results.

Whenever other choices than z are more convenient for the quantization axis, the spin operators must be submitted to the corresponding rotation [13].

Some applications of these methods to typical paramagnetic ions can be found in the book by Al'tschuler and Kozyrev [17]. Effects of covalency between the magnetic ion and the ligands are taken into account in the approach of *ligand field theory* [16].

1.5   Quenching of L

We are now almost in a position to discuss this effect, which has been already mentioned before. We need, however, to digress and discuss the time-reversal transformation properties of the electronic wave functions involved. The latter can be constructed as a product of a many electron orbital wave function times a many electron spinor.

We prove in Appendix B that:

1. Any antilinear operator A satisfies the identity:

$$(< v \mid A) \mid u >= (\ < v (\mid A \mid u >) \ )^*$$

   for arbitrary $< v \mid$ and $\mid u >$

2. The time reversal operator K_0 for the orbital part of the electron wave function is an antiunitary operator. This means: $K_0 K_0^\dagger = 1$.

3. If $\mid E >$ is an eigenstate of the stationary Schrödinger equation, with eigenvalue E, and K is the time reversal operator, then $K \mid E >$ is also an eigenstate, degenerate with $\mid E >$.

4. For systems without spin, or as long as we consider only the orbital part of an electron state, the time reversal operator K can be taken identical to the complex conjugation operator K_0 with the properties $K_0 = K_0^\dagger = K_0^{-1}$, or $K_0^2 = 1$.

Let us now act with K_0 on an orbital state which we assume is not degenerate. In this particular case, the resulting state must be linearly dependent upon the original one, since otherwise the latter would be degenerate. Then,

$$K_0 \mid n >= c \mid n > \qquad (1.37)$$

with $c =$ some complex constant. Since $K_0^2 = 1$, applying K_0 again on both sides of Eq. (1.37) yields:

$$\mid n >= K_0 \left(c \mid n > \right) \qquad (1.38)$$

Since K_0 is antilinear,

$$K_0 c = c^* K_0$$

so from Eq. (1.38) we get

$$\mid n >= c^* K_0 \mid n >=\mid c^2 \mid\mid n >$$

which implies $c = e^{i\phi}$, with ϕ real.

We consider now a matrix element of $\mathbf{L}$, $< n \mid \mathbf{L} \mid m >$, and insert K_0^2 inside the bracket:

$$< n \mid \mathbf{L} \mid m >=< n \mid K_0 \left(K_0 L K_0^\dagger \right) K_0 \mid m > \qquad (1.39)$$

But, from the very definition of time reversal we have:

$$K_0 \mathbf{L} K_0^\dagger = -\mathbf{L} \qquad (1.40)$$

Let us now call $\mid v >=\mid n >$, $\mid u >= \mathbf{L} K_0 \mid m >$ and apply the first property of antilinear operators enumerated above:

$$< v \mid (K_0 \mid u >) = [(< v \mid K_0) \mid u >]^*$$

or:

$$< n \mid (K_0 \mathbf{L} K_0 \mid m >) = [(< n \mid K_0)\,(\mathbf{L} K_0 \mid m >)]^*$$

Since we have assumed that $< n \mid$ and $< m \mid$ are both non-degenerate, we finally have:

$$< n \mid \mathbf{L} \mid m >= - < n \mid \mathbf{L} \mid m >^* e^{i(\phi_n - \phi_m)} \qquad (1.41)$$

We have $< n \mid \mathbf{L} \mid m >^* =< m \mid \mathbf{L}^\dagger \mid n >$, but, since $\mathbf{L}$ is an observable, $\mathbf{L}^\dagger = \mathbf{L}$. In the special case $n = m$ Eq. (1.41) yields

$$< n \mid \mathbf{L} \mid n >= 0 \qquad (1.42)$$

This is then the effect of quenching of the orbital angular momentum which we already mentioned before. This phenomenon is effective if the crystal-field symmetry is low enough that the ground state of the paramagnetic ion be a singlet.

Let us discuss as an example the quenching of the orbital angular momentum of the Cr^{++} ion in the salt $CrSO_4.5H_2O$, where the ion is at the centre of a square of water molecules, while two oxygen ions are located exactly above and below it, in the apical positions [15]. We shall use the international symbols for the point groups throughout. The six oxygens in this arrangement form approximately a regular octahedron. The crystalline field is the sum of a pre-dominantly cubic component (group $m3m$) and a smaller tetragonal component (group $4/mmm$ in international notation) [15]. The tetragonal contribution can be thought of as analogous to that produced by charges arranged on a tetra-hedron. A small distorsion of the square of water molecules gives rise to an additional, still smaller, orthorhombic term in the potential, with symmetry group mmm , of magnitude similar to the spin-orbit coupling. The Hamilto-nian without the spin-orbit interaction is invariant, in a first approximation, under the rotations and reflections of the space coordinates belonging to the cubic group $m3m$, and besides, being spin-independent, under any spin rota-tion. The chromous ion has the configuration $3d^4$, with a ground state term $^5D_0$ (verify this). As we have seen before, the $L = 2$ orbital wave functions trans-form according to the irreducible representation $D^{(2)}$ under the full rotation group, and this representation, which has five dimensions, reduces into the sum of the representations E_g (twofold) and T_{2g} (threefold) of the cubic group, the doublet having the lowest energy [15]. The tetrahedral distorsion splits further the doublet into two orbital singlets, since the tetragonal group has only one dimensional irreducible representations. They are, in this case, A_{1g}, which is the identity representation, with symmetry basis z^2 and B_{1g}, with symmetry basis $x^2 - y^2$. We shall assume, following Ref. [15] that B_{1g} is the lowest in en-ergy. B_{1g} becomes (i.e., it has the same characters as) A_g of the rhombic group mmm, and then this is the lowest level when all the crystal-field contributions have been included.

Let us now consider the spin variables. The spin-orbit interaction is invariant under all simultaneous spin and coordinate rotations because it is a scalar. The spin representation is in this case that corresponding to $S = 2$, which is $D^{(2)}$, but due to the restrictions imposed by the crystal-field point group, the only symmetry operations are again those of the group mmm. Therefore, the lowest spin multiplet, which contains 5 spin states, transforms under spin and orbital rotations as the external product

$$A_g \times D^{(2)} = A_g + A_g + B_{1g} + B_{2g} + B_{3g} \qquad (1.43)$$

(see appendix A), which are five spin-orbital singlets. Eq. (1.43) states that the external product considered of the subspaces which are the basis for representations A_g and $D^{(2)}$ is a *reducible* representation space of the lower symmetry group, which decomposes into the external sum of the irreducible basis subspaces of the 5 representations which appear on the r. h. s. of the equation. As we shall see in next section, where we analyze the effect of spin reversal on spin, in this case, since the ionic configuration contains an even number of electrons, Kramers theorem allows a singlet ground state. As we proved above, the expectation value of any component of the angular momentum in one orbital singlet like A_g vanishes, and we have complete quenching, just as in the transition element compounds we mentioned in the previous section.

1.6   Time reversal and spin

Under time reversal, spin angular momentum reverses sign, just as the orbital one. For systems with spin we must therefore look for a time reversal operator K such that

$$K\mathbf{s}K^{\dagger} = -\mathbf{s} \qquad (1.44)$$

For one electron, the standard Pauli matrices transform under complex conjugation as

$$
\begin{aligned}
K_0 s_x K_0 &= s_x \\
K_0 s_z K_0 &= s_z \\
K_0 s_y K_0 &= -s_y
\end{aligned}
\qquad (1.45)
$$

The basic commutation relations like

$$[\, x, p_x \,] = i\hbar \qquad (1.46)$$

must change sign under K, so that K must be anti-unitary like K_0. We then introduce a unitary operator U in spin space:

$$K = UK_0 \qquad (1.47)$$

Then, since $K_0^2 = 1$, we also have $U = KK_0$ and $U^{\dagger} = K_0 K^{\dagger}$. Under U, which is an operator on spin space, $\mathbf{r}$ and $\mathbf{p}$ remain unaltered:

$$
\begin{aligned}
U\mathbf{r}U^{\dagger} &= KK_0\mathbf{r}K_0 K^{\dagger} = K\mathbf{r}K^{\dagger} = \mathbf{r} \\
U\mathbf{p}U^{\dagger} &= KK_0\mathbf{p}K_0 K^{\dagger} = -K\mathbf{p}K^{\dagger} = \mathbf{p}
\end{aligned}
\qquad (1.48)
$$

The Pauli matrices transform accordingly under U as:

$$
\begin{aligned}
U\, s_x\, U^\dagger &= -s_x \\
U\, s_y\, U^\dagger &= s_y \\
U\, s_z\, U^\dagger &= -s_z
\end{aligned}
\tag{1.49}
$$

These transformations in spin space can be recognized as the result of a rotation of π around the y axis, which can be represented by the operator

$$
Y = e^{-i\pi s_y/\hbar}
\tag{1.50}
$$

Therefore Y and U can only differ by an arbitrary phase factor, which we choose $= 1$, and consequently

$$
K = Y K_0
\tag{1.51}
$$

For one electron, upon expanding the exponential in Eq. (1.50) we find

$$
K = -i\sigma_y K_0
\tag{1.52}
$$

and for a system with N electrons

$$
K = \prod_{j=1}^{N} (\, -i\sigma_y(j)\,)\ K_0
\tag{1.53}
$$

The square of this operator is

$$
K^2 = (-1)^N
\tag{1.54}
$$

1.6.1 Kramers degeneracy

We have seen that the (degenerate or not) subspace $S(E)$ belonging to the eigenvalue E is invariant under time reversal when this is a symmetry operation of the Hamiltonian. We must consider separately two cases:

- $K^2 = +1$ (even number of electrons). Here we can choose the phases of the eigenstates in such a way that

$$
K_0 \mid n \,\rangle = \mid n \,\rangle \ , \quad \forall n
$$

  and we can use the complete orthonormal set of eigenstates as the basis vectors of the representation. This is called a *real basis*.

- $K^2 = -1$ (odd number of electrons). A consequence of the negative square of K is that $K^\dagger = -K$, and since K is anti-unitary we have

$$
\begin{aligned}
&\langle\, n \mid (K \mid n \,\rangle) = \\
&\langle\, n \mid (K^\dagger \mid n \,\rangle\,) = -\langle\, n \mid (K \mid n \,\rangle) = 0
\end{aligned}
\tag{1.55}
$$

  Then all state vectors can be arranged in degenerate orthogonal pairs, whose orbital parts are mutually complex conjugate. This twofold degeneracy cannot be lifted by any time-reversal-invariant perturbation.

We have just proved Kramers theorem, which can be stated as:

Theorem 1.1 (Kramers)
The energy levels of a system with an odd number of electrons are at least two-fold degenerate in the absence of magnetic fields.

1.7   Effective spin Hamiltonian

We have seen that for H^{LS}, L and S are good quantum numbers. As shown in the previous section, the crystal-field potential maintains L as a good quantum number, but it mixes states with different values of M_L according to the point group of the ligand charge distribution. Since the crystal-field potential does not depend on spin, an eigenstate of the Hamiltonian :

$$H_0 = H^{LS} + V_{cryst} \tag{1.56}$$

can be constructed as the external product of kets

$$\mid \Psi > = \mid L\,\Gamma\,\gamma > \mid S\,M_S > \tag{1.57}$$

where Γ, γ indicate respectively the representation and the particular basis state chosen.

Let us consider a case as the Cr salt of previous section, in which the crystal-field has a sufficiently low symmetry that all degeneracies of the ground state are lifted, resulting in an orbital singlet which we denote $\mid \Gamma_0 >$:

$$H_0 \mid \Gamma_0 > = E_0 \mid \Gamma_0 > \tag{1.58}$$

while for the excited states $\mid \Gamma\,\gamma >$:

$$H_0 \mid \Gamma\,\gamma > = E_{\Gamma\,\gamma} \mid \Gamma\,\gamma > \tag{1.59}$$

We now add to H_0 as perturbation the Zeeman term:

$$H_z = -\mu_B(\mathbf{L} + \mathbf{2S}) \cdot \mathbf{B}$$

describing the interaction with an external field, and the spin-orbit term:

$$H_{s-o} = \lambda(L\,S)\mathbf{L} \cdot \mathbf{S}$$

Let us now calculate, by the use of perturbation theory, the correction to the ground state due to the total perturbation

$$V = H_z + H_{s-o}$$

within the orbital manifold consisting of the ground state and the excited states denoted above. In this process, we shall consider the spin operators as c - numbers, since the calculation will be done entirely within the orbital space. As a result we shall obtain an expression for the perturbed energy of the ground state which will depend upon the spin operators, thereby becoming an effective spin Hamiltonian $\mathcal{H}_{eff}$, which eventually must be diagonalized within the spin $2S + 1$-dimensional subspace.

By direct application of perturbation theory formalism up to second order in V, we have:

$$\mathcal{H}_{eff} = \mathcal{H}^{(1)} + \mathcal{H}^{(2)}$$

where:

$$
\begin{aligned}
\mathcal{H}^{(1)} &= -\mu_B \mathbf{B} \cdot < \Gamma_0 \mid \mathbf{L} \mid \Gamma_0 > \\
&\quad -2\mu_B \mathbf{B} \cdot \mathbf{S} + \lambda \mathbf{S} \cdot < \Gamma_0 \mid \mathbf{L} \mid \Gamma_0 >
\end{aligned}
\tag{1.60}
$$

and:

$$\mathcal{H}^{(2)} = -\sum_{\Gamma\gamma} \frac{< \Gamma_0 \mid V \mid \Gamma\gamma >< \Gamma\gamma \mid V \mid \Gamma_0 >}{(E_{\Gamma\gamma} - E_0)} \tag{1.61}$$

Exercise 1.9
Prove that $\mathcal{H}_{eff}$ can be written as:

$$\mathcal{H}_{eff} = -\mu_B B_\alpha g_{\alpha\beta} S_\beta + \lambda^2 S_\alpha \Lambda_{\alpha\beta} S_\beta - \mu_B^2 B_\alpha \Lambda_{\alpha\beta} B_\beta \tag{1.62}$$

where the second rank tensors $\mathbf{g}$ and $\mathbf{\Lambda}$ are:

$$
\begin{aligned}
\Lambda_{\alpha\beta} &= \sum_{\Gamma\gamma} \frac{< \Gamma\gamma \mid L_\alpha \mid \Gamma_0 >< \Gamma_0 \mid L_\beta \mid \Gamma\gamma >}{(E_{\Gamma'\gamma'} - E_0)} \\
\\
g_{\alpha\beta} &= 2\delta_{\alpha\beta} - \Lambda_{\alpha\beta}
\end{aligned}
\tag{1.63}
$$

Let us now discuss some consequences of Eq. (1.63).

1.7.1 Effective gyromagnetic ratio

There will be some contribution of the orbital angular momentum to $\mathbf{g}$ in the ground state -although by assumption $\mathbf{L}$ was quenched in $\mid \Gamma_0 >$- due to the admixture of the higher energy states. This correction is in general anisotropic, since it reflects the symmetry of the crystal-field. In the case in which there is one axis of highest symmetry (uniaxial symmetry), which is usually chosen as the z axis, the tensor $\mathbf{\Lambda}$ is diagonal in such a coordinate system, with equal eigenvalues $\Lambda_\perp$ along the perpendicular directions to z, and in general a different one $\Lambda_\parallel$ along z. According to Eq. (1.63) the same is true for the $\mathbf{g}$ tensor.

1.7.2 Single-ion anisotropy energy

In the case that the crystal-field is cubic, the $\mathbf{\Lambda}$ tensor is a scalar. If instead we have uniaxial symmetry, Eq. (1.62) becomes:

$$
\begin{aligned}
\mathcal{H}_{eff} \ = \ & -g_{\parallel}\mu_B B_z S_z - g_{\perp}\mu_B \left(B_x S_x + B_y S_y\right) \\
& + D\left[S_z^2 - \frac{S\left(S+1\right)}{3}\right] \\
& + \frac{1}{3}S\left(S+1\right)\left(2\Lambda_{\perp} + \Lambda_{\parallel}\right) \\
& - \mu_B^2 \left\{\Lambda_{\perp}\left[\left(B_x\right)^2 + \left(B_y\right)^2\right] + \Lambda_{\parallel}\left(B_z\right)^2\right\}
\end{aligned}
\tag{1.64}
$$

where we introduced the single-ion anisotropy constant D:

$$
D = \lambda^2 \left(\Lambda_{\parallel} - \Lambda_{\perp}\right)
$$

Observe that the quadratic anisotropy term vanishes identically in the case $S = 1/2$. The last term in Eq (1.64) yields at low temperatures a constant contribution to the static susceptibility, called *Van Vleck susceptibility*. The term in $\mathcal{H}_{eff}$ quadratic in the magnetic field is much larger in magnitude than the quadratic diamagnetic contribution which we are neglecting.

Exercise 1.10
Neglect the spin-orbit term in Eq. (1.64), and assume for simplicity that the system is cubic. Prove that if both the thermal energy β^{-1} and the eigenvalues of $\mathcal{H}_{eff}$ are small compared to the orbital excitation energies appearing in the denominators of the expression defining Λ in Eq. (1.63), then the static susceptibility is

$$
\chi = \chi^0 + 2\mu_B^2 \Lambda
\tag{1.65}
$$

where χ^0 is the Brillouin susceptibility as calculated before.

As an example of the level splittings of the ground state spin multiplet produced by the anisotropy and the Zeeman terms, let us consider the case of a spin $S = 3/2$ in a uniaxial crystal-field, with the external magnetic field applied along the z axis, for simplicity. Then, the spin dependent part of the effective Hamiltonian is

$$
\mathcal{H}_{eff} = -g_{\parallel}\mu_B B S_z + D\left[S_z^2 - \frac{1}{3}S\left(S+1\right)\right]
$$

which is diagonal in the $\mid S, M_S >$ basis. The eigenvalue structure is depicted in Fig. 1.3.

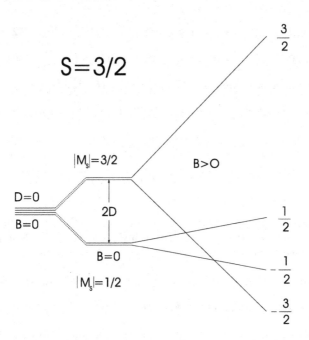

Figure 1.3: *Zeeman splittings for an S=3/2 atom in a uniaxial crystal-field.*

References

1. Langevin, P. (1905) *J.de Phys.* **4**, 678.

2. Langevin, P. (1895) *Ann. Chem. Phys.* **5**, 289.

3. A.Messiah, A. (1960) *"Mécanique Quantique"*, Dunod, Paris.

4. R.E.Trees, R. E. (1951) *Phys. Rev.* **82**, 683; Marvin, H. H. (1947) *Phys. Rev.* **71**, 102; R.G.Breene, R. G. (1960) *Phys. Rev* **119**, 1615.

5. W.E.Henry, W. E. (1952) *Phys. Rev.* **88**, 559.

6. Kittel, C. (1971) *"Introduction to Solid State Physics"*, John Wiley & Sons, Inc.

7. Becquerel, J. (1929) *Z. Phys.* **58**, 205.

8. Bethe, H. A. (1929) *Ann. Phys.* **3**, 135.

9. Kramers, H. A. (1930) *Proc. Acad. Sci. Am.* **33**, 959.

10. Van Vleck, J. H. (1952) *The Theory of Electric and Magnetic Susceptibilities*, Oxford University Press.

11. Newman, D. J. and Ng, Betty (1989) *Rep. Prog. Phys.* **52**, 699.

12. Runciman, W. A. (1956) *Phil. Mag.* **1**, 1075.

13. Hutchings, M. T. (1964) *Sol. State Phys.* **16**, 227.

14. Bleaney, B. and Stevens, K. W. (1953) *Rep. Prog. Phys.* **16**, 108.

15. Heine, Volker (1977) *"Group Theory in Quantum Mechanics"*, Pergamon Press, International Series in Natural Philosophy, Vol. 91, Chap. IV.

16. Ballhausen, J. C. (1962) *"Introduction to Ligand Field Theory"*, Mc Graw-Hill.

17. Al'tshuler, S. A. and Kozyrev, B. M. (1974) *"EPR in Compounds of Transition Elements"*, Halstead Press, John Wiley & Sons (Keter Publishing House, Jerusalem, Ltd).

18. Larsen, E. M. (1965) *"Transitional Elements"*, W. A. Benjamin, Inc.

19. Mulliken, R. S. (1932) *Phys. Rev.* **40**, 55.

20. Jackson, J. D. (1975) *"Classical Electrodynamics"*, John Wiley & Sons, Inc.

Chapter 2

Interacting Spins

2.1 Weiss model of ferromagnetism

Before considering any microscopic model for ordered magnetic systems, it is instructive to review Weiss phenomenological model [1] of ferromagnetism, which introduces the concept of a local molecular magnetic field. The very existence of materials which exhibit spontaneous magnetization in equilibrium demands, of course, some kind of interaction which tends to align the atomic magnetic dipolar moments, in such a way that they display a coherent pattern in space which, for a hypothetically infinite sample, can extend indefinitely. The simplest possible example of a long range ordered system is a single-domain ferromagnet. If we start from the ideas developed in the previous chapter on paramagnetic systems, we should assume that, since spins in a ferromagnet spontaneously align parallel at sufficiently low T, there must exist some local field acting on each of them (or at least most of them) which, in the absence of an external field, can only result from the presence of the other spins. If the system is completely homogeneous and has a total spontaneous moment $\mathbf{M}$, we expect a uniform magnetization $\mathbf{m} = \mathbf{M}/V$, where V is the total volume. Then the natural assumption to make is that the local field acting on each spin is proportional to $\mathbf{m}$:

$$\mathbf{B}_{loc} = \lambda \mathbf{m} \qquad (2.1)$$

which is Weiss local molecular field, introducing a phenomenological constant λ.

We have hitherto disregarded the magnetic dipolar forces, but they are certainly present in every case, so they would appear as the first candidate for the spin-spin interaction responsible for the magnetic molecular field. Applying to this case the Lorentz theory of electric dipolar systems, which are completely analogous [2] to the magnetic ones, we would find that for a macroscopic system with external spherical shape, the magnetic dipolar forces yield, for a uniform magnetization, a local field

$$\mathbf{B}_{loc} = \frac{4\pi}{3} \mathbf{m}$$

The minimum potential energy of a spin J in such a field is:

$$W = -\frac{4\pi}{3}mg\mu_B J \sim -\frac{4\pi}{3}(Jg\mu_B)^2/a^3$$

where a is the average separation between neighbouring spins. Temperature, as we have seen for paramagnets, has the effect of counteracting the aligning torques produced by the external field. The critical temperature $T_c \sim W/k_B$ turns out to be in the range of $\sim 10^{-2}K$. So, in order to explain critical temperatures of $100 - 1000\,K$ of known ferromagnetic substances we need a factor λ in Eq. (2.1) of the order $\sim 10^4 - 10^5$ instead of ~ 4 from the Lorentz theory, which rules out dipole-dipole forces as the source of the spin-spin interactions in ferromagnets of practical interest. Weiss left λ undetermined, and proceeded to apply this idea of a local molecular field to Langevin theory of paramagnetism, adequate for classical magnetic dipolar moments. We shall now profit from the quantum theory of paramagnetism developed in the previous chapter, and obtain a version of Weiss theory which incorporates the space quantization of atomic angular momentum. As we have seen already, the passage to the classical Langevin limit is simply obtained in the limit $J \to \infty$. Upon appeal to Brillouin's formula for the paramagnetic magnetization, Eq. (1.25), we add the local molecular field to the external one, to follow Weiss, so that we can write for the average magnetic moment per atom m, the modified Brillouin equation:

$$m(T,B) = g\mu_B J B_J \left(\beta g\mu_B J(B + \lambda m) \right) \tag{2.2}$$

2.1.1   Critical behaviour of Weiss model

We shall now verify that this model exhibits a phase transition. We must find a spontaneous (i.e., in the absence of an external field) finite magnetization at low T, which must vanish above a given temperature. We take the external field $B = 0$. Let us call $x = \beta g J \mu_B m$, and rewrite Eq. (2.2) as:

$$x = (gJ\mu_B)^2 \beta B J_J(\lambda x) \tag{2.3}$$

Near the critical temperature where the phase transition occurs, the magnetization is small, by hypothesis, so that we can expand the r. h. s. of Eq. (2.3) in a power series in x. It can be proved that for any x,

$$B'_J(x) \leq B'_J(0) \tag{2.4}$$

Exercise 2.1
Prove Eq. (2.4). Using this property, prove that, whenever

$$T \leq T_c = \frac{J(J+1)}{3k_B}(g\mu_B)^2\lambda \tag{2.5}$$

there are two solutions, $\pm m_s(T)$ of Eq. (2.3), while if $T > T_c$ the only solution is $m(T) = 0$.

This defines T_c as the critical (Curie) temperature for the transition from ferromagnetic (at low $T < T_c$) to the paramagnetic ($T > T_c$) phase.

We can also calculate the static longitudinal susceptibility, by partial differentiation of the equation of state, Eq. (2.2) with respect to B.

Exercise 2.2
Obtain the static longitudinal susceptibility for $T > T_c$ and $B = 0$:

$$\chi_{\parallel}(T,0) = \frac{\partial m}{\partial B} = \frac{(gJ\mu_B)^2 \beta B'_J(0)}{1 - (gJ\mu_B)^2 \beta \lambda B'_J(0)} = \frac{(gJ\mu_B)^2(J+1)}{3Jk_B(T - T_c)} \qquad (2.6)$$

Equation (2.6) is called the *Curie-Weiss law*. [3] We can also study the behaviour of $m(T,0)$ as a function of T in the neighbourhood of the critical temperature, or *Curie temperature*, T_c. This is easily done by expanding m in powers of the *reduced temperature* $\tau = (T_c - T)/T_c$:

Exercise 2.3
For $B = 0$, obtain the expansion

$$\frac{m(T,0)}{gJ\mu_B} = \frac{kT_c}{(gJ\mu_B)^2} \frac{\lambda}{\sqrt{A}} \tau^{1/2} + O(\tau^{3/2}) \qquad (2.7)$$

Equation (2.7) is compatible with some characteristics of a second order phase transition, namely:

- the long range order parameter m, which is finite at low T, tends continuously to zero at a finite temperature, in this case the Curie temperature, and in the neighbourhood of T_c, $m \sim \tau^{\beta}$, where β is one of the *critical exponents* which characterize the phase transitions [4];

- the partial derivative of m with respect to T is singular at the transition temperature, which implies that $\beta < 1$.

The Weiss model of ferromagnetism yields $\beta = 1/2$, which is the typical value for this exponent in the mean field approximation.

2.2 Microscopic basis of magnetism

As we have seen in our discussion of the Weiss model of ferromagnetism, the purely magnetic (dipole-dipole) forces are completely insufficient to provide a quantitative explanation for the high values of the Curie temperatures of the relevant magnetic materials, like Fe, Ni and Co, which are in the range of several hundred K, corresponding in the energy scale to $0.1\ eV$, characteristic of electronic excitations in atomic systems. The recognition that the electronic interactions, essentially of electrostatic nature, bring about the phenomena related to collective magnetic behaviour, has been one of the consequences of the quantum theory of the electronic structure of matter. Heisenberg removed the difficulties encountered by Weiss. Such large values of the molecular field as are needed to reproduce the measured Curie-Weiss temperatures, could be justified when he attributed the interactions between different spins in a magnetic system to electron exchange effects [7]. The idea, further developed by Dirac [8] and Van Vleck [9], was that the combined effect of the Coulomb interelectronic interactions and Pauli's exclusion principle between two atoms with spins $\mathbf{S}_1$ and $\mathbf{S}_2$ would produce an effective interaction potential of the form

$$V = -J(R_{12})\mathbf{S}_1 \cdot \mathbf{S}_2 \tag{2.8}$$

where $J(R)$ is the exchange energy, a function of the interatomic distance R. If J is positive in Eq. (2.8), this effective spin-spin interaction provides a model of ferromagnetic behaviour. We could now postulate this interaction and go onwards to calculate the Weiss field on this basis. However, we prefer to give first the derivation of the exchange interaction for a variety of situations, which will essentially cover the most important families of insulators with collective magnetic behaviour.

Let us discuss first the case of *direct exchange*. This is a situation in which the ground state orbital wave functions of neighbouring atoms have a non negligible overlap. There are however very few cases in which direct exchange, as described below, is the basis of ferromagnetism. CrO_2 and $CrBr_3$ are examples of this behaviour. In the majority of systems which show ordered magnetic behaviour, other exchange mechanisms must be invoked as the basic interactions, and we shall review these alternatives in the following sections. We leave the discussion of itinerant magnetism of transition metals for a later chapter.

2.2.1   The direct exchange interaction

Let us then consider two atoms a and b at a distance R. We shall disregard the states of the system in which two electrons are occupying orbitals on the same site, since the large Coulomb repulsion between them in this case will make such configurations very unlikely in the low energy states involved in the magnetic interactions, which, as previously mentioned, are in the $0.1eV$ range, while the on-site Coulomb repulsion energies are typically of the order of several eV. This is the so called Heitler-London approximation to the electronic structure of molecules. The Hamiltonian we shall consider is:

$$H = H_1 + H_2 + e^2/R \tag{2.9}$$

where H_1 includes one electron terms, H_2 two electron ones and R is the internuclear distance. We are considering monovalent atoms for simplicity, and

$$H_1 = \frac{p_1^2}{2m} + \frac{p_2^2}{2m} + v_a(\mathbf{r}_1) + v_b(\mathbf{r}_1) + v_a(\mathbf{r}_2) + v_b(\mathbf{r}_2) \tag{2.10}$$

$$H_2 = e^2/\mid \mathbf{r}_1 - \mathbf{r}_2 \mid \tag{2.11}$$

where the atomic core potential of atom a is

$$v_a(\mathbf{r}) = e^2 / \mid \mathbf{r} - \mathbf{R}_a \mid \tag{2.12}$$

with $\mathbf{R}_a$ the position vector of atom a, and correspondingly for the atom b.

We recall now briefly the second quantized representation of the many-electron states, since we shall find it convenient to make a slight generalization of the usual one to include the non-orthogonality of the atomic states which is the basis of the direct exchange. Let us then consider one-electron spin-orbitals, which are products of an orbital wave function of atom a or b times the corresponding spinor. The internal product of the orbital wave functions centered on different atoms, like

$$\int \phi_a^*(\mathbf{r})\phi_b(\mathbf{r}) \, \mathrm{d}^3 \mathbf{r} \equiv S_{ab} \equiv < a \mid b > \tag{2.13}$$

is in general non vanishing. We shall use sometimes Dirac bras and kets notation to denote the electron states as above. In the following we call $S_{ab} \equiv l$. This overlap integral in general could be a complex number. The matrix $\mathbf{S}$ is hermitian, by definition. Let us now consider the representation of the matter field operator associated with the electrons [10] in a non-orthogonal basis. We shall assume that we have a complete, non-orthogonal set of states, which satisfies the completeness relation

$$\sum_{m,n} \phi_m(\mathbf{x}) \, \mathbf{S}_{mn}^{-1} \, \phi_n^*(\mathbf{x}') = \delta^3(\mathbf{x} - \mathbf{x}') \tag{2.14}$$

where the matrix $\mathbf{S}$ is the overlap matrix :

$$\mathbf{S}_{mn} \equiv < m \mid n >$$

and we can always assume that the original basis states are normalized, so that the diagonal elements of $\mathbf{S}$ are unity.

Exercise 2.4
Consider the linear transformation

$$\mid \nu > = \sum_n \mid n > \left(\mathbf{S}^{-1/2} \right)_{n\nu} \tag{2.15}$$

where $\mathbf{S}$ is the same overlap matrix defined above. Verify that the new set $\{\nu\}$ is orthonormal. Prove that if one assumes this set is also complete, then Eq. (2.14) follows.

Let us now consider the matter field operator $\psi_\sigma(\mathbf{x})$ that annihilates an electron of spin σ at point $\mathbf{x}$. This operator can be expanded in the original basis set as:

$$\psi_\sigma(\mathbf{x}) = \sum_{n,m} \phi_{n\sigma} \left(\mathbf{S}^{-1} \right)_{nm} c_{m\sigma} \tag{2.16}$$

with an obvious notation including spin, and where the operator $c_{m\sigma}$ annihilates an electron of the corresponding spin in state m and satisfies the anticommutation relations

$$\left\{ c_{n\sigma}, c_{m\sigma'}^\dagger \right\} = S_{nm}\delta_{\sigma\sigma'}$$

$$\left\{ c_{n\sigma}, c_{m\sigma'} \right\} = \left\{ c_{n\sigma}^\dagger, c_{m\sigma'}^\dagger \right\} = 0$$

Let us review briefly some of the properties of the second quantized field operators. As a consequence of Pauli's principle they must satisfy the anticommutation relations:

$$\left\{ \psi_\sigma(\mathbf{x}), \psi_{\sigma'}^\dagger(\mathbf{x}') \right\} = \delta_{\sigma\sigma'}\,\delta^3(\mathbf{x} - \mathbf{x}') \tag{2.17}$$

$$\left\{ \psi_\sigma(\mathbf{x}), \psi_{\sigma'}(\mathbf{x}') \right\} = \left\{ \psi_{sigma}^\dagger(\mathbf{x}), \psi_{\sigma'}^\dagger(\mathbf{x}') \right\} = 0 \tag{2.18}$$

Exercise 2.5
Prove that Eq. (2.16) and the anticommutation relations of the $c, c^\dagger$ operators lead to the correct anticommutation relations (2.17) and (2.18) of the field operators.

Let us disregard spin for the moment to simplify the notation, and look into other properties of the field operators. By its very definition, $\psi^\dagger$ creates, upon acting on the vacuum state $\mid 0 >$, an electron at the point determined by its argument:

$$\psi^\dagger(\mathbf{x}) \mid 0 >= \mid \mathbf{x} >$$

On the other hand, the one electron state $\mid a >$ is created from the vacuum as

$$\mid a >= c_a^\dagger \mid 0 >$$

Then, consider the internal product

$$< \mathbf{x} \mid a >=< 0 \mid \psi(\mathbf{x})\, c_a^\dagger \mid 0 > \tag{2.19}$$

Substituting the expansion of ψ from (2.16) into (2.19) we find:

$$< \mathbf{x} \mid a >= \sum_{n,m} \phi_n(\mathbf{x}) S_{nm}^{-1} < 0 \mid c_m\, c_a^\dagger \mid 0 > \tag{2.20}$$

where

$$< 0 \mid c_m\, c_a^\dagger \mid 0 >= S_{ma} \tag{2.21}$$

So finally we find that

$$< \mathbf{x} \mid a >= \phi_a(\mathbf{x})$$

is the wave function of state a.

We can also evaluate internal products of two electron states. Consider

$$\mid ab >\equiv c_b^\dagger c_a^\dagger \mid 0 >$$

and

$$| 12 >\equiv \psi^\dagger(\mathbf{x}_2)\psi^\dagger(\mathbf{x}_1) \mid 0 >$$

Exercise 2.6
Prove that

$$< 12 \mid ab >= \phi_a(1)\phi_b(2) - \phi_a(2)\phi_b(1) \tag{2.22}$$

which is the as yet un-normalized Slater determinant.

Exercise 2.7
Show that

$$< ab \mid ab >\equiv< c_a c_b c_b^\dagger c_a^\dagger \mid 0 >= 1- \mid S_{ab} \mid^2 \tag{2.23}$$

In the Hilbert space of two electron states, the identity operator is:

$$\mathbf{1} = \frac{1}{2} \int \int dx_1 dx_2 \mid 12 >< 12 \mid$$

so that

$$\int \int dx_1 dx_2 < ab \mid 12 >< 12 \mid ab >= 2 < ab \mid ab >$$

Therefore the normalized Slater determinant for orbitals a and b is:

$$\Phi(12) = \frac{\phi_a(1)\phi_b(2) - \phi_a(2)\phi_b(1)}{\sqrt{2 < ab \mid ab >}}$$

We are now ready to study the diatomic molecule in the Heitler-London approximation (HLA). This approximation consists in neglecting completely the states with double occupancy of one of the atomic orbitals (which of course could only occur if both electrons have different spins). This approximation excludes the ionic states, in which there is charge transfer to one of the atoms, and considers only the covalent configurations, in which electrons are shared between both atoms. The application of this approximation to the Hydrogen molecule within a variational approach is a standard example in the Quantum Mechanics of molecules. An extension to include ionic configurations called the AMO (Alternant Molecular Orbitals) approximation [11] provides a much better description of the Hydrogen molecule at short distances than the HLA [12], which in general is expected to be reasonable at a typical internuclear separation, but not at very small or very large distances.

At this point we must re-introduce spin. We consider a state

$$\psi_1(12) = \frac{1}{\sqrt{2 < ab \mid ab >}} < 12 \mid a \uparrow b \uparrow> \tag{2.24}$$

and the time reversed state $\mid \psi_4 >$, with $\uparrow$ changed into $\downarrow$ for both electrons, which is automatically orthogonal to the first because so are the spinors.

Their norm squared is

$$< \psi_1 \mid \psi_1 >=< \psi_4 \mid \psi_4 >= 1- \mid l \mid^2$$

and we denote the corresponding normalized states as $\mid 1 >$ and $\mid 4 >$.

We can also consider states with antiparallel spins:

$$\mid \Phi_1 >=\mid a \uparrow b \downarrow>$$

$$\Phi_2 >=\mid a \downarrow b \uparrow>$$

These states are not orthogonal:

Exercise 2.8
Prove that

$$< \Phi_1 \mid \Phi_2 >= - \mid S_{ab} \mid^2 \equiv -l^2$$

The linear combinations $\Phi_1 \pm \Phi_2$ are mutually orthogonal, so that the states

$$\mid 2 >= \frac{1}{\sqrt{2 - 2l^2}} (\mid \Phi_1 > + \mid \Phi_2 >)$$

and

$$\mid 3 >= \frac{1}{\sqrt{2 + 2l^2}} (\mid \Phi_1 > - \mid \Phi_2 >)$$

are normalized and mutually orthogonal, besides being obviously orthogonal in spin space to states $\mid 1 >$ and $\mid 4 >$.

Let us denote by E_a and E_b the eigenvalues of the ground states of atoms a and b when at an infinite distance apart:

$$\left(\frac{p^2}{2m} + v_\alpha(\mathbf{r}) \right) \phi_\alpha(\mathbf{r}) = E_\alpha \phi_\alpha(\mathbf{r}) \tag{2.25}$$

where $\alpha = a$ or b.

We can now calculate the matrix elements of H within the four dimensional manifold we are considering. In the first place, since the Hamiltonian is spin-independent, it does not mix states $1, 4$ with $2, 3$. By the same token, it does not connect 1 with 4, so that within the Heitler-London approximation states 1 and 4 are eigenstates.

Exercise 2.9
Prove that $< 1 \mid H \mid 1 >=< 4 \mid H \mid 4 >$ *and that*

$$< 1 \mid H \mid 1 > = E_a + E_b + \frac{1}{1 - l^2} (< a \mid v_b \mid a > + < b \mid v_a \mid b >)$$

$$- \frac{l^* < a \mid v_a \mid b > + l < b \mid v_b \mid a >}{1 - \mid l \mid^2} + \frac{k - j}{1 - \mid l \mid^2} \tag{2.26}$$

where we use the definitions of the direct (k) and exchange (j) integrals as:

$$k = \int d^3\mathbf{r_1}\, d^3\mathbf{r_2}\, \mid \phi_a(1) \mid^2 \frac{1}{r_{12}} \mid \phi_b(2) \mid^2 \tag{2.27}$$

and

$$j = \int d^3\mathbf{r_1}\, d^3\mathbf{r_2}\, \phi_a^*(1)\phi_b^*(2)\frac{1}{r_{12}}\phi_b(1)\phi_a(2) \tag{2.28}$$

It is clear that $k \geq 0$. One can also prove that $j \geq 0$:

Exercise 2.10
Express the Coulomb potential in the definition of j above in terms of its Fourier transform and prove that $j \geq 0$.

It is convenient to add together some one-electron and two-electron terms as:

$$K_{ab} \equiv k + \frac{<a \mid v_b \mid a> + <b \mid v_a \mid b>}{1- \mid l \mid^2} \tag{2.29}$$

and

$$J_{ab} \equiv j + \frac{l^* <a \mid v_a \mid b> +l <b \mid v_b \mid a>}{1- \mid l \mid^2} \tag{2.30}$$

We can now consider the remaining two dimensional subspace:

Exercise 2.11
Prove that

$$< \Phi_1 \mid H \mid \Phi_1 >= E_a + E_b + K_{ab} =< \Phi_2 \mid H \mid \Phi_2 > \tag{2.31}$$

For the non-diagonal elements:

Exercise 2.12
Show that
$$< \Phi_1 \mid H \mid \Phi_2 >= -(E_a + E_b) \mid l \mid^2 -\tilde{J}_{ab} \tag{2.32}$$

where we have defined
$$\mid l \mid^2 J_{ab} \equiv \tilde{J}_{ab} \tag{2.33}$$

It is easy to see that the orthonormalized linear combinations $\mid 2 >$ and $\mid 3 >$ are eigenstates of H. In fact, one finds that states $\mid 1 >$, $\mid 2 >$ and $\mid 4 >$ are degenerate, and the corresponding triplet energy E_t is (2.31), while the remaining singlet eigenvalue E_s is

$$E_s = E_a + E_b - \frac{K_{ab}}{1+ \mid l \mid^2} - \mid l \mid^2 \frac{J_{ab}}{1+ \mid l \mid^2} \tag{2.34}$$

We must add to the eigenvalues obtained so far the Coulomb mutual repulsion energy of both positively charged cores, $\frac{e^2}{R_{ab}}$.

The splitting between the triplet and the singlet is

$$E_t - E_s = \frac{2 \mid l \mid^2 (K_{ab} - J_{ab})}{1 - \mid l \mid^2} \qquad (2.35)$$

Exercise 2.13

Show that for the triplet states the total spin $S = 1$, while M_S is 1 for $\mid 1 >$, 0 for $\mid 2 >$ and -1 for $\mid 4 >$, and that for the singlet $\mid 3 >$, $S = M_S = 0$;

Show also that within the 4-dimensional subspace considered, one can represent the Hamiltonian by an effective spin Hamiltonian

$$H_{eff} = E_0 - J_{eff}(R_{ab})\mathbf{S_1} \cdot \mathbf{S_2} \qquad (2.36)$$

where

$$E_0 = \frac{3E_t + E_s}{4}$$

and

$$J_{eff}(R_{ab}) = E_s - E_t$$

Heisenberg [7], Dirac [8] and Van Vleck [9] remarked that the Hamiltonian 2.36 favours a ferro(antiferro)-magnetic pairing of the electrons if $J_{eff} > 0$ ($<$ 0) respectively. On the other hand, if J_{eff} scales with the overlap squared, which is expected to decrease exponentially with the distance between the spins, this would imply that the range of the exchange interaction be limited to a small number of neighbouring atoms. A spin Hamiltonian with the Heisenberg form (2.36) can in fact be found in situations more general than the ones we assumed before in applying the Heitler-London approach, and as we shall see its application is fairly wide. As to order of magnitude, the highest critical temperatures of ferromagnets are around a few hundreds to around a thousand K, so that the exchange interaction constant J_{eff} between n. n. atoms should be around 10^{-2} to 10^{-1} eV, a relatively small energy in atomic scale, which makes its quantitative determination a difficult task.

2.2.2   The superexchange mechanism

The previous results, particularly (2.36) are relevant as long as there is some non-zero overlap $\mid l \mid$ of the ground state wave functions of atoms a and b. We expect that the asymptotic dependence of l on R_{ab} be exponential, so that direct exchange should not be expected to have a strong influence in substances where the nearest neighbour spins are situated at a distance R which is large compared to the mean radius of the atomic ground state wave functions of a and b. One could think that direct exchange might be an important ingredient in transition metals, in which the d atomic orbitals would have an appreciable overlap. However, in metals the Heitler-London approximation, which excludes interatomic charge transfer, is certainly inappropriate, and we shall see further on that for metals other hamiltonians have been proposed than the simple direct exchange Heisenberg one, although within certain approximations one can reconcile the Heisenberg Hamiltonian with the magnetism of narrow band metals. At any rate, one would be led to expect that collective magnetic behaviour in insulating materials could be well described by the Hamiltonian (2.36). It turns out, however, that many insulating compounds, particularly many oxides and halogenides, show collective magnetic behaviour, although the magnetic ions have ground state wave functions fairly localized in space, and are not first n. n. of each other, so that no appreciable direct overlap is to be expected in these cases. One typical example is provided by the extensively studied family of transition metal fluorides MF_2, with $M = Fe, Co$ or Mn, in which the transition metal ions occupy a body centered tetragonal lattice, each metal ion M^{2+} being surrounded by a distorted octahedron of F^-, as shown in Fig. 2.1. The electronic wave functions of the metal cations do not have any overlap so that direct exchange is excluded, which led Kramers [13] in 1934 to propose that a strong admixture of the cation and anion wave functions could be invoked to couple the cations indirectly, thus providing a mechanism for effective exchange between the transition metal spins. The mathematical treatment involves fourth order perturbation expansion of the total ground state energy. Anderson [14] revised Kramers perturbation approach, and showed that it is poorly convergent. Besides, several electron transfer processes can be simultaneously important, and it becomes necessary to have recourse to methods of ligand field theory. Let us now, however, discuss as an introduction to the main processes in action, the simple minded perturbative model of what has become known as *superexchange*.

The simplest possible system which will exhibit superexchange behaviour is a molecule consisting of one diamagnetic anion surrounded symmetrically by two paramagnetic cations, as shown schematically in Fig. 2.2, which displays both the ferromagnetic (FM) and the antiferromagnetic (AFM) configurations of the electronic spin distribution.

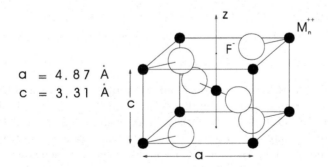

$a = 4.87 \text{ Å}$

$c = 3.31 \text{ Å}$

Figure 2.1: *Crystal structure of* MnF_2

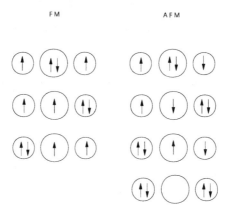

Figure 2.2: *Electron configurations of cations and anions in the superexchange process*

We shall now simplify drastically the atomic hamiltonian, reducing it to the minimum elements which still contain the basic physics we need to describe.

Let us consider the relevant states.

FM configuration

In this case we have one electron occupying the ground orbital state of each cation, and they both have parallel spins, while two electrons, naturally with opposite spins, occupy the ground orbital state of the ligand atom, which shall be denoted p, as a reference to its probable s, p character. The corresponding state is:

$$| E_0 > = | c_{b\uparrow}^{\dagger} \, c_{p\downarrow}^{\dagger} \, c_{p\uparrow}^{\dagger} \, c_{a\uparrow}^{\dagger} \, | 0 \rangle$$

and has total spin 1. Let us consider, to simplify, the symmetric case in which atoms a and b are equal. In the tight-binding approximation, in which one neglects the effects of overlap, the energy of this state is

$$E_0 = 2E_a + 2E_p + U_p$$

where U_p contains the effects of the Coulomb repulsion between both electrons occupying the ground state orbital, assumed non degenerate, of the ligand, naturally with opposite spins to satisfy Pauli exclusion principle. Due to the Coulomb electron interaction, there is the possibility of electron transfer from the ligand anion to the transition cations and viceversa, so one is led to consider states with double occupation of the cation orbitals, to either side of the anion. We shall assume that the Coulomb repulsion term U_a of the cation, is much larger than that of the anion, because the corresponding ground state orbitals are more localized. Since we need to consider singly occupied states of the anion, we must allow however for doubly occupied cation states. We include then in the relevant subspace the state

$$| a > = | c_{b\uparrow}^{\dagger} c_{p\uparrow}^{\dagger} c_{a\downarrow}^{\dagger} c^{\dagger} a \uparrow | 0 >$$

and the symmetric one $| b >$, with two electrons on cation b. These two excited states have unperturbed energy $E_1 = E_p + 3E_a + U_a$. The Hamiltonian in the second quantized Fock representation that will contain all the terms described above is:

$$H = H_0 + V$$

where

$$H_0 = E_a \sum_{\sigma} (n_{a\sigma} + n_{b\sigma}) + E_p \sum_{\sigma} n_{p\sigma} +$$
$$U_a (n_{a\uparrow}n_{a\downarrow} + n_{b\uparrow}n_{b\downarrow}) + U_p n_{p\uparrow} n_{p\downarrow} \qquad (2.37)$$

where the number operator $n_\alpha = c_\alpha^{\dagger} c_\alpha$, and

$$V = t \sum_{\sigma} \left(c_{a\sigma}^{\dagger} c_{p\sigma} + c_{b\sigma}^{\dagger} c_{p\sigma} \right) + h.c. \qquad (2.38)$$

is a simplified expression for the terms responsible for the electron transfer between the anion and the cations. One usually assumes that the two-body contribution to the hopping potential V can be substituted by a one-body potential in the Hartree–Fock approximation, and t is the sum of all possible contributions to the hopping matrix element.

The AFM configuration

In this case, in which, say, ion a has spin up, and b spin down, the unperturbed energy is the same as for the FM configuration. The excited states $| a \rangle$ and $| b \rangle$ are the same as before, but now there is the possibility of transferring both electrons from the central ligand anion to the cations, which can be accomplished by the action of the perturbing hopping potential on one of the excited states, but not directly on the unperturbed ground state. The new state to consider is

$$| ab \rangle = | c_{b\downarrow}^{\dagger} c_{b\uparrow}^{\dagger} c_{a\downarrow}^{\dagger} c_{a\uparrow}^{\dagger} \rangle$$

with unperturbed energy

$$E_{ab} = 2E_a + 2U_a$$

Perturbation Calculation

We can now follow the standard perturbation procedure. Let us then substitute $V \to \lambda V$, where λ is a dimensionless small parameter:

$$H(\lambda) = H_0 + \lambda V$$

and we look for solutions of the eigenvalue equation

$$H(\lambda) \mid \Psi\rangle = \tilde{E}_0 \mid \Psi\rangle$$

Now we expand the exact ground state vector as a linear combination of all the states compatible with each configuration, as discussed above, with coefficients which are a series in λ:

$$\mid \Psi\rangle = \mid E_0\rangle + \sum_{n=1}^{\infty}\sum_{\alpha} \lambda^n a_n(\alpha) \mid \alpha\rangle \qquad (2.39)$$

and expand the ground state energy $\tilde{E}_0$ in a power series in λ:

$$\tilde{E}_0 = E_0 + \sum_{n=1}^{\infty} \lambda^n \epsilon_n \qquad (2.40)$$

Exercise 2.14
Prove that

1. *$\epsilon_1 = \epsilon_3 = 0$ for the AFM and FM configurations, while the second order terms are equal for both.*

2. *The first different contribution between both configurations is the fourth order one.*

3. *The difference of the ground state energies to this order is*

$$\Delta = \tilde{E}_{AFM} - \tilde{E}_{FM} = \frac{4t^4(E_a - E_{ab})}{(E_a - E_0)^3(E_{ab} - E_0)} \qquad (2.41)$$

With this result, we can immediately write an effective-exchange spin Hamiltonian:

$$H_{eff} = -J\,\mathbf{S}_a \cdot \mathbf{S}_b + const. \qquad (2.42)$$

with $J = \Delta$ from (2.41). Under the assumptions we made regarding the unperturbed energies, we would have in general $E_{ab} - E_0 > 0$ and $E_a - E_{ab} \leq 0$, so $J < 0$, and the superexchange interaction is generally antiferromagnetic. For more detailed information on superexchange including many references see Refs. [5] and [6].

2.2.3 The RKKY interaction

Some systems have both localized spins and electrons in delocalized, or itinerant, states. The main examples are the rare-earth metals and the so-called magnetic semiconductors. In these systems there is an important contribution to the spin-spin interaction due to the polarization of the itinerant electrons by interaction with the localized spins. Zener [16] propposed a mechanism for polarizing the s and p conduction electrons in transition metals by exchange interaction with the d electrons of the paramagnetic ions. This idea was based on previous work by Frölich and Nabarro [17], in which they proposed a model of a long-range interaction between nuclear spins. In that model, the hyperfine interaction between a nucleus and the conduction electrons induces a polarization of the latter, which can in turn polarize another nucleus at a distance R, resulting in an effective exchange interaction between the nuclear spins. This is, accordingly, an effect obtained by considering second order perturbation contributions to the ground state energy of the system. The equivalent process, as applied to localized spins interacting with conduction electrons, was later developed by Ruderman and Kittel [15] and Bloembergen and Rowland [20]. This idea was applied by Kasuya [18] to metallic ferromagnetism and by Yosida [19] to alloys of the kind of a transition metal and a diamagnetic one, like $Cu - Mn$. In this case an indirect interaction is obtained between local electronic spins interacting with the Fermi sea of the conduction electrons. We shall derive this interaction in Chap. 9. For the time being, we just mention the relevant results in connection with the present discussion of exchange.

Suppose that we consider a dilute alloy, in which a small percentage of magnetic ions have been dissolved in, for instance, a noble metal. Under reasonable assumptions, one can obtain an effective interaction between two spins at a distance R by second order perturbation theory. Let us consider a metal in which electrons in the conduction band have an effective-mass m^*. Assume as well that the exchange interaction between a conduction electron of spin $\mathbf{s}$ and a localized spin $\mathbf{S}_n$ at $\mathbf{r}_n$ is local, and has the simple form

$$H_{int} = J_{sd}\mathbf{s} \cdot \mathbf{S}_n \delta^3(\mathbf{x}_{el} - \mathbf{r}_n) \tag{2.43}$$

Then the effective interaction of spins n and m is

$$H_{eff} = J_{sd}^2 \mathbf{S}_n \cdot \mathbf{S}_m \Phi(R_{nm}) \tag{2.44}$$

where the *range function* Φ of the distance between the localized spins is

$$\Phi(R) = \frac{m^* \mu_B^2}{h^3} \frac{1}{R^4} \left(\sin 2k_F R - 2k_F R \cos 2k_F R \right) \tag{2.45}$$

The indirect exchange mechanism is considered to be responsible for the colllective magnetic behaviour of the rare earth metals and of the magnetic semiconductors. In these cases, instead of an alloy with a dilute distribution of local spins, every atom, or at least every lattice cell, in the case of the magnetic semiconductors, has an unpaired resultant spin arising from electrons in fairly localized orbitals, while other electronic states are itinerant, and a local exchange interaction between these two families of electrons gives rise, as seen above, to an indirect long-range effective-exchange interaction between different local spins.

References

1. Weiss, P. (1909) *J. de Phys.* **6**, 667.

2. Kittel, C. (1971) *"Introduction to Solid State Physics"*, John Wiley and Sons, Inc., New York.

3. Curie, P. (1895) *Ann. Chim. Phys.* **5**, 289.

4. Stanley, H. Eugene (1987) *Oxford University Press, Inc.*, New York, Oxford.

5. Anderson, P. W. (1964) in *"Magnetism"*, Rado, G. and Suhl, H. (editors), Academic Press, New York, vol. I, Chap.2.

6. J.B.Goodenough, J. B. (1963) *"Magnetism and the Chemical Bond"*, Interscience, New York, Chap. 3.

7. Heisenberg, W. (1928) *Z. Phys.* **49**, 619.

8. Dirac, P. A. M. (1929) *Proc. Roy. Soc. (London)* **A123**, 714.

9. Van Vleck, J. H. (1934) *Phys. Rev.* **45**, 405.

10. March, N. H., Young, W. H. and Sampanthar, S. (1967) *"The many -body problem in Quantum Mechanics"*, Cambridge, at the University Press.

11. Lödin, P.-O. (1962) *J. Appl. Phys.* **33**, 251.

12. Chao, K. A., Oliveira, F. A., De Cerqueira, R. O. and Majlis, N. (1977) *Int. J. Quant. Chem.* **XII**, 11.

13. Kramers, H. (1934) *Physica* **1**, 182.

14. Anderson, P. W. (1950) *Phys. Rev.* **79**, 350.

15. Ruderman, M. A. and Kittel, C.(1954) *Phys. Rev.* **96**, 99.

16. Zener, C. (1951) *Phys. Rev.* **87**, 440.

17. Frölich, F. and Nabarro, F. R. N. (1940) *Proc. Roy. Soc. (London)* **A175**, 382.

18. Kasuya, T. (1956) *Prog. Theoret. Phys.* **16**, 45.

19. Yosida, K. (1957) *Phys. Rev.* **106**, 893.

20. Bloembergen, N. and Rowland, T. J. (1955) *Phys. Rev.* **97**, 1679.

Chapter 3

Mean Field Approximation

3.1 Helmholtz free energy

We start by considering an inequality satisfied by the thermodynamic Helmholtz free energy [1]. If a system is represented by a Hamiltonian H the Helmholtz free energy $F\{H\}$ is a functional of H and a function of the thermodynamic variables T and V [2]. In the case of magnetic systems we can neglect in general the small variations of volume involved, so that we take T and the external field B as the variables defining F [3]. For T and B fixed, one can try to develop a perturbative approach to calculate F, similar to the usual scheme for calculating the ground state energy $E_0 = \langle H \rangle_{T=0}$, by writing the identity

$$H \equiv H_0 + (H - H_0) \tag{3.1}$$

where H_0 is some Hamiltonian simple enough to be exactly solved. We can expand $F\{H\}$ as a series in the difference $H - H_0$, around $F_0 = F\{H_0\}$, with the hope that the correction terms are small, and that H_0 has been chosen wisely enough that the main physical properties of the original system are well described by the first few terms in the series. R. E. Peierls obtained a perturbative expansion for the free energy which starts like

$$F = F_0 + \langle H - H_0 \rangle_0 + \cdots \tag{3.2}$$

where $F_0 = -\frac{1}{\beta} \log Z_0$, $Z_0 = Tr\rho_0$, $\rho_0 = \exp -\beta H_0$, and we use the notation

$$\langle A \rangle_0 \equiv Tr(\rho_0 A)/Tr(\rho_0) \tag{3.3}$$

for any operator A. The general result proved in Ref. [1] is that the remainig terms in the series 3.2 add to a *non-positive* correction (which, eventually, can be infinite if the perturbation series diverges!) irrespectively of the choice of H_0. Therefore, we have a variational principle at our disposal:

$$F \leq F_0 + \langle H - H_0 \rangle_0 \equiv \phi \tag{3.4}$$

One can now obtain variational upper bounds for F by minimizing ϕ with respect to some parameters which are included in the definition of H_0. The minimum of ϕ gives accordingly the best upper bound of F which can be obtained with the particular form chosen for H_0. Let us now apply inequality (3.4) to the Heisenberg Hamiltonian with a Zeeman term:

$$H = -J \sum_{<ij>} \vec{S}_i \cdot \vec{S}_j - \gamma B \sum_i S_i^z \qquad (3.5)$$

Assume only first nearest neighbour (n. n.) interactions on a spin lattice. Let us choose for H_0 the Hamiltonian we know how to diagonalize exactly, namely the Zeeman paramagnetic one. Following Weiss' idea of a molecular field, we substitute the external field by an effective local field on each atom i,

$$\mathbf{B}_i^{eff} = \mathbf{B} + \mathbf{B}_i^{mol} \qquad (3.6)$$

In an ordered ferromagnet we assume that the local field is the same on all sites and that it is parallel to the direction of the external field, which is chosen as the spin quantization axis. Then,

$$H_0 = -\gamma B^{eff} \sum_{i=1}^{N} S_i^z \qquad (3.7)$$

as in Sect. 2.1. Then, just as in Sect. 1.1 we have an unperturbed free energy per spin

$$f_0(\beta, B^{eff}) = -\frac{1}{\beta} \log Z_0(\beta, B^{eff}) \qquad (3.8)$$

where, for $S = 1/2$,

$$Z_0(\beta, B^{eff}) = 2 \cosh \left(\frac{1}{2} \beta \gamma B^{eff} \right) \qquad (3.9)$$

In order to evaluate $\langle H \rangle_0$ and $< H_0 >_0$ we need the average magnetic moment of each site

$$m \equiv \langle S_i^z \rangle_0 = \frac{1}{Z_0} Tr(\rho_0 S^z) = -\frac{1}{\gamma} \left(\frac{\partial f_0(\beta, B)}{\partial B} \right)_{B=B^{eff}} \qquad (3.10)$$

where S^z is the corresponding operator of a generic site (since all sites are equivalent). Then we have:

$$\langle H \rangle_0 / N = -\frac{J}{2} \nu m^2 - \gamma B m \qquad (3.11)$$

where ν is the number of first nearest neighbours of a site. On the other hand, $\langle H_0 \rangle_0$ involves the unknown B^{mol} :

$$\langle H_0 \rangle_0 / N = -\gamma (B^{mol} + B) m \qquad (3.12)$$

Exercise 3.1
Verify Eqs. (3.11) and (3.12).

For $S = 1/2$ the variational function ϕ in (3.4) is :

$$\phi = -\frac{1}{\beta} \log \left(2 \cosh \left[\frac{\beta \gamma}{2} \left(B^{mol} + B \right) \right] \right) - J\nu m^2 + \gamma B^{mol} m \qquad (3.13)$$

Both variables B^{mol} and m are variational parameters. The condition

$$\left(\frac{\partial \phi}{\partial B^{mol}} \right)_m = 0 \qquad (3.14)$$

leads to Eq. (3.10), but we must still impose the second condition for an extremum of ϕ:

$$\left(\frac{\partial \phi}{\partial m} \right)_{B^{mol}} = 0 \qquad (3.15)$$

Exercise 3.2
Verify that Eq. (3.15) yields

$$\gamma B_{mol} = 2J\nu m \qquad (3.16)$$

Upon substituting (3.16) into (3.10) we get, for $S = 1/2$:

$$m = \frac{1}{2} \tanh \left[\frac{\beta}{2} \left(2J\nu m + \gamma B \right) \right] \qquad (3.17)$$

One can easily solve Eq. (3.17) analytically for B as a function of m. We see that the Weiss molecular field parameter λ can be related to microscopic variables:

$$\lambda = \frac{2J\nu}{\gamma}$$

This solves the puzzle created by the impossibility of explaining the high values of typical Curie temperatures of the known ferromagnetic substances within the classical Weiss-Langevin model of paramagnetism, since the molecular field in a transition metal can be as large as 10^6 *gauss*.

Let us now discuss in more detail the consequences of the molecular field approximation for ferromagnetic systems in the $S = 1/2$ case, with only first n. n. interactions. The spontaneous behaviour of this system, corresponding to the case $B = 0$ is described by the self-consistent solution of (3.17). We shall verify that the parameter

$$K = \beta J\nu$$

controls the kind of solutions to be found. Let us measure $k_B T$ in the natural unit provided by J. For zero applied field, $m = 0$ is always a solution, at any temperature. Besides this paramagnetic solution there are two other non-zero ones, $\pm m_s$, where s stands for 'spontaneous', iff $K > 2$. Therefore the critical, or Curie-Weiss temperature

$$T_c = \frac{J\nu}{2k_B}$$

separates the ferromagnetic low T phase from the paramagnetic phase which exists for $T > T_c$. For zero external field, as T increases from 0 to T_c, the spontaneous magnetization given by the non trivial solution of (3.17) decreases from $m(0) = S$ to zero, as indicated in Fig. 3.1. The thermodynamic average

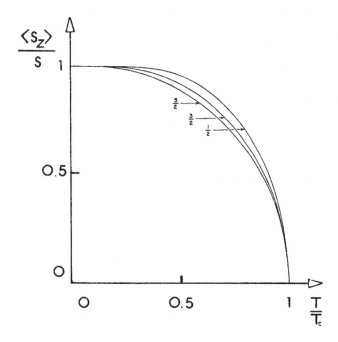

Figure 3.1: *Mean field results for the spontaneous magnetization of a ferromagnet as a function of T/T_c for several values of S.*

$S^z \equiv m$ is the adequate long range order parameter. The continuous vanishing of m and the discontinuity in its first derivative as $T \nearrow T_c$ are signatures of a second order phase transition. We already found in Chap. 2 that $m(T, B = 0)$ has precisely these properties, as described in particular, by Eq. (2.7). This is the same critical behaviour predicted by the Van der Waals theory of the gas-liquid phase transition [4] in which the order parameter is $\Delta v \equiv v_g - v_l$, the difference of the specific volumes of the gas and the liquid phase, which plays the role of m_s, while the pressure p is the analogue of B. The critical temperature of the Van der Waals fluid corresponds to the Curie temperature of the ferromagnetic-paramagnetic transition [4]. One disturbing feature of the mean field results is that the critical behaviour is independent on the dimensionality of the system, in the sense that it yields a finite Curie temperature — that is, long range order at finite temperatures — for any physical dimension of the system. However, we shall see that the isotropic Heisenberg model with short range interactions cannot sustain long range order at any finite temperature for dimension less than three. This contradiction is a consequence of the absence of fluctuations in the mean field description. Even the simplest harmonic approximation to the transverse spin fluctuations in a Heisenberg system with short range interactions leads to the absence of long range order in less than three dimensions, as we shall find upon studying the effect of spin waves on the order parameter. In terms of the reduced temperature $\tau = (T - T_c)/T_c$ we found in Chap. 2 (Eq. (2.7)) that for $\tau \nearrow 0$

$$m_s(\tau) = C \mid \tau \mid^{\beta} \qquad (3.18)$$

where β is one of the so called *critical exponents* [5] and C is a constant. We found that the mean field approximation yields $\beta = 1/2$. Experimental values for β [5] are: $0.32 - 0.36$ for the liquid-vapour critical point, and $0.32 - 0.39$ for the magnetic phase transitions in different systems. For the $3D$ Heisenberg model, numerical calculations yield 0.37 ± 0.04 [9]. Just as in the Van der Waals case, the present results describe thermodynamic equilibrium states. Within the mean field approximation, Eq. (3.17) is the equation of state of the ferromagnet. For $\tau < 0$, i.e in the ferromagnetic phase, we shall now discuss in more detail the mean field results. In Fig. 3.2 the magnetization, as given by (3.17) is plotted as a function of the dimensionless external field $b = \gamma B/J$. The example corresponds to the coordination number $\nu = 12$, the number of first n. n. in an f.c.c. lattice. We have taken $T/T_c = 1/2$ and $S = 1/2$.

For large $\mid B \mid$ the system tends to saturation, that is $\mid m \mid \to S$. From Eq. (3.13) we see that for $S = 1/2$ we can write the variational approximation to the Helmholtz free energy as:

$$\frac{\phi}{J\nu} = -\frac{1}{2}\frac{T}{T_c} \log\left(2 \cosh 2(T/T_c)(m + b/2)\right) + m^2 \qquad (3.19)$$

Let us start from $b > 0$, and then imagine that the field is slowly (i.e. reversibly) removed so that the system evolution is a succession of equilibrium states, which in our case are represented precisely by Fig. 3.2. As b decreases further and it becomes negative, we find a Van der Waals kind of isotherm, similar to a $p(v)$ isotherm for a gas below T_c. As in that case, we should expect that the section $\overline{BCDEF}$ of the curve be a series of metastable or unstable states. The analogue of the two separate phases of the gas-liquid system is in our case the pair of different solutions $m(b)$ of the mean field equations for given b. Since Fig. 3.2 represents a succession of equilibrium states, then at all points on the curve

$$\phi' \equiv \left(\frac{\partial \phi}{\partial m}\right)_b = 0$$

At an equilibrium configuration, on the other hand, we must have a positive second partial derivative: $\phi''|_b > 0$. Upon differentiating Eq. (3.13) twice with respect to m one obtains:

$$\phi''|_b = -J\nu \left[\beta J\nu \left(1 - 4m^2\right) - 2\right] \qquad (3.20)$$

which has two roots, $\pm m_0$, where m_0 depends on the parameters. For positive m we have $\phi''|_b > (<)0$ if $m > (<)m_0$.

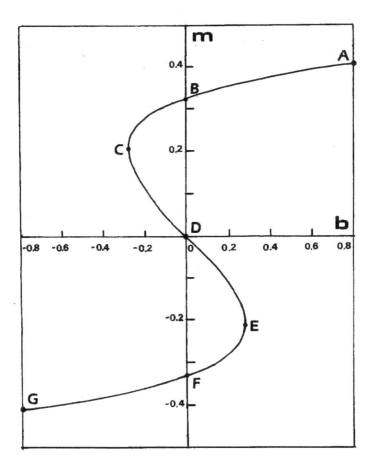

Figure 3.2: *Equilibrium isotherm of an $S = 1/2$ f.c.c. ferromagnet for $k_B T/J = 5/2$ in the MFA. The average $m = \langle S_z \rangle$ is plotted vs. the dimensionless external field $b = \gamma B/J$.*

Exercise 3.3
Prove that:

- ϕ *as function of m for constant b has* local minima *(stable configurations) on the mutually symmetric portions $\overline{ABC}$ and $\overline{EFG}$ of the curve in Fig. 3.2, while it has local maxima (unstable configurations) on the portion $\overline{CDE}$.*

- *the partial derivative $(\partial m/\partial b)_\phi$ on the curve is infinite at points C and E where $b = -(+)b_0$ and b_0 depends on the parameters of the system.*

Figure 3.3 shows $\phi(m, b)$ in a 3D view. Sections of $\phi(m)_{b=const.}$ are shown. The Helmholtz free energy obtained in Eq. (3.13) is invariant under the simul-

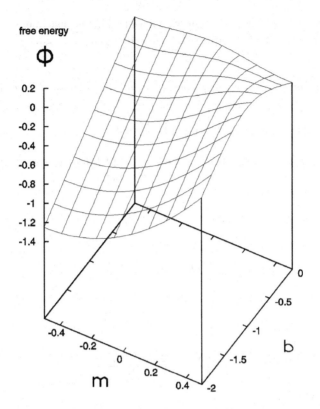

Figure 3.3: *Surface of the variational Helmholtz free-energy ϕ on the m, b plane for an $S = 1/2$ f.c.c. ferromagnet.*

taneous inversion of m and b. For $b = 0$ there are two symmetric solutions $\pm m_s$ and the two points B and F on the isotherm are symmetric local minima of ϕ. We choose $b < 0$ for convenience with the graphical representation. For $b < 0$ points on portion $\overline{BC}$ of the isotherm in Fig. 3.3 are local minima. The points on the $\overline{CD}$ portion are local maxima (unstable configurations). At fixed b if the system is on some point on the piece $\overline{BC}$ it would relax to the lower ϕ states on portion $\overline{FG}$, were it not for the barrier represented by the intermediate maximum along $\overline{CD}$, so that we have metastability on $\overline{BC}$. As seen in Fig. 3.3 the barrier disappears for $\mid b \mid > b_0$ at point C, where the maximum and minimum coincide and the second differential of ϕ vanishes. One would predict a hysteresis magnetization cycle if the field oscillates slowly in time between positive and negative values. However, let us remark that a typical barrier height, as the one for $b = 0$ is $\Delta\phi \approx 1.3J$, while the thermal energy in this example is $k_B T \sim 2.5J$. Therefore, one concludes that, had the thermal fluctuations been correctly taken into account, they would have destroyed the metastability along $\overline{BC}$ and correspondingly the hysteresis, since the system could, through thermal activation, traverse the free energy barrier. This is just one example of the limitations of the mean field approximation due to its disregard of fluctuations.

Exercise 3.4
Sketch the hysteresis loop corresponding to the example considered above.

3.2   Mean field susceptibility

In the paramagnetic regime the zero field susceptibility is isotropic: the longitudinal and transverse components are equal. Let us calculate the longitudinal component:

$$\chi = \frac{\gamma N}{V} \left(\frac{\partial \langle S^z \rangle}{\partial B} \right)_{B=0} \tag{3.21}$$

From Eq. (3.17),

$$\frac{\partial \langle S^z \rangle}{\partial B} = S B_S'(x) \frac{\partial x}{\partial B} \tag{3.22}$$

where

$$x = \beta S(\gamma B + 2J\nu \langle S^z \rangle) \tag{3.23}$$

Finally, we find

$$\chi = \frac{\gamma N}{V} \frac{\gamma S^2 B_S'(x)}{1 - 2\beta S^2 J\nu B_S'(x)} \tag{3.24}$$

where x is evaluated at $B = 0$.

We shall presently require the limiting expansions of $B_S(x)$ for small and large x:

For small x,

$$B_S(x) = \frac{S+1}{3S} \left(x - Ax^3 + \mathcal{O}(x^5) \right) \tag{3.25}$$

where

$$A = \frac{(2S+1)^2 + 1}{60S^2} \tag{3.26}$$

For large x,

$$B_S(x) = 1 - (1/S)e^{-x/S} + \mathcal{O}(e^{-(2x+x/S)}) \tag{3.27}$$

The form (3.24) for χ shows the enhancement of the paramagnetic (Curie) susceptibility due to the interaction among spins. We expect that in the limit $B \to 0$, $x \to 0$ aswell if $T > T_c$. We substitute the derivative of (3.25) into (3.24), and we let T decrease from $T \gg T_c$, looking for the highest temperature T (or the smallest β) at which the denominator of (3.24) vanishes - and the susceptibility diverges in consequence. This determines β_c :

$$2\beta_c J\nu S(S+1)/3 = 1 \tag{3.28}$$

One verifies that for $S = 1/2$, $k_B T_c = J\nu/2$ as found before. At low T the system should be in the FM phase, and therefore $\langle S^z \rangle \neq 0$. We must bear in mind that, since the Heisenberg Hamiltonian is completely isotropic, any orientation can be chosen for the ground state magnetization. If one direction in particular is selected, this breaks this continuous symmetry. Upon the application of a small field the magnetization will align along the applied field, no matter how small that be. We can assume that we apply a field proportional to $N^{-\alpha}$ with $\alpha > 0$, so that in the thermodynamic limit the Zeeman energy per spin tends to 0 and the ground state energy in this limit is not affected by this fictitious field. Then we obtain at low T a Weiss ground state with all spins aligned in the direction z of the external field.

Let us calculate the longitudinal susceptibility $\chi_\parallel$ in the $S = 1/2$ case, where

$$\sigma \equiv \langle S^z \rangle / S = \tanh(\gamma \beta B / 2 + \nu \beta J \sigma / 4) \tag{3.29}$$

Then

$$\chi_\parallel(T) = \frac{\gamma^2 \beta \rho}{4(\cosh^2(x) - T_c/T)} \tag{3.30}$$

where x is the argument of tanh in Eq. (3.29) and $\rho = N/V$, the spin volume-concentration.

Exercise 3.5
Verify Eq. (3.30).

For $T \to 0$ we obtain

$$\chi_\parallel(T) \to \frac{\gamma^2 \rho}{k_B T} \exp(-\nu J / 2 k_B T) \tag{3.31}$$

Finally, for $T \nearrow T_c$ we can expand $cosh^2(x)$ in powers of $(T - T_c)/T_c$ to obtain:

$$\chi_\parallel(T) \to \frac{\gamma^2 T T_c \beta \rho / 4}{(T_c - T)(3T - T_c)} \simeq \frac{\beta \gamma^2 T_c \rho}{8(T_c - T)} \tag{3.32}$$

On the other side of T_c, for $T \searrow T_c$, we have

$$\chi_\parallel(T) \to \frac{\beta \gamma^2 \rho / 4}{1 - \nu \beta J / 2} \tag{3.33}$$

We conclude that for $T \sim T_c$

$$\chi_\parallel(T) \to \frac{\gamma^2 \rho}{4 \mid T - T_c \mid} (1 - \frac{1}{2} \theta(T_c - T)) \tag{3.34}$$

where $\theta(x)$ = Heaviside step function. Then, in the neighbourhood of T_c

$$\chi_\parallel^{-1}(T) = C \mid T - T_c \mid \tag{3.35}$$

with a different coefficient C below and above T_c.

The exponential decrease of $\chi_\parallel(T)$ as $T \to 0$ is not observed experimentally. One important reason for this is that any macroscopic FM is never found in a single domain, uniformly magnetized state at zero applied field. Both re-orientation of domains magnetization and domain-walls displacement change the bulk magnetic moment. These processes yield a non-zero susceptibility even at very small T. Besides, each specimen exhibits a particular behaviour, which depends on the detailed structure of domains when the magnetization is being measured, which reflects in a history dependent response.

3.3   Specific heat of ferromagnet

In the mean field approximation each spin i contributes to the total internal energy with the same average amount

$$U/N = -\gamma \langle S_i^z \rangle B_{mol} = -\gamma \lambda m^2 \tag{3.36}$$

The specific heat per atom at zero external field is then

$$c_B = -2\gamma \lambda m (\partial m/\partial T)|_{B=0} \tag{3.37}$$

Let us now obtain for given spin S the limiting forms of the specific heat in different temperature ranges. For $T \to 0$, we find for m the form

$$m|_{T\to 0} \to S - \exp\left(-\frac{2J\nu S}{k_B T}\right) \tag{3.38}$$

so that

$$c_B = \frac{2S^2 J\nu}{k_B T^2} \exp\left(-\frac{2J\nu S}{k_B T}\right) \tag{3.39}$$

which tends exponentially to 0 as $T \to 0$.

For $T \to T_c$ we need to consider the expansion of $B_S(x)$ up to cubic terms.

Let us define the reduced temperature $\tau \equiv T/T_c$, and the reduced magnetic moment $\sigma \equiv \langle S^z \rangle/S$. For $B = 0$, we have

$$x = \frac{3S\sigma}{(S+1)\tau}$$

(see Eq. (3.23)). We find, for $T \nearrow T_c$

$$\sigma = A\left(\frac{T_c - T}{T_c}\right)^{1/2} \tag{3.40}$$

and

$$c_B(T_c^-) = 5/2\frac{(2S+1)^2 - 1}{(2S+1)^2 + 1}\, k_B \tag{3.41}$$

where use has been made of the Brillouin function expansion (3.25).

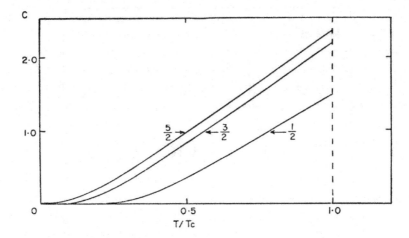

Figure 3.4: *Specific heat per magnetic ion for a Heisenberg ferromagnet, in units of k_B, as obtained in the mean field approximation. The value of spin is indicated on each curve.[6]*

Exercise 3.6
Verify Eq. (3.41), obtain the classical limit $S \to \infty$ and give an explanation of the result in terms of classical kinetic theory.

Above T_c in the MFA one finds $c_B(T) \equiv 0$, since both the order parameter $m = \langle S_z \rangle$ and the molecular field vanish identically. This is physically not satisfactory, since we expect some contribution to the specific heat even at $T \gg T_c$. The reason for this un-physical result of the MFA lies in the complete neglect of the short range correlations. The only order parameter in the theory, by construction, is the macroscopic long-range order parameter σ, which is an average over the whole (infinite) system. It is completely reasonable to expect that some short range order survives above the phase transition. The average $\langle S_i^z S_{i+\delta}^z \rangle$ for a small distance δ should not be identically zero in the paramagnetic phase. Some refinements of the MFA incorporate explicitly one or more short range order parameters in the theory from the start, as we shall discuss presently. We found in the MFA a finite discontinuity of the specific heat at the critical temperature, as shown in Fig. 3.4. This is a characteristic feature of a second order phase transition [7]. In such a process Gibbs' free energy per spin g_F and g_P of the ferromagnetic (F) and paramagnetic (P) phases in equilibrium at T_c satisfy the two conditions:

$$g_F(T_c)|_{B=0} = g_P(T_c)|_{B=0}$$
$$\frac{\partial g_F}{\partial T}(T_c)|_{B=0} = \frac{\partial g_P}{\partial T}(T_c)|_{B=0} \qquad (3.42)$$

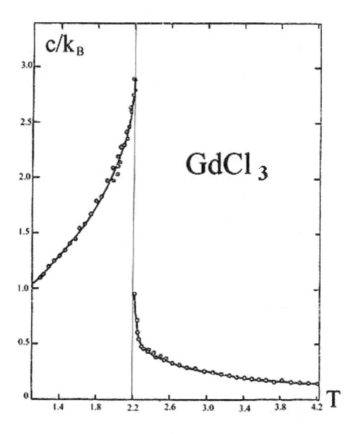

Figure 3.5: *Specific heat per spin of GdCl₃ for $B = 0$, in units of k_B, as a function of T in K.*

while the second derivatives of g are different, resulting in a discontinuity of the specific heat.

For real systems, when the symmetry of the system is known to belong to the Heisenberg (isotropic) class in $3d$, experiments [5] yield a *divergence* of the specific heat at the transition point like

$$c \sim |\tau|^{-\alpha} \tag{3.43}$$

where $\tau = (T - T_c)/T_c$, and

$$0 < \alpha < 0.11 - 0.17 \quad \tau > 0$$
$$0 < \alpha < 0.13 - 019 \quad \tau < 0 \tag{3.44}$$

Experimental results for the ferromagnetic compound $GdCl_3$ [8] are shown in Fig. 3.5. Numerical calculations [9] yield for α the values:

- -0.14 ± 0.06 for the $3d$ Heisenberg FM;

- 0 exactly, which in fact corresponds to a logarithmic singularity, for the $2d$ Ising FM, and

- $\alpha \approx 1/8$ for the $3d$ Ising FM.

3.4 The Oguchi method

We shall now consider corrections to the simple MFA by introducing the contribution of the short range static correlations. A convenient short range order (SRO) parameter is

$$\Delta \equiv \langle \mathbf{S}_i \cdot \mathbf{S}_{i+\delta} \rangle / S^2 \tag{3.45}$$

for nearest neighbour sites.

Within the MFA, since all spins are independent, the average of the product above factorizes, and

$$\Delta^{MFA} = \sigma^2 / S^2 \tag{3.46}$$

so that Δ vanishes above T_c. If instead we explicitly obtain the exact partition function of the isolated pair of spins, the SRO parameter will not vanish above the transition. Upon trying to improve the MFA Van Vleck [10] already in 1937 used a method which was later developed by Oguchi [11]. Bethe and Peierls incorporated several improvements to the self-consistent MFA by considering several convenient SRO parameters [12] which are obtained through the exact calculation of the partition function of a small cluster of spins, and the substitution of the rest by a self-consistent mean effective field. Let us briefly discuss the Oguchi method. To this end we shall first solve exactly the Hamiltonian for a pair of n.n. spins:

$$H^{ij} = -J\mathbf{S}_i \cdot \mathbf{S}_j - \gamma(S_i^z + S_j^z)B_{eff} \tag{3.47}$$

where B_{eff} is in the direction of the magnetization, chosen as z axis. Consider the sum of both spins

$$\mathbf{S}_p = \mathbf{S}_i + \mathbf{S}_j$$

with square

$$(\mathbf{S}_p)^2 = S'(S' + 1) \tag{3.48}$$

where

$$0 \leq S' \leq 2S$$

and a total z component $S_i^z + S_j^z$ with eigenvalues M', where $-2S \leq M' \leq 2S$. The eigenvalues of the pair Hamiltonian (3.47) are

$$E_p = -J \left[\frac{S'(S' + 1)}{2} - S(S + 1) \right] - \gamma M' B_{eff} \tag{3.49}$$

and the SRO parameter is

$$\Delta = \frac{\langle S'(S'+1)\rangle - 2S(S+1)}{S^2} \tag{3.50}$$

To simplify the notation we define $j \equiv \beta J$ and $b \equiv \gamma\beta B_{eff}$. The partition function for the two-spin cluster is

$$Z_{pair}(j,b) = \sum_{S'=0}^{2S} \sum_{M'=-S'}^{S'} \exp\left[(j/2)S'(S'+1) + bM' \right] \tag{3.51}$$

Exercise 3.7
Show that the partition function for the pair of spins with $S = 1/2$ is

$$Z_{pair}(j,b) = 1 + e^j(2\cosh b + 1) \tag{3.52}$$

and obtain the average pair magnetization

$$\langle M'\rangle/S = \frac{4e^j \sinh b}{1 + e^j(2\cosh b + 1)} \tag{3.53}$$

For consistency, we require that $< S_i >^{MFA}$ satisfy the condition

$$\langle M'\rangle^{pair}/S = 2\langle S_i^z\rangle^{MFA}/S \equiv 2\sigma \tag{3.54}$$

To complete the self-consistency conditions we demand that each spin in the pair be acted upon by the same effective molecular field as any other one from the rest of the system not in the pair, which we obtained in previous section as

$$B_{mol} = 2J\nu S\sigma \tag{3.55}$$

with the only change that now one nearest neighbour has to be excluded since its contribution to the interaction has been already included in the pair Hamiltonian, so that for the spins of the pair the effective field in the present units is

$$b = 2j(\nu - 1)S\sigma + b_0 \tag{3.56}$$

with $b_0 = \gamma B\beta$ and B is the external field. We now substitute (3.56) into Eq. (3.53). From Eq. (3.54) we find

$$\sigma^{Oguchi} = \frac{2\sinh(b_0 + (\nu-1)j\sigma)}{e^{-j} + 1 + 2\cosh(b_o + (\nu-1)j\sigma)} \tag{3.57}$$

The transition temperature can be obtained from (3.57) choosing $B = 0$ and assuming that σ vanishes at some T_c. The values of T_c obtained from the numerical solution of the Oguchi equations turn out a few percents smaller than the MFA results [12]. One can compare expression (3.57)for $b_0 = 0$ to the one previously obtained within the MFA,

$$\sigma^{MFA} = \frac{\sinh\left(j\nu\sigma^{MFA}\right)}{1 + \cosh\left(j\nu\sigma^{MFA}\right)} \tag{3.58}$$

Exercise 3.8

a) Show that one finds again the critical behaviour

$$\sigma^{Oguchi} \sim (T_c^{Oguchi} - T)^{1/2} \tag{3.59}$$

b) Obtain the longitudinal susceptibility below T_c in the Oguchi approximation :

$$\chi = \lim_{B\to 0} \frac{N\gamma\sigma}{VB} = \left(\frac{C}{T}\right)\frac{4}{e^{-2j} + 3 - 2(\nu - 1)j} \tag{3.60}$$

where C = Curie constant, and show that it has the same critical behaviour as in the MFA:

$$\chi^{-1} \sim |T_c - T| \tag{3.61}$$

We show now that the SRO parameter Δ defined in (3.45) does not vanish in the Oguchi approximation at any finite T. The cluster (in this case, pair) average we need is

$$\langle S'(S'+1)\rangle = \frac{2(1 + 2\cosh b)}{e^{-j} + 1 + 2\cosh b} \tag{3.62}$$

Exercise 3.9

Verify Eq. (3.62).

If $j > 0$ we expect $\Delta > 0$. We can verify this for $T \to 0$ and for $T \nearrow T_c$. In fact, as $T \to 0$, $\Delta \to 1$. In the limit $T \nearrow T_c$ we get

$$\Delta \to \frac{3(1 - e^{-j})}{3 + e^{-j}} \tag{3.63}$$

which is positive for $j > 0$. At the transition temperature Δ is continuous but its slope is not, due to the discontinuity in the slope of σ. The specific heat per spin at constant B can be calculated from the SRO parameter, since Δ is proportional to the contribution to the internal energy of each pair of spins in the absence of an external field:

$$c = C/N = -\frac{J\nu S^2}{2}\frac{\partial\Delta}{\partial T} \tag{3.64}$$

For $T > T_c$, where $\sigma \equiv 0$, we obtain

$$c = 6\nu k_B \frac{j^2 e^{-2j}}{(3 + 2e^{-2j})^2} \tag{3.65}$$

This specific heat has a discontinuity at T_c and it does not vanish above T_c, but instead decreases monotonically. For very large T (very small j) it decreases like T^{-2} . This behaviour agrees qualitatively with experiments on FM materials, except very near the Curie temperature.

3.5   Modulated phases

The MFA can yield a great variety of equilibrium phases besides the ferromagnetic one. Suppose that the applied static field oscillates in space. Then the Zeeman term becomes

$$-\gamma B \sum_i S_i^z \cos\left(\mathbf{q_0} \cdot \mathbf{R_i}\right) \tag{3.66}$$

We introduce now the Fourier transformed (FT) spin operators

$$\mathbf{S_q} = \sum_i e^{-i\mathbf{q} \cdot \mathbf{R_i}} \mathbf{S_i} \tag{3.67}$$

Since we are assuming translational invariance of H, the effective exchange integrals J_{ij} depend only on the difference $(\mathbf{R_i} - \mathbf{R_j})$, and we obtain for the Hamiltonian in wave vector space

$$H = -\frac{1}{N} \sum_q [\, J(\mathbf{q})\mathbf{S_q} \cdot \mathbf{S_{-q}} - 1/2\gamma B(S^z(\mathbf{q}) + S^z(-\mathbf{q})) \,] \tag{3.68}$$

We now linearize H to get a mean field approximation equivalent to the static case, by use of the identity

$$\mathbf{S_q} \equiv (\mathbf{S_q} - \langle \mathbf{S_q} \rangle) + \langle \mathbf{S_q} \rangle \tag{3.69}$$

The first term in brackets above is the fluctuation of the spin operator. We now substitute (3.69) into (3.68) and neglect all terms which are of second order in the fluctuations, thereby obtaining

$$H \approx -\frac{1}{N} \sum_q [\, J(-\mathbf{q}) (\mathbf{S_q} \cdot \langle \mathbf{S_{-q}} \rangle + \langle \mathbf{S_q} \rangle \cdot \mathbf{S_{-q}}) -$$

$$\gamma B_0 / 2 \, (\, S^z(\mathbf{q_0}) + S^z(-\mathbf{q_0}) \,) \,] \tag{3.70}$$

If we assume that only the wave vectors $\pm \mathbf{q_0}$ are important in the response to the modulated field, then we find an effective molecular field

$$\mathbf{B}_{eff}(\mathbf{q_0}) = 1/2\mathbf{B} + \frac{J(-\mathbf{q_0})}{N}\langle S(-\mathbf{q_0})\rangle/\gamma \tag{3.71}$$

The linearization procedure leads exactly to the equations of the MFA, so it is equivalent to the variational approximation we developed at the beginning of this chapter for the Helmholtz free energy. Let us remind ourselves that the response function of a non-interacting spin system to a uniform static (that is, $\omega = 0$, $\mathbf{q} = 0$) external field is Curie's susceptibility

$$\chi_0 = C/T \tag{3.72}$$

This is the response of independent spins, so if we apply a local external field at at lattice site $\mathbf{R_0}$, that is if $\mathbf{B(R)} = \mathbf{B}\delta^3(\mathbf{R} - \mathbf{R_0})$ we must obtain a magnetization which is also local, of the form $\mathbf{M(R)} = \mathbf{m_0}\delta^3(\mathbf{R} - \mathbf{R_0})$. This implies that the site-dependent non-interacting response is

$$\chi_0(\mathbf{R}) = (C/T) \, \delta^3(\mathbf{R} - \mathbf{R_0}) \tag{3.73}$$

whose FT is a constant:

$$\chi_0(\mathbf{q}) = C/T \tag{3.74}$$

The average magnetization is

$$\mathbf{M(r)} = \frac{\gamma}{V} \sum_i \delta^3(\mathbf{r} - \mathbf{R_i})\langle \mathbf{S}_i \rangle \tag{3.75}$$

Therefore the FT of the magnetization for independent spins is aligned in the direction z of the applied field and its magnitude is

$$M^z(\mathbf{q}) = \frac{\gamma}{V}\langle S^z(\mathbf{q}) \rangle = \chi_0(\mathbf{q})B^z(\mathbf{q}) \tag{3.76}$$

We notice that the non vanishing Fourier components of the response (magnetization) are only those present in the applied field. Returning now to the case of interacting spins, we substitute the effective field for the external one in (3.76), obtaining

$$\langle S^z(\mathbf{q_0}) \rangle = \frac{V\chi_0 B_0/(2\gamma)}{1 - J(-\mathbf{q_0})\chi_0 V/(N\gamma^2)} \tag{3.77}$$

and consequently

$$\chi(\mathbf{q_0}) = \frac{\chi_0}{1 - J(-\mathbf{q_0})\chi_0 V/(N\gamma^2)} \tag{3.78}$$

or, substituting χ_0 by the expression (3.74),

$$\chi(\mathbf{q_0}) = \frac{C}{T - J(-\mathbf{q_0})CV/(N\gamma^2)} \tag{3.79}$$

We consider the case of positive $J(\mathbf{R})$, and a fixed wave vector $\mathbf{q_0}$. Suppose that we start from a very large T, where we regain from (3.78) Curie's law, and start decreasing T. There will be a value of T for which the denominator in (3.78) vanishes. If the lattice has inversion symmetry, $J(\mathbf{q}) = \mathbf{J(-q)}$. Then the maximum T for which (3.78) diverges will be proportional to the maximum value of $J(\mathbf{q})$, so that the critical temperature for the transition from the paramagnetic to the ordered phase is

$$T_c = C(V/N\gamma^2) \max\{J(\mathbf{q})\} \tag{3.80}$$

Exercise 3.10
Show that if $\max\{J(\mathbf{q})\} = J(0)$ *we return to the previous results for a FM.*

Consider now $J(\mathbf{R}) < 0$, i.e. AFM interactions. In the case that we limit the range of J to first nearest neighbours, we have

$$J(\mathbf{q_0}) = -\mid J \mid \sum_{\{d\}} e^{i\mathbf{d}\cdot\mathbf{q_0}} \tag{3.81}$$

where $\{\mathbf{d}\}$ is the set of all translations from a given site to the nearest neighbours. If

$$\mathbf{d} \cdot \mathbf{q_0} = \pm \pi \quad \forall \mathbf{d} \tag{3.82}$$

then

$$\max \{J(\mathbf{q})\} = J(\mathbf{q_0}) \tag{3.83}$$

and we find

$$\chi(\mathbf{q_0}) = \frac{C}{T - \mid J \mid \nu CV/(N\gamma^2)} \tag{3.84}$$

for the AFM, so that the Curie and the Néel temperature have the same expression

$$T_c = T_N = CV \mid J \mid \nu/(N\gamma^2)$$

in both cases . We see that for the AFM the longitudinal susceptibility diverges at T_N for $\mathbf{q} = \mathbf{q_0}$, with $\mathbf{q_0}$ defined by (3.82), while it remains finite for $\mathbf{q} = \mathbf{0}$. If we depart from the case in which the range of the exchange interaction is limited to the star of first n. n. we find a great variety of phases. For instance, consider a s. c. lattice in which each spin has interactions J_1 with its 6 first n. n. and $-J_2$ with the 12 second n. n., with $J_{1,2} > 0$. Then

$$\begin{aligned} J(\mathbf{q_0}) &= 2J_1 \left(\cos q_{0x}a + \cos q_{0y}a + \cos q_{0z}a \right) \\ &- 4J_2[(\cos q_{0x}a)(\cos q_{0y}a) + (\cos q_{0y}a)(\cos q_{0z}a) \\ &+ (\cos q_{0z}a \cos q_{0x}a)] \end{aligned} \tag{3.85}$$

Exercise 3.11
Show that the extrema of (3.85) satisfy

$$a\mathbf{q_0} = \left(\cos^{-1} \left(J_1/4J_2 \right) \right) (1, 1, 1) \tag{3.86}$$

There are real solutions of Eq. (3.86) if $J_1 < 4J_2$. If this is so, in general the paramagnetic phase will be unstable against the spontaneous generation of a non-commensurate magnetization wave, with wave vector as defined by (3.86), at the critical temperature determined by (3.80). The direction of the magnetization is fixed in this phase. The amplitude of the longitudinal component varies periodically as we move in the crystal in the direction of $\mathbf{q_o}$, so that this configuration is called a *longitudinal (spin) wave* (LW).

Exercise 3.12
Show that the transition temperature from the paramagnetic to the LW phase is

$$T_c = \frac{S(S+1)J_1^2}{4k_B J_2} \tag{3.87}$$

and that the uniform susceptibility for this phase is

$$\chi(0)^{LW} = \frac{C}{T - 8J_2(J_1 - 2J_2)T_c/J_1^2} \tag{3.88}$$

The coefficient of T_c in the denominator of (3.88) is always < 1, and this guarantees that the uniform susceptibility is finite at T_c. The high T behaviour is FM like if $J_1 > 2J_2$ and AFM like in the opposite case. In the family of the rare earths, Tm and Er display this kind of magnetic order within definite temperature intervals [13].

Static transverse standing waves can also exist. Both transverse and longitudinal waves are known as examples of *Helimagnetism*.

3.6 MFA for antiferromagnetism

As we mentioned before, the AFM instability occurs if Eq. (3.82) is satisfied. Then the average z components nearest neighbour spins tend to have opposite sign. Those lattices that can be subdivided into two interpenetrating sublattices, in such a way that all first n. n. of a site on a given sublattice belong to the other one, are called *bipartite*. One expects such a lattice with $J < 0$ to display AFM order. Going back to Eq. (3.2) we choose as H_0 a sum of mean field Hamiltonians for both sublattices A and B:

$$H_0 = -\gamma \sum_{i_A} B_{i_A}^{mol} S_{i_A}^z - \gamma \sum_{i_B} B_{i_B}^{mol} S_{i_B}^z \tag{3.89}$$

where $\{i_\alpha\}$ are sites on sublattice α, while H is the Heisenberg Hamiltonian of Eq. (3.5) with $J < 0$. We shall assume that the external field B and the sublattice magnetization are parallel to z. In general, there will be some anisotropy present that will favour the orientation of the magnetization along special crystallographic axes, and we assume that z is along one of these axes.

The partition function Z_0 for the non-interacting H_0 can be obtained immediately:

Exercise 3.13
Show that

$$Z_0 = Z_0(A) Z_0(B) \tag{3.90}$$

where $Z_0(\alpha)$ for $\alpha = (A, B)$ can be explicitly written for each sublattice as in Sect. 3.1.

For $S = 1/2$ we already obtained:

$$Z_0(\beta, B) = 2 \cosh\left(\frac{1}{2}\beta\gamma B\right) \tag{3.91}$$

Then we write

$$Z_0^{A,B} = 2\cosh\left(\frac{1}{2}\beta\gamma B_{A,B}^{mol}\right) \tag{3.92}$$

If N is the total number of sites in the whole lattice, we calculate the paramagnetic free energy for each sublattice as

$$f_0^{A,B} = \frac{F_0^{A,B}}{N/2} = -\frac{1}{\beta}(2/N)\log Z_0^{A,B} \tag{3.93}$$

and of course $f_0 = f_0^A + f_0^B$. It is convenient to add to the Hamiltonian a *staggered field* B_a which can be the result of a single site anisotropy term in the energy or can be considered as an artificially constructed external field, with the property that $B_a(i_A)\sigma_A > 0$ and $B_a(i_B)\sigma_B > 0$ where

$$\sigma_{A,B} = \langle S_{A,B}^z \rangle$$

Eventually, we shall add a Zeeman term to include a uniform external field B. We now calculate the variational function ϕ as defined in section 1:

$$\phi = F_0 + \langle H - H_0 \rangle_0 \tag{3.94}$$

To this end we calculate the averages of H and H_0 in the non-interacting ensemble:

$$\frac{\langle H \rangle_0}{N/2} = |J|\,\nu\sigma_A\sigma_B - \gamma\,|B_a|\,(|\sigma_A| + |\sigma_B|) - \gamma B(\sigma_A + \sigma_B) \tag{3.95}$$

and

$$\frac{\langle H_0 \rangle_0}{N/2} = \gamma B_A\sigma_A - \gamma B_B\sigma_B - \gamma\,|B_a|\,(|\sigma_A| + |\sigma_B|) - \gamma B(\sigma_A + \sigma_B) \tag{3.96}$$

Upon equating to zero the partial derivatives of ϕ with respect to $\{B_\alpha\}$ and $\{\sigma_\alpha\}$ we find

$$\sigma_{A,B} = -(1/\gamma)\frac{\partial f_0^{A,B}}{\partial B_{A,B}} \tag{3.97}$$

and

$$B_{A,B} = -|J|\,\nu\sigma_{B,A} \tag{3.98}$$

We have finally:

$$\sigma_{A,B} = S B_S(S\gamma\beta B_{A,B}) \tag{3.99}$$

For $S = 1/2$ we have

$$\sigma_{A,B} = (1/2)tanh(\beta\gamma B_{A,B}/2) \tag{3.100}$$

Let us now find the critical temperature for $S = 1/2$. As $T \nearrow T_N$, $\sigma_{A,B} \searrow 0$ and we expand $tanh$ for small argument. Let us call $a = \beta\gamma \mid J \mid \nu$. The pair of equations (3.100) reduces to the linear system:

$$\left. \begin{array}{l} 2\sigma_A + a\sigma_B = 0 \\ a\sigma_A + 2\sigma_B = 0 \end{array} \right\} \tag{3.101}$$

The eigenvalues of the secular 2×2 determinant of system (3.101) are $\pm a$, and the corresponding eigenvectors are

$$(\sigma_A/\sigma_B)^{\pm} = \mp 1$$

where the choice $(+)$ yields an unphysical ferromagnetic order solution, with negative critical temperature, while the $(-)$ solution corresponds to the AFM order with antiparallel sublattices and with Néel temperature

$$T_N = \frac{\nu \mid J \mid}{2k_B} \tag{3.102}$$

which coincides with our previous expression for the ferromagnetic T_c if one makes the substitution $J \rightarrow \mid J \mid$.

3.6.1   Longitudinal susceptibility

Equation (3.98) contains implicitly the generalization of Weiss molecular field for anti-ferromagnets. To make this more explicit we introduce as many parameters λ_α as sublattices. The bipartite AFM with only two Weiss molecular field parameters is the next simplest case after the single domain FM:

$$B_\alpha = -\lambda_{\alpha\beta}M_\beta \tag{3.103}$$

where $\alpha, \beta = A, B$. We have

$$\lambda_{AB} = 2\frac{\mid J \mid \nu_B}{\gamma^2 \rho_A} \tag{3.104}$$

where $\rho_A = \rho_B = N/(2V)$ volume concentration of each sublattice spins and $\nu_{A,B}$ = number of $B(A)$ spins surrounding an AB spin. The total field on sites $\alpha = (A, B)$ is

$$B_\alpha = B - \lambda_{\alpha\beta}M_\beta \tag{3.105}$$

Let us call

$$b_\alpha = \frac{\gamma B_\alpha}{k_B T} \tag{3.106}$$

and

$$b_0 = \frac{\gamma B}{k_B T} \tag{3.107}$$

Then,

$$\sigma_\alpha \equiv \langle S_{i_\alpha}^z \rangle = SB_S(S\,(b_\alpha + b_0)) \tag{3.108}$$

and the sublattice magnetization can be written as

$$M_\alpha = \gamma \rho \sigma_\alpha \tag{3.109}$$

Let us consider first the paramagnetic phase, so that $M_\alpha = \chi_0^{(\alpha)} B$. Then

$$b_\alpha = b_0 - \lambda_{\alpha\beta} \chi_0^\beta b_0 \tag{3.110}$$

We assume both sublattices are atomically identical, so that we have $\lambda_{\alpha\beta} = \lambda_{\beta\alpha} = \lambda$ and $\rho_\alpha = \rho$.

Let us now calculate the susceptibility as the response of the total magnetization $M = M_A + M_B$ to the external field in the zero field limit, that is

$$\chi = \left(\frac{\partial M}{\partial B} \right)_{B=0} \tag{3.111}$$

Now we expand the Brillouin function in (3.108) for small B, and after some simple algebra we obtain:

$$\chi_\| = \frac{\chi_0}{1 + \chi_0 \lambda} = \frac{\gamma^2 S(S+1)\rho}{3k_B} \frac{1}{T + \theta} \tag{3.112}$$

where

$$\theta = \frac{\gamma^2 S(S+1)\rho\lambda}{3k_B} \tag{3.113}$$

and we substituted in Eq. (3.112) the expression for the paramagnetic susceptibility

$$\chi_0 = \frac{\gamma^2 S(S+1)\rho}{3k_B T} \tag{3.114}$$

We verify that for $S = 1/2$, $\theta = | J | \nu/(2k_B) = T_N$.

The curve of $\chi^{-1}(T)$ has the negative intersect $-\theta$ with the T axis. In the special case of only first n. n. interactions θ coincides with T_N, but in general this is not necessarily the case: if the range of the interaction is greater than first n. n., the ordered phases are more complicated and one may have to deal with more than two sublattices. Depending on the ratios J_n/J_1 one can obtain $\theta/T_N > 1$, as found experimentally for many systems.

We remark that if only $J_1 = - | J | \neq 0$, in which case $\theta = T_N = \lambda C/2$, we obtain

$$\chi(T)|_{T \searrow T_N} = 1/\lambda \tag{3.115}$$

Now let us turn to the ordered Néel phase at low temperatures with an external field B_0 applied along the magnetization of the A sublattice. Then for each sublattice

$$M^{A,B} = M_0^{A,B} + \chi_\| B_0/2 \tag{3.116}$$

where $M_0^{A,B}$ is the spontaneous magnetization of each sublattice for $B_0 = 0$.

The response in (3.116) is proportional to $\chi_\|/2$, since there are only $N/2$ spins in each sublattice and they occupy the total volume of the system. Call

$x_0 = S\gamma\beta\lambda M_0$ the argument of the Brillouin function in the absence of the external field. M_0 is the spontaneous sublattice magnetization. One obtains

$$\chi_{\parallel} = \frac{\rho^{at}\gamma^2\beta S^2 B'_S(x_0)}{1 + \lambda\rho^{at}\gamma^2\beta S^2 B'_S(x_0)} \qquad (3.117)$$

Eq. (3.117) is valid at all temperatures, as one verifies by substituting $x_0 = 0$ and obtaining the susceptibility in the paramagnetic phase. We have used $2\rho_{A,B} = \rho^{at}$, the atomic volume concentration of the magnetic ions in the lattice.

As $T \to 0$, $\chi_{\parallel}$ tends exponentially to 0. At $T \to T_N$, it has a finite limit, but the derivative $\frac{\partial\chi_{\parallel}}{\partial T}$ is finitely discontinuous, due to the discontinuity of the temperature derivative of the spontaneous sublattice magnetization M_0:

Exercise 3.14
Show that the derivative of $\chi_{\parallel}$ with respect to temperature is discontinuous at $T = T_N$.

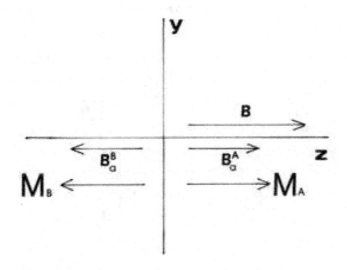

Figure 3.6: *Configuration of magnetization and anisotropy field of both sublattices for an applied field B $\parallel$ (z , $\mathbf{M_{A,B}}$) in an AFM.*

3.6.2   Transverse susceptibility

Let us find the average energy of the AFM in the presence of an external magnetic field perpendicular to the spin quantization axis z, which we assume coincides with the high symmetry axis of the system. Let us include in the model Hamiltonian an anisotropy energy term of the form

$$V_a = -(1/2)D\sum_{i,\alpha}(S_{i_\alpha}^z)^2 \tag{3.118}$$

so that now the total Hamiltonian $H + V_a$ has uniaxial symmetry. In Fig. 3.7 we show schematically the situation considered. In equilibrium, we expect both sublattices to orient symmetrically with respect to the y axis, at an angle ϕ with the symmetry axis. We neglect in first order the possible change of the magnitude of the sublattice magnetization. We have

$$\frac{\langle H_0\rangle_0}{V} = (N/2V)2Jv\vec{\sigma}_A \cdot \vec{\sigma}_B - \gamma\vec{B}\cdot(\vec{\sigma}_A + \vec{\sigma}_B)$$
$$-1/2D[(\sigma_A^z)^2 + (\sigma_B^z)^2] \tag{3.119}$$

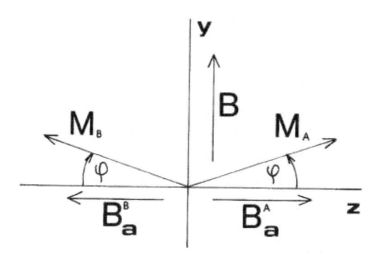

Figure 3.7: *Configuration for transverse field on antiferromagnet.*

We are assuming that to first order the average component of spin along the equilibrium orientation of the magnetization is the same, for both sublattices, as it was before the application of the field. Therefore,

$$\vec{\sigma}_{A(B)} = \sigma_0(0, \sin\phi, +(-)\cos\phi) \tag{3.120}$$

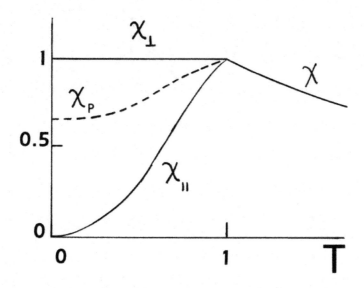

Figure 3.8: *Longitudinal* (∥) *and transverse* (⊥) *static susceptibility of antiferromagnet for zero field, in MFA. T in units of T_N and $\chi(T)$ in units of $\chi(T_N)$. The dotted curve is the expected average susceptibility χ_P for a powder, in which all orientations are equally probable. For $T > T_N$, χ is isotropic.*

where σ_0 is the sublattice magnetization without field as defined in Eq. (3.108) and

$$\frac{\langle H_0 \rangle_0}{N/2} \equiv u(\phi) = -2Jv\sigma_0^2 \cos(2\phi)$$

$$2 - \gamma B_0 \sigma_0 \sin\phi - D\sigma_0^2 \cos^2(\phi) \tag{3.121}$$

The minimum of $u(\phi)$ occurs for

$$\sigma_0 \sin(\phi) = \frac{\gamma B}{4Jv + D} \tag{3.122}$$

which yields for the transverse susceptibility

$$\chi_\perp = \frac{1}{\lambda + D/(\rho\gamma^2)} \tag{3.123}$$

Notice that the temperature dependent σ_0 was cancelled out of the expresion above, so that the transverse susceptibility turns out to be independent on T. The results of the MFA for the longitudinal and the transverse AFM susceptibil-

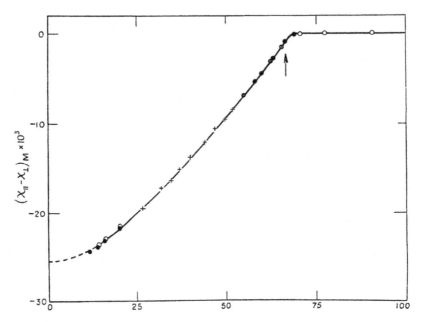

Figure 3.9: $(\chi_\perp - \chi_\parallel)$ per 10^{-3} mol for MnF_2 vs. T in K. The arrow indicates the temperature of the maximum in the specific heat [14].

ities are summarized in Fig. 3.8. Experimental values of the difference $(\chi_\perp - \chi_\parallel)$ for MnF_2 are shown in Fig. 3.9.

We shall return to the calculation of $\chi_\perp$ with Green's functions methods in Chap. 5.

3.6.3   Spin-flop and other transitions

When an external field is applied along the easy axis of an antiferromagnet, the Néel phase can become unstable in relation to the configuration in which the sublattice magnetization aligns approximately perpendicular to the field, which is why the new phase is called *flopped*. To verify this let us calculate the MFA energy in the spin-flopped configuration, represented in Fig. 3.10. Now the field B is along the z axis , and we assume that both sublattices align symmetrically almost normal to this axis. Then we have (SF stands for spin-flop):

$$\frac{u(\phi)^{SF}}{N/2} = -2J\nu\sigma_0^2 - 2\gamma B\sigma_0 \sin\phi - D\sigma_0^2 \sin^2\phi \qquad (3.124)$$

The minimum of u^{SF} occurs for

$$\sigma_0 \sin\phi = \frac{\gamma B}{4J\nu - D} \qquad (3.125)$$

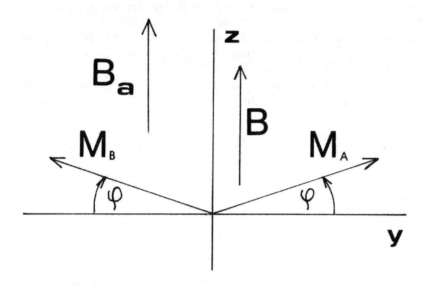

Figure 3.10: *Spin-flop configuration of antiferromagnet under applied field in an easy direction.*

(compare with Eq. (3.122)). Let us repeat the calculation of the minimum energy for the AF phase with the applied field parallel to the anisotropy axis and to the magnetization, this time including the anisotrpy energy. We call $\delta\sigma$ the induced change of the sublattice order parameter due to the field, and we assume that, if the field is applied in the direction of the A sublatice magnetization, the new equilibrium values are $\sigma_{A(B)} = \sigma_0 + (-)\,\delta\sigma$. Then we obtain

$$
\frac{u^{AF}}{N/2} = -2J\nu[\,\sigma_0^2 - (\delta\sigma)^2\,]
$$

$$
-\frac{D}{2}[\,(\sigma_0 + \delta\sigma)^2 + (\sigma_0 - \delta\sigma)^2)\,] - 2\gamma B\delta\sigma \qquad (3.126)
$$

Now the minimum u^{AF} occurs for

$$
(2J\nu - D)\delta\sigma = 2\gamma B \qquad (3.127)
$$

and we obtain for the energy

$$
u^{AF} = -2J\nu\sigma_0^2 - D\sigma_0^2 - (1/2)\frac{\chi_\| B^2}{\rho/2} \qquad (3.128)
$$

We remark here that since we assume that B and $\delta\sigma$ are both positive, Eq. (3.127) shows that there is a maximum anisotropy compatible with the AF phase in this configuration, namely

$$
D = 2J\nu \qquad (3.129)
$$

For larger anisotropies the combined anisotropy and Zeeman energies overcome the exchange energy, and the system aligns with the field, just as in the paramagnetic (P) phase, without undergoing the transition to the flop phase. A material with such a large anisotropy is called a *metamagnet*. Paradigms of metamagnets are $FeCl_2$ and $DyPO_4$.

Let us now study the boundary curves in the (B, T) plane between different phases. The difference of the minimum energies of the AF and SF phases is

$$u^{AF} - u^{SF} = -D\sigma^2 + (1/2)\frac{(\chi_\perp - \chi_\parallel)B^2}{\rho/2} \tag{3.130}$$

Since $\chi_\perp - \chi_\parallel \geq 0$, $\exists B_1$ such that for $B > B_1$ the spin-flopped configuration is more stable, and this field is

$$B_1 = \left(\frac{D\sigma_0^2}{1/2(\chi_\perp - \chi_\parallel)}\right)^{1/2} \tag{3.131}$$

Of course the transition only exists for $T < T_N$, because the AF phase disappears at that temperature. We see that the critical field B_1 for the $AF \rightarrow SF$ transition depends on the equilibrium magnetization in the absence of field, which in turn depends upon T. Clearly in the absence of anisotropy the critical field is zero at any $T < T_N$, so that even an infinitesimal applied field along the spin quantization axis z will produce the spin-flop transition.

In summary, we found that the mean-field energy of the AF and SF phases coincide when the applied field has the value B_1 given by Eq. (3.131). If $B > B_1$ the system will stay in the SF phase. As B increases above B_1 we see from (3.125) that there is a second critical field B_2 such that

$$1 = \sin\phi = \frac{\gamma B_2}{\sigma_0(4J\nu - D)} \tag{3.132}$$

For $B > B_2$ both sublattices are aligned parallel to the field, so that this phase is also called *paramagnetic* (P): upon increase of T it goes over continuously to the disordered phase under a field. The anisotropy energies are in most cases smaller than the exchange energies, so in general $B_2 > B_1$ at low temperatures. We must still determine the boundary curve $AF \leftrightarrow P$ for $T < T_N$, which requires to find the value B_3 of the applied field such that the energies of the AF and the P phases are equal. In order to obtain the energy for the P phase we must take into account the change $\delta\sigma$ induced by the applied field, which as before can be obtained by minimizing the corresponding energy with respect to $\delta\sigma$.

Exercise 3.16
Show that the boundary curve $AF \leftrightarrow P$ is

$$(2J\nu + D)\sigma_0 - \gamma B_3 = 0 \tag{3.133}$$

Exercise 3.17
Show that if $B = B_1 = B_2$, then the same value of B satisfies (3.133) as well.

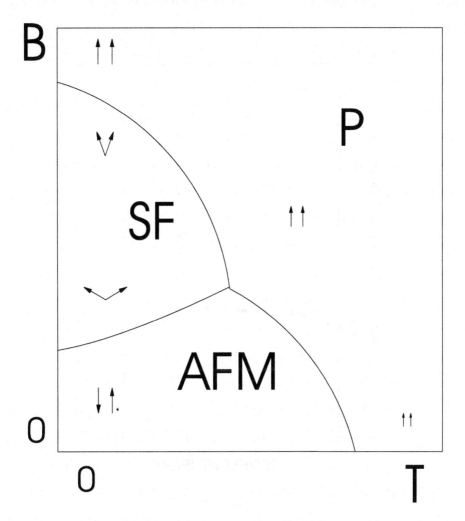

Figure 3.11: *Phase diagram of antiferromagnet in the (B,T) plane within the MFA.*

The conditions $B_1 = B_2 = B_3$ determine the coordinates of the triple point, at which the three boundary curves meet. These phase boundaries can be calculated in the spin wave approximation [15] and with Green's function methods [16, 17] (See Chap. 5). A very detailed classical reference on the application of MFA to the phase transitions in antiferromagnets is the review by Nagamiya et al. [18]. Keffer's [19] monograph on spin waves contains a general review of the subject.

Phase boundaries obtained in the mean field approximation agree qualitatively with experimental data on typical AF compounds [20, 21]. We show in Fig. 3.12 the experimental results for the phase boundaries of $NiCl_2.4H_2O$ [21].

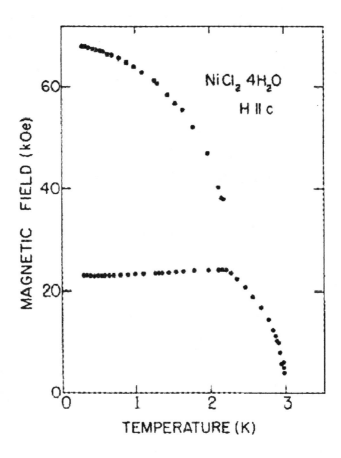

Figure 3.12: *Phase diagram of $NiCl_2.4H_2O$ antiferromagnet (from [21]).*

3.7   Helimagnetism

Let us consider now the Heisenberg Hamiltonian on a $3d$ lattice *with first and second* nearest neighbour exchange integrals, and let us assume that the intra-plane interactions are ferromagnetic on all lattice planes belonging to the family perpendicular to a given direction, say the x axis, so that spins on each plane align parallel at low T. We assume that interactions among different planes could be either positive or negative, so that we expect to find the magnetization of different planes not necessarily parallel to each other. It is convenient to Fourier analyze the local spin operators. This allows to take advantage of translation invariance on the planes. We assume that each plane is ferromagnetically ordered with uniform magnetization. Let us specify a lattice site by the two component position vector $\mathbf{n} = (n_1, n_2)$ on its plane and the index m of the plane:

$$H = -\sum_{m,m'} \sum_{\mathbf{n},\mathbf{n}'} J^{m,m'}(\mathbf{n} - \mathbf{n}') \mathbf{S}_{\mathbf{n},m} \mathbf{S}_{\mathbf{n}',m'} \qquad (3.134)$$

Here we must include $m = m'$ in the sum. We assume that the successive planes along the x direction are identical $2d$ lattices, although not necessarily in registered positions with each other. Now we Fourier transform the local spin operators:

$$\mathbf{S}_{\mathbf{n},m} = \frac{1}{\sqrt{N_s}} \sum_{\mathbf{k}} e^{-i\mathbf{k}\cdot\mathbf{n}} \mathbf{S}_{\mathbf{k},m} \qquad (3.135)$$

where $\mathbf{k}$ is the $2d$ wave vector along the plane and $N_s = $ the number of sites on each plane. The Hamiltonian in terms of the Fourier transformed operators is now

$$H = -\sum_{m,m'} \sum_{\mathbf{k}} J^{m,m'}(\mathbf{k}) \; \mathbf{S}_{\mathbf{k},m} \cdot \mathbf{S}_{-\mathbf{k},m'} \qquad (3.136)$$

where

$$J_{m,m'}(\mathbf{k}) = \sum_{\mathbf{n}} J^{m,m'}(\mathbf{n}) \; e^{i\mathbf{k}\cdot\mathbf{n}}$$

Exercise 3.18
Prove that the Fourier transformed operators satisfy the commutation relations

$$\left[S^\alpha_{\mathbf{k},m}, S^\beta_{\mathbf{k}',m'} \right] = \frac{1}{\sqrt{N_s}} \, i \, \epsilon_{\alpha\beta\gamma} \, \delta_{m,m'} \, S^\gamma_{\mathbf{k}+\mathbf{k}',m} \qquad (3.137)$$

Since we have assumed that in the ground state of the system each plane is uniformly magnetized, we can limit ourselves to the $\mathbf{k} = \mathbf{0}$ term in (3.136), and take

$$\mathbf{S}_{\mathbf{k},m} = \delta_{\mathbf{k},0} \mathbf{S}_{0,m} \qquad (3.138)$$

We consider the case where spins are aligned parallel to the planes.
      Our effective Hamiltonian is

$$H_{eff} = -\sum_{m,m'} J_{m,m'}(\mathbf{0}) \; \mathbf{S}_{\mathbf{k}=0,m} \cdot \mathbf{S}_{\mathbf{k}=0,m'} \qquad (3.139)$$

where

$$J_{m,m'}(\mathbf{0}) = \sum_{\mathbf{h}} J^{m,m'}(\mathbf{h})$$

We drop the 0 wave vector index in (3.139) from the FT of the exchange integrals to simplify the notation, and rewrite the effective Hamiltonian as

$$H_{eff} = -2J_1 \sum_m \mathbf{S}_m \cdot \mathbf{S}_{m+1} - 2J_2 \sum_m \mathbf{S}_m \cdot \mathbf{S}_{m+2} \qquad (3.140)$$

which can be interpreted as describing a Heisenberg chain. Let us first look at the semi-classical limit $S \to \infty$. The equilibrium magnetization of each plane is aligned along a given direction, which we take as the quantization axis for spins on that plane, but since we allow the inter-plane exchange interactions to be either FM or AFM the directions of the magnetization of different planes might be not parallel, and we should rotate the spin operators in each plane accordingly so as to make the local quantization axis to coincide with the equilibrium magnetization of the plane. Since all planes are parallel, we can define this direction for each plane m by specifing θ_m, the angle the magnetization at that plane makes with a given fixed z axis. Then the semi-classical energy corresponding to (3.140) is

$$\frac{E}{S(S+1)} = -J_1 \sum_m \cos\left(\theta_m - \theta_{m+1}\right) -$$

$$J_2 \sum_m \cos\left(\theta_m - \theta_{m+2}\right) \qquad (3.141)$$

In an infinite chain all sites are equivalent, so

$$\theta_{m+1} - \theta_m \equiv \theta$$

is m independent, and

$$\theta_{m+2} - \theta_m = 2\theta$$

In Fig. 3.13 we show the assumed spin configuration, where θ is the helix turn angle. If M is the total number of parallel planes (eventually $M \to \infty$) we can now rewrite the semi-classical energy per plane as

$$\frac{E}{S(S+1)M} = -2J_1 \cos\theta - 2J_2 \cos 2\theta \qquad (3.142)$$

the factor 2 on the r.h.s. coming from the neighbours to the left and right of each plane. Border effects are neglected, and they should be negligible as $M \to \infty$, at least as regards the bulk equilibrium phases. One finds the following extrema solutions:

1. $\theta = 0$ (ferromagnetic order);

2. $\theta = \pi$ (antiferromagnetic order);

3. $J_1 + 2J_2 \cos\theta = 0$ which leads to the helimagnetic ordering.

We call the first two solutions *collinear*.

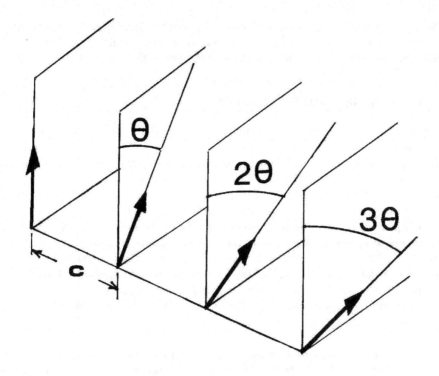

Figure 3.13: *Spin configuration of helimagnet. The lattice constant in the direction of k is denoted c, and the helix turn angle is θ.*

Exercise 3.19
Prove that

$$\Delta \equiv \frac{E^{coll} - E^{heli}}{8J_2} =$$

$$-\frac{S(S+1)M}{4J_2} \left(8J_2 \mid J_1 \mid +16J_2^2 + J_1^2\right) \qquad (3.143)$$

If $J_2 < 0$, the helical phase is stable except at the particular case

$$\mid J_2 \mid = (1/4) \mid J_1 \mid .$$

If $J_2 > 0$, the collinear phase is stable except at the same special case above, where both types of phases have the same energy. This conclusions do not depend on the sign of J_1, which favours either the F or the AF collinear phase. Helical magnetic (HM) structures were predicted independently by Yoshimori [13], Villain [22] and Kaplan [23], and they were found afterwards in several materials. One can mention the ionic compound MnO_2 [13] and the alloy $MnAu_2$ [24] as examples. Also $FeCl_3$ shows a complicated HM structure

[25]. Some heavy rare earths show spiral configurations in some temperature ranges, vith a very rich variety of structures and phase transitions. We refer the interested reader to Smart's book [12] for further detail.

3.8   Goldstone's theorem

We have found within the mean field approximation a low T solution of the variational equations for the free energy F of a Heisenberg FM which consists of two pure states, with spontaneous magnetization $\pm m_s$. We observe that the symmetry group of the Hamiltonian is the complete rotation group in $3d$, a continuous orthogonal group with three real parameters (the Euler angles). In the Weiss ground state, one particular direction is space has been selected for the magnetization, so this is a ground state with broken symmetry. Since all directions in space are equivalent, there is an infinite manifold of states with the same free energy which are obtained simply by a rigid rotation of all spins, and we find soft modes associated with the transverse fluctuations of long wave-lenght. The reason for this is that there is no energy change associated with a global rotation of spins over large distances. This is, for this case, the contents of Goldstone's theorem [26], which states that:

If there is a manifold of spontaneously-broken-symmetry degenerate ground-states mutually related by transformations which belong to a continuous group, there are excitations which in the long-wave limit have zero energy.

We check this property in the case of a FM in the ordered state, magnetized along the z axis, where the transverse susceptibility χ_{xx} is (see Chap. 5):

$$\chi_{xx}(\omega, k) = \frac{\gamma^2 \rho}{2\pi} \frac{E_k}{(\omega + i\epsilon)^2 - E_k^2} \tag{3.144}$$

and $E_k = 2\nu J(1 - \gamma_k) + \gamma B$. We see that for vanishing ω and B, $\chi_{xx} \propto k^{-2}$. On the other hand, we remind that the longitudinal susceptibility χ_{zz} of a FM in the MFA (Sec. 3.) is instead finite in the $k \to 0$ limit for $0 < T < T_c$.

References

1. Feynman, R. P. (1972) *Statistical Mechanics*, W. A. Benjamin, Inc.

2. Sommerfeld, A. (1956) *Thermodynamics and Statistical Mechanics*, Lectures on Theoretical Physics, Vol.V, Academic Press.

3. Stanley, Eugene H. (1971) *Introduction to Phase Transitions and Critical Phenomena*, Oxford University Press.

4. Reichl, L. E. (1980) *A Modern Course in Statistical Physics*, University of Texas Press, Austin, Texas, U.S.A.

5. Patashinskii, A. Z. and Pokrovskii, V. L. (1980) *Fluctuation Theory of Phase Transitions*, Pergamon Press, International Series in Natural Philosophy, vol. 98.

6. Lidiard, A. B. (1954) *Rep. Prog. Phys.* **17**, 201.

7. Landau, L. D. and Lifshitz, E .M. (1059) *Statistical Physics*, Pergamon Press, London-Paris, p. 434.

8. Garton, G., Leask, M. J., Wolf, W. P. and Wyatt, A. F. G. (1963) *J. Appl. Phys.* **34**, 1083.

9. Ferer, M., Moore, M. A. and Wortis, M. (1971) *Phys. Rev.***B4**, 3954.

10. Van Vleck, J. H. (1937) *Phys. Rev.***52**, 1178.

11. Oguchi, T. (1955) *Prog. Theoret. Phys. (Kyoto)***13**, 148.

12. Smart, J. Samuel (1966) *Effective Field Theories of Magnetism* ,W.B.Saunders Company, Philadelphia and London.

13. Yoshimori, A. (1959) *J.Phys. Soc.(Japan)***14**, 807.

14. Griffel, M. and Stout, J. W. (1950) *J. Am. Chem. Soc.* **72**, 4351.

15. Feder, J. and Pytte, E. (1968) *Phys. Rev.* **168**, 640.

16. Anderson, F. B. and Callen, H. B. (1964) *Phys. Rev.* **136** , A1068.

17. Arruda, A. S. de, et al., (1996) *Sol. State. Comm.* **97**, 329.

18. Nagamiya T., Yosida K. and Kubo, R. (1955) *Adv. Phys.* **4**, 1.

19. Keffer, F. (1966) *Handb. der Phys.* **2 Aufl., Bd. XVIII** Sect. 49, Springer -Verlag.

20. J.N.McElearney, J.N. et al. ,(1973) *Phys. Rev.* **B7**, 3314.

21. Paduan Filho, A. (1974) *Phys.Lett.* **50A**, 51.

22. Villain, J. (1959), *J. Phys. Chem.Solids* **11**(1961), 303.

23. Kaplan, T.A. (1960), *Phys. Rev.* **116**, 888.

24. Herpin, A. *et al.* (1958), *Compt. rend.* **246**, 3170 ; *id.* (1959), *Compt. rend.* **249**, 1334.

25. Cable, J.W. *et al.* (1962), *Phys. Rev.* **127**, 714.

26. Goldstone, J. (1961) *Nuovo Cimento* **19**, 154.

Chapter 4

Spin Waves

4.1 Introduction

We have just seen that the Heisenberg Hamiltonian

$$H = -\sum_{ij} J_{ij} \mathbf{S}_i \mathbf{S}_j - \gamma B \sum_{i=1}^{N} S_i^Z \tag{4.1}$$

is a reasonable model for the description of the magnetic properties of systems which contain isolated spins. The derivation of (4.1) was based on the Heitler-London approximation applied to a pair of atoms. We can extend the application of that Hamiltonian to multi-electron ions, if we assume that the effective exchange interactions between any pair of electrons belonging one to the un-filled shells of ion i, the other to the corresponding shell of j, are the same for all the possible pairs. Clearly in this case one can add independently the spins of each ion, and obtain the Hamiltonian (4.1), where now the spins are the total spins of the ions, as obtained by applying Hund's rule to all unpaired spins in the ground state of each ion.

The dimensionless spin operators in (4.1) obey the commutation relations:

$$[\, S_i^\alpha, S_j^\beta \,] = i\epsilon_{\alpha\beta\gamma}\, \delta_{ij} S_i^\gamma \tag{4.2}$$

where the indices refer to the cartesian coordinates and $\epsilon_{\alpha\beta\gamma}$ is the Levi-Civita tensor (totally antisymmetric in the three indices). (4.2) can also be written as

$$\mathbf{S} \wedge \mathbf{S} = i\mathbf{S} \tag{4.3}$$

The circular components of $\mathbf{S}$, defined as $S^\pm = S^x \pm iS^y, S^z$, satisfy:

$$[\, S^+, S^- \,] = 2S^z$$

$$[\, S^\pm, S^z \,] = \mp S^\pm$$

If the spin quantization axis is z,

Exercise 4.1
Prove that

$$\sum_i S_i^z = (\mathbf{S}_{tot})^z$$

commutes with the Hamiltonian (4.1).

Accordingly the eigenvalue of the z-component of the total angular momentum operator is one of the good quantum numbers one can use to label the eigenstates of the Hamiltonian. Of course, if $B = 0$, we can choose any component of $\mathbf{S}_{tot}$ as a constant of motion, which is a reflection of the isotropic structure of Eq. (4.1).

Let us now consider the Hamiltonian (4.1), with $B > 0$ and assume

$$J_{ij} \geq 0\,,\; \forall i,j \;. \tag{4.4}$$

Since the electron charge is negative, so is μ_B , and the minimum potential energy configuration in an external field corresponds to the most negative $S_{tot}^z = -NS$, N being the number of spins in the system. One can prove that the state $\mid -NS > \equiv \mid 0 >$ is an eigenstate of H, with energy eigenvalue

$$E_0/N = -\gamma B - S^2 \sum_{<i,j>} J_{ij} \tag{4.5}$$

It is easy to see that under the assumption above that all $J_{ij} \geq 0$, this state has the minimum *classical* energy. This is precisely Weiss ferromagnetic phase for a single domain sample. Weiss state is also the exact ground state of the quantum Heisenberg ferromagnet if condition (4.4) is satisfied [4].

One obtains N linearly independent states upon introducing in $\mid 0 >$ one spin deviation at a given site. That is, states in which the spin at site n has its z component increased by one unit: $(-S \rightarrow -S+1)$, and which have consequently a total z component of spin $-NS + 1$. We shall denote these states $\mid n >$. They can be obtained from the Weiss state by acting on it with the local spin raising operator S_n^+. We recall the matrix elements of the circular components of the angular momentum operator:

$$
\begin{aligned}
S^- \mid S, m > &= \sqrt{(S - m + 1)(S + m)} \mid S, m - 1 > \\
S^+ \mid S, m > &= \sqrt{(S - m)(S + m + 1)} \mid S, m + 1 > \\
S^z \mid S, m > &= m \mid S, m >
\end{aligned}
\tag{4.6}
$$

We construct the Weiss state ket for N down spins $\mid 0 >$ as the direct product of N individual spin kets (spinors), each of which we shall denote by just the m eigenvalue since all spinors have spin S:

$$\mid 0 > \equiv \prod_{i=1}^{N} \mid -S >_i$$

Then, the ket with one spin deviation with respect to the Weiss state is:

$$| \, n > \equiv | -S + 1 >_n \prod_{i \neq n}^{N} | -S >_i$$

Of course, the quantum Weiss state is a singlet in an applied field, but a different situation arises if the external field vanishes, for in that case there is no special quantization direction, and the Hamiltonian is invariant under the complete rotation group in three dimensions, because the exchange energy is a scalar. This implies an infinite degeneracy, as was mentioned in the previous chapter. We can construct orthonormalized states with one spin flip on site n by acting on the Weiss ground state with the S^+ operators and using Eqs. (4.6):

$$| \, n >= \frac{1}{\sqrt{2S}} S_n^+ \, | \, 0 > \tag{4.7}$$

We assume that the exchange integrals J_{ij} depend only on the difference $\mathbf{R_j} - \mathbf{R_i}$. Then, if spins are on sites of a perfect infinite lattice in any dimensions, the Hamiltonian has translational symmetry. One conseqence thereof is that each exact eigenstate must be a basis for an irreducible representation of the translation group of the lattice (appendix A) . This statement is nothing but Bloch's theorem , which in this case establishes that the state

$$| \, \mathbf{k} >= \frac{1}{\sqrt{N}} \sum_n e^{i \mathbf{k \cdot n}} \, | \, n > \tag{4.8}$$

transforms as the basis of a symmetry type characterized by the wave vector $\mathbf{k}$.

Exercise 4.2
Prove that (4.8) is an eigenstate of H, and that

$$H \, | \, \mathbf{k} >= (E_0 + \epsilon_k) \, | \, \mathbf{k} > \tag{4.9}$$

with

$$\epsilon_k = 2S \left(J(0) - J(\mathbf{k}) \right) + \gamma B \tag{4.10}$$

where we have defined the Fourier transform of the exchange interaction:

$$J(\mathbf{k}) = \sum_{\mathbf{h}} J_{\mathbf{h}} \; e^{i \mathbf{k \cdot h}} \tag{4.11}$$

and $E_0 = -2NS^2 J(0) - NS\gamma B$ is the ground (Weiss) state energy.

The vectors $\{\mathbf{h}\}$ above are those connecting each spin to all those interacting with it through an exchange integral $J \neq 0$. In the particular case in which only the nearest neighbour J is non vanishing, we have, with a simplified notation:

$$\epsilon(\mathbf{k}) = 2SJ\nu(1 - \gamma_k) + \gamma B \tag{4.12}$$

where

$$\gamma_k = \frac{1}{\nu} \sum_{\mathbf{h}} e^{i\mathbf{k} \cdot \mathbf{h}}$$

is the form factor of the star of first n. n. of each spin and ν = the number of first n. n. If the lattice has inversion symmetry both ϵ_k and γ_k have that symmetry as well.

In the more general case in which the range of J is greater one verifies that as long as (4.4) is satisfied the magnon dispersion relation (4.10) is positive semi-definite, vanishing at $\mathbf{k} = 0$ in the absence of an external field, in which case the $\mathbf{k} = 0$ mode is a Goldstone boson as mentioned in the previous chapter. With an applied external field, the excitation spectrum corresponding to exactly one spin deviation has a gap

$$\Delta = \gamma B \qquad (4.13)$$

which is precisely the Larmoor precession frequency. The $\mathbf{k} = 0$ mode is a uniform (i.e., all spins in phase) precession of the system.

Exercise 4.3
Calculate ϵ_k of Eq.(4.10) for (a) the linear chain, (b) the square lattice and (c) the three primitive cubic lattices, assuming first n.n. interactions only.

The rotational degeneracy of the isotropic Heisenberg hamiltonian, makes it necessary, when one considers a single-domain *spontaneously magnetized* ferromagnet, to apply an infinitesimal magnetic field, which vanishes in the thermodynamic limit at least as $N^{-1+\epsilon}$, with $0 < \epsilon < 1$, such that the energy per spin is not affected in this limit, but the direction of the field defines the spin quantization axis. This field breaks the rotational symmetry of the ground state. The Weiss state, as defined by this limiting process, is therefore a broken-symmetry ground state. In real samples, the rotational symmetry is broken by anisotropic terms which must be added to the Heisenberg isotropic Hamiltonian, like the spin-orbit interaction and the crystal field — sources of single-site anisotropy terms, spin-anisotropy of the exchange interactions and the dipole-dipole interaction, among others.

4.2   Holstein-Primakoff transformation

Suppose we choose the Weiss ground state with all spins ↑, following the common use of disregarding the negative sign of Bohr's magneton. The applied field is assumed to point also up. Let us now introduce the local spin-deviation operator

$$\hat{n} = S - S^z \qquad (4.14)$$

with the eigenvalues

$$n = S - m$$

So, increasing m decreases n and viceversa. For fixed S we label the $\mid m >$ states in terms of n. For example:

$$S^- \mid n >= \sqrt{(2S - n)(1 + n)} \mid n + 1 >=$$

$$\sqrt{2S}\sqrt{1 - \frac{n}{2S}}\sqrt{1 + n} \mid n + 1 > \qquad (4.15)$$

We remind here that the creation operator $a^\dagger$ for an energy quantum of a harmonic oscillator satisfies the relation

$$a^\dagger \mid n >= \sqrt{n + 1} \mid n + 1 > \qquad (4.16)$$

where n is the number of energy quanta in the state $\mid n >$. Therefore we can rewrite (4.15) in terms of harmonic oscillator creation and annihilation operators:

$$\begin{aligned} S^- &= \sqrt{2S}a^\dagger \hat{f} \\ S^+ &= \sqrt{2S}\hat{f}a \end{aligned} \qquad (4.17)$$

where the non linear operator $\hat{f}$ is defined as

$$\hat{f} = \sqrt{1 - \hat{n}/2S} \qquad (4.18)$$

It is easy to verify that definition (4.17) is compatible with (4.15), that is, so long as

$$n \leq 2S \qquad (4.19)$$

The latter is a very stringent condition, although it is easier to satisfy for large S. At any rate this inequality is incompatible with true harmonic oscillators, since the spectrum of $\hat{n}$ is unbounded. We expect however that at low temperatures the statistical average of $\hat{n}/2S$ be small compared to one, so that the square root in (4.20) be well defined. Then we follow Holstein and Primakoff (HP) [2] and represent the local spin operators at point i as

$$\begin{aligned} S_i^- &= \sqrt{2S}a_i^\dagger \hat{f}_i \\ S_i^+ &= \sqrt{2S}\hat{f}_i a_i \\ S_i^z &= S - \hat{n}_i \end{aligned} \qquad (4.20)$$

Exercise 4.4
Prove that if the operators $a, a^\dagger$ satisfy boson commutation relations, namely

$$\begin{aligned} [a_i, a_j^\dagger &= \delta_{ij} \\ [\hat{n}_i, a_j^\dagger] &= \delta_{ij}a_j^\dagger \\ [\hat{n}_i, a_j] &= -\delta_{ij}a_i \end{aligned}$$

with $\hat{n}_i = a_i^\dagger a_i$, then the transformation (4.20) leaves invariant the commutation relations of the spin operators.

4.3   Linear spin-wave theory

The operators which create or destroy a state with the required transformation properties under the translation symmetry group of an infinite lattice are
the Fourier transforms of the local spin flip operators, as already indicated by
Eq. (4.8):

$$a_k^\dagger = \frac{1}{\sqrt{N}} \sum_n e^{i\mathbf{k}\cdot\mathbf{R_n}} a_n^\dagger \qquad (4.21)$$

with the corresponding definition for the hermitian conjugate annihilation operator. The new harmonic oscillator operators satisfy:

$$[\,a_k\,,\,a_{k'}^\dagger\,] = \delta_{kk'}$$

$$[\,a_k\,,\,a_{k'}\,] = [\,a_k^\dagger\,,\,a_{k'}^\dagger\,] = 0$$

$$[\,a_k\,,\,\hat{n}_{k'}\,] = a_k \delta_{kk'}$$

$$[\,a_k^\dagger\,,\,\hat{n}_{k'}\,] = -a_k^\dagger \delta_{kk'}$$

with $\hat{n}_k \equiv a_k^\dagger a_k$ We see then that the operators a_k, when acting on the vacuum
state $\mid 0 >$ create the one-deviation states described in Eq. (4.8). These excitations are called " magnons". The transformation equations of HP, contain
nonlinear terms which describe interactions between magnons. To see this, let
us expand the square root in (4.20) in a power series in $\hat{n}/2S$, and let us also
substitute on the r. h. s. the local operators by the magnon operators, by using
the inverse Fourier transform of (4.21). Then we obtain:

$$
\begin{aligned}
S_i^+ &= \sqrt{\frac{2S}{N}} \sum_k e^{i\mathbf{k}\cdot\mathbf{R}_i} a_k \\
&\quad - \frac{1}{N\sqrt{8NS}} \sum_{k\,k'\,k''} e^{i(\mathbf{k}-\mathbf{k'}+\mathbf{k''})\cdot\mathbf{R}_i} a_{k'}^\dagger\, a_{k''} a_k - \cdots \\
S_i^z &= S - \frac{1}{N} \sum_{k\,k'} e^{i(\mathbf{k'}-\mathbf{k})\cdot\mathbf{R}_i} a_k^\dagger a_{k'} \qquad (4.22)
\end{aligned}
$$

Upon substituting (4.22) for the spin operators in the exchange term of (4.1)
one obtains, up to quadratic order in the operators, the harmonic oscillator
Hamiltonian:

$$H^{(2)} = -N\nu JS^2 + 2\nu JS \sum \epsilon_k a_k^\dagger a_k \qquad (4.23)$$

where the dispersion relation ϵ_k coincides with that obtained in (4.10). We
shall call the restriction to this Hamiltonian the *"free spin wave approximation"*
(FSWA) for ferromagnets.

For the present choice of the Weiss state with all spins up, the one magnon
states $\mid k >$ have one quantum less of angular momentum than the Weiss state,
i.e. they belong to the subspace of the states with

$$S_{tot}^z = NS - 1$$

If we consider a one-magnon eigentate (4.8) we find that the spin deviation *is not localized on any particular site:*

Exercise 4.5
Calculate the expectation value of the local spin deviation $\hat{n}_i$ in a one magnon state $\mid k >$ and prove that

$$< k \mid \hat{n}_i \mid k >= \frac{1}{N}$$

The exact solution of the Heisenberg Hamiltonian obtained by Bethe [3] in $1d$ contains in particular the one-magnon excited states. J. C. Slater [5] proved the existence of spin-wave-like excitations in a FM ring of N atoms. The application of the spin-wave theory to $3D$ is due to F. Bloch [6] for spin $1/2$ and to C. Möller [7] for arbitrary spin. The magnon dispersion relation does not depend upon the orientation of the wave vector relative to the direction of the magnetization, due to the perfect isotropy assumed for the exchange interactions. In Fig. 4.1 we show the instantaneous configurations of spins as a spin-wave propagates through the lattice, in a semiclassical picture in which the spin operators are represented as ordinary vectors.

4.4 Semiclassical picture

In the large S limit the quantum description goes over to the classical vector model for spin. In this limit, a spin operator can be substituted by an ordinary vector of length $\sqrt{S(S+1)} > S$. This can be interpreted as meaning that even in the Weiss ground state, where for all spins in the lattice $S^z = S$, there is still some contribution from the transverse components to $\mathbf{S}^2$. The averages $< S^\pm >= 0$, but the average of the squares of these operators is positive, thus contributing the remainder of the length of $\mathbf{S}$. In the semi-classical picture, one can think of the spin vectors in the ground state as if they were precessing around the quantization axis z with random phases. [8]

Klein and Smith [9] show that one can express the difference in energy between the classical and quantum ground state of an FM in terms of the energy of the zero point fluctuations of the quantum oscillators corresponding to the independent spin waves. To see this, consider the limit of large S. Then the length of the classical spin vector is $S_0 \simeq S + 1/2$. We substitute each spin by such a vector, and calculate the energy of the Weiss state, including a Zeeman interaction with an external field B. Then the energy of the classical system is:

$$E_0^{class} = -\gamma BN(S + 1/2) - N(S^2 + S) \sum_\delta J_\delta \qquad (4.24)$$

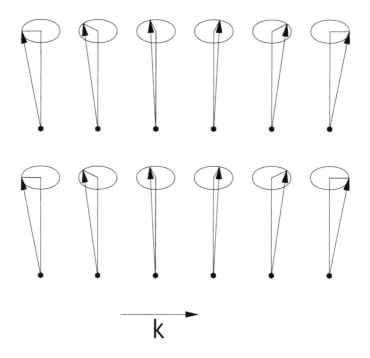

Figure 4.1: *Pictorial representation of spin wave propagating along a direction perpendicular to the quantization z axis.*

Exercise 4.6
Prove that the difference between the quantum and classical ground state energies in the Weiss state is:

$$E_0^{quant} - E_0^{class} = 1/2 \sum_k \epsilon(k) \qquad (4.25)$$

We can also write the classical equations of motion of the angular momentum and verify that the corresponding precession frequency coincides with the spin wave excitation frequency [10]. To this end let us calculate the torque exerted on a magnetic moment **m** by an effective magnetic field:

$$\mathbf{T} = \mathbf{m} \wedge \mathbf{B}_{eff} = \gamma \mathbf{S} \wedge \mathbf{B}_{eff} \qquad (4.26)$$

From Euler's equation of motion of rotating bodies, we can equate the rate of change of angular momentum with the applied torque (4.26):

$$\hbar \frac{d\mathbf{S}}{dt} = \gamma \mathbf{S} \wedge \mathbf{B}_{\text{eff}} \qquad (4.27)$$

In the case of a linear chain with only n. n. interactions, the effective field acting on spin l is:

$$\mathbf{B}_{eff} = \mathbf{B} + (J/\gamma)(\mathbf{S}_{l-1} + \mathbf{S}_{l-1}) \qquad (4.28)$$

so that the equation of motion (4.27) turns into:

$$\hbar \frac{d\mathbf{S}_l}{dt} = \gamma \mathbf{S}_1 \wedge [\mathbf{B} + (J/\gamma)(\mathbf{S}_{l-1} + \mathbf{S}_{l+1})] \tag{4.29}$$

We verify that the same equation is obtained quantum mechanically by calculating the commutator of $\mathbf{S}_l$ with the Hamiltonian [13]. For the general case of arbitrary FM exchange interactions in any dimensions, Heisenberg equation of motion of the spin operator is

$$\hbar \frac{d\mathbf{S}_l}{dt} = \hbar \gamma \mathbf{S}_l \wedge \left[\mathbf{B} + (1/\gamma) \sum_j J_{lj} \, \mathbf{S}_j \right] \tag{4.30}$$

which reduces to Eq. (4.29) for the chain. We get the conservation law for the z component of the total spin if we sum (4.30) over all sites l. We find that all exchange terms cancel out:

Exercise 4.7
Prove that

$$\hbar \frac{d\mathbf{S}_{tot}}{dt} = \mathbf{S}_{tot} \wedge \gamma \mathbf{B} \tag{4.31}$$

where $\mathbf{S}_{tot} = \sum_i \mathbf{S}_i$. Show that this implies

$$\mathbf{S}_{tot}^z = const. \tag{4.32}$$

This result, as we mentioned before (Exercise 4.1), can be directly obtained by calculating the commutator of $\mathbf{S}_{tot}$ with the Hamiltonian. Equations (4.31) and (4.32) can be interpreted as describing a precession movement of $\mathbf{S}_{tot}$ around the fixed vector $\mathbf{B}$ with the Larmoor frequency $\omega_0 = \gamma B$.

Let us now consider solutions of (4.27) which describe a normal mode where all spins precess at the same frequency. A precession motion with constant frequency corresponds to a solution of (4.27) in which the transverse component of $\mathbf{S}$ rotates at a constant angular frequency. Then, we look for a solution satisfying

$$\frac{ds_x}{dt} = \lambda S_y$$
$$\frac{ds_y}{dt} = -\lambda S_x \tag{4.33}$$

with some real λ. Let us assume that successive spins precess with a phase difference ka as in Fig. 4.1.

Exercise 4.8

Prove that for the linear chain with only first n.n. interactions,

$$\lambda = \omega(k) = \gamma B + \frac{4J}{\hbar} S(1 - \cos ka) \tag{4.34}$$

which coincides with the spin wave energy (4.10).

4.5   Macroscopic magnon theory

Let $\mathbf{S}_0$ be the spin located at the point chosen as the origin of the coordinate system in the lattice. Let $\{\delta\}$ be the star of its ν first n.n. and assume that J only connects $\mathbf{S}_0$ with $\{\mathbf{S}_\delta\}$. We want to analyze the long-wave semi-classical limit and we substitute the spin operators by c-vectors. In the small k limit the central vector and its neighbours in the star considered above are approximately parallel, even at high T, in a FM system ($J > 0$). That is,

$$| \mathbf{S}_\delta - \mathbf{S_0} | \ll (| \mathbf{S}_\delta |, \ | \mathbf{S_0} |) \ .$$

This assumption is consistent with experimental results on the specific heat (see Chap. 3) and with a vast evidence from inelastic neutron scattering, which show the survival of SRO even at elevated temperatures, both in FM and AFM systems.

Then it is natural to expand $\mathbf{S}_\delta$ in a Taylor series:

$$\mathbf{S}_\delta = \mathbf{S_0} + (\delta \cdot \nabla)\mathbf{S_0} + \frac{1}{2}(\delta \cdot \nabla)^2 \mathbf{S_0} + \cdots \tag{4.35}$$

If the crystal has inversion symmetry we find

$$\sum_\delta \mathbf{S}_\delta = \nu \mathbf{S_0} + a^2 \nabla^2 \mathbf{S_0} + \cdots \tag{4.36}$$

Expansion (4.35) can be performed for the neighbours of each spin in the crystal, so that the exchange Hamiltonian can be written as

$$H = -J\nu \sum_i \mathbf{S}_i^2 - a^2 J \sum_i \mathbf{S}_i \cdot \nabla^2 \mathbf{S}_i \tag{4.37}$$

Let us disregard the constant in the r. h. s. above, substitute the sum in the subsequent term by an integral and neglect derivatives higher than the second, on the basis that the magnetization only varies appreciably over distances much greater than a lattice constant. Then

$$H = -\frac{J}{a} \int \mathrm{d}^3 \mathbf{r} \ \mathbf{S}(\mathbf{r}) \cdot \nabla^2 \mathbf{S}(\mathbf{r}) + constant \tag{4.38}$$

We can express the spin field above in terms of the magnetization:

$$H = -\frac{JS^2}{M_0^2 a} \int \mathrm{d}^3 \mathbf{r} \ \mathbf{M}(\mathbf{r}) \cdot \nabla^2 \mathbf{M}(\mathbf{r}) \tag{4.39}$$

which leads to the definition of the exchange effective field $\mathcal{H}_{eff}$ through

$$H = -\frac{1}{2} \int \mathrm{d}^3 \mathbf{r} \ \mathcal{H}_{eff}(\mathbf{r}) \cdot \mathbf{M}(\mathbf{r}) \tag{4.40}$$

where

$$\mathcal{H}_{eff} \equiv \frac{2Js^2}{M_0^2 a} \nabla^2 \mathbf{M}(\mathbf{r}) \tag{4.41}$$

and M_0 is the saturation magnetization at $T = 0$.

In order to obtain the total effective field we add to (4.41) the external applied field, the demagnetization field, and eventually the anisotropy field. We shall have occasion to use this form of the exchange effective field further on.

4.6   Thermal properties

4.6.1   Total spin deviation

Let us now calculate some low T properties of the Heisenberg ferromagnet within the FSWA in which magnons act as independent quantum harmonic oscillators. Then an arbitrary excited state is approximately described as a linear superposition of many one-magnon states. This picture makes sense insofar as one can neglect the effect of the interaction of two or more such excitations . We shall treat states of two magnons in Chap. 14. It seems in order, however, to remark that in the $3D$ case, the correction to the energy of a pair of independent magnons is very small, $O(1/N)$. Bound states of two magnons do exist, but they require a threshold of the single magnon energies which is of the order of the magnon band-width, and as a consequence of no importance except at very high temperatures.

The FSWA then leads us to consider an ensemble of boson-type excitations which should accordingly obey Bose-Einstein statistics. Let us start the discussion of the effects of temperature by calculating the total spin deviation. Since

$$S_m^z = S - \hat{n}_m$$

at site m, we can obtain the magnetization at a temperature T as:

$$VM(T) = \gamma NS - \gamma \sum_m \langle \hat{n}_m \rangle_T \tag{4.42}$$

where N is the total number of spins and V is the volume of the whole system.

Exercise 4.9
Verify that the transformation from the site operators $\hat{n}_m$ to spin wave operators $\hat{n}_k$ is unitary and that, accordingly,

$$\sum_m \hat{n}_m = \sum_k \hat{n}_k \tag{4.43}$$

Substituting (4.43) into (4.42) we find:

$$\hbar \gamma NS - VM(T) \equiv V\Delta M(T) = \hbar \gamma \sum_k \langle \hat{n}_k \rangle_T \tag{4.44}$$

Eq. (4.44) gives the decrease of the thermodynamic average of the total magnetic moment of the sample at temperature T from the $T = 0$ saturated value $\hbar \gamma NS$.

Let us now consider the thermodynamic limit of (4.44), that is $(V, N) \to \infty$, with $\lim_{N,V \to \infty} N/V = \rho$, the volume concentration of spins in the sample. The summation in (4.44) must be limited to the interior of the first Brillouin zone (BZ). We consider now that the magnons are running waves with periodic boundary conditions in the volume V. In this case the density of points in k space is $V/(2\pi)^d$, where V is the total volume and d the dimension of the lattice. The k-space summation can be substituted, in the thermodynamic limit, by an integration over the volume of the first BZ in d-dimensions:

$$\frac{1}{N} \sum_k \cdots = \frac{v_0^{(d)}}{(2\pi)^d} \int d^d k \cdots$$

where $v_0^{(d)}$ is the volume of the lattice unit cell in d dimensions. For a hypercubic lattice the dispersion relation behaves, in the small k limit, as

$$\epsilon_k \simeq \epsilon_0 + Dk^2 \tag{4.45}$$

with a parameter D , which is called the *spin stiffness*, adequate for each case.

Exercise 4.10
Prove (4.45) and obtain the parameter D for the lattices considered in Exercise 4.3.

The statistical average of the number of excited quanta of a harmonic oscillator is the Bose-Einstein probability distribution, so that the total number of spin deviations at temperature T in thermodynamic equilibrium for the approximation of free spin-waves, is:

$$\Delta N = \frac{v_0^{(d)}}{(2\pi)^d} \int_{BZ} \frac{d^d k}{e^{\beta \epsilon_k} - 1} \tag{4.46}$$

Let us first consider the zero field case: $\epsilon_0 = 0$. At low T, we shall find with reasonable probability only excitations of small energy, of the order of $k_B T$, and this corresponds to small k, so that the quadratic approximation (4.45) is applicable. Then the integral in (4.46) is approximately

$$\int \frac{d^d k}{e^{\beta \epsilon_k} - 1} \simeq (\beta D)^{-d/2} \Omega_d \int_0^\infty \frac{\xi^{(d-2)/2} \, d\xi}{e^\xi - 1} \tag{4.47}$$

where the constant Ω_d is the solid angle subtended by the sphere in d dimensions. We find that the number of spin deviations at any non-zero temperature diverges if $d \le 2$. This is the consequence of Mermin-Wagner theorem [1]. For more than two dimensions this integral can be evaluated in terms of the Riemann zeta function, defined as:

$$\zeta(s) = \frac{1}{\Gamma(s)} \int_0^\infty \frac{x^{s-1} dx}{e^x - 1} = \sum_{p=1}^\infty \frac{1}{p^s} \tag{4.48}$$

For $d = 3$ we obtain:

$$\Delta M(T) = \left(v_0/8\pi^3\right) 4\pi(\beta D)^{-3/2}\Gamma(3/2)\zeta(3/2) \tag{4.49}$$

Then, we have for the magnetization of a simple cubic magnetic lattice with lattice constant a_0 :

$$M_0 - M(T) = \gamma\frac{N}{V}\frac{\Gamma(3/2)\zeta(3/2)}{2\pi^2}\left(\frac{k_B T a_0^2}{D}\right)^{3/2} \tag{4.50}$$

and this is *Bloch's $T^{3/2}$ law* [6]. This law is valid, with suitable definitions of D and a_0, for any lattice, and even for a continuous, structureless magnetic medium [8]. Another way of writing (4.50) is:

$$M(T)/M_0 \simeq 1 - \zeta(3/2)\Theta^{3/2} \tag{4.51}$$

The specific heat per unit volume at zero external field B and constant volume is given by:

$$c_V = \frac{1}{V}\left(\frac{\partial E}{\partial T}\right)_V = \frac{15 M_0 k_B}{4\gamma}\zeta(3/2)\,\Theta^{3/2} \tag{4.52}$$

In (4.51) and (4.52) Θ is the dimensionless parameter

$$\Theta = \left(\frac{\gamma}{M_0}\right)^{3/2}\frac{k_B T}{4\pi D} \tag{4.53}$$

Exercise 4.11
Prove (4.52).

The internal energy of the ensemble of magnons in equilibrium, referred to the ground state energy is

$$E(T) = <H>_T - <H>_0 = const. \cdot T^{5/2} \tag{4.54}$$

4.6.2 Non-linear corrections

Let us write the Hamiltonian as

$$H = -\sum_{<i,j>} J_{i,j}(S_i^- S_j^+ + S_i^z S_j^z) \tag{4.55}$$

and substitute the local spin operators, according to Eq. (4.22), in terms of the boson operators $a, a^\dagger$, in such a way that the latter appear in normal order (all creation operators on the left). Then we recover the quadratic Hamiltonian, and we find higher order corrections. If we collect all quartic terms we find the correction

$$H^{(4)} = \frac{1}{2N}\sum_{k,k',k''}(J(\mathbf{k}) + J(\mathbf{k + k' + k''}) - 2J(\mathbf{k - k''})) \times$$

$$a_\mathbf{k}^\dagger a_{\mathbf{k}'}^\dagger a_{\mathbf{k}''} a_{\mathbf{k+k'-k''}} \tag{4.56}$$

as the first higher order correction which has a non-vanishing average.

Exercise 4.12
Prove (4.56).

One can extract from this expression a quadratic form by substituting all pairs of operators of the form $a^\dagger a$ by their thermal average with the density matrix corresponding to the unperturbed harmonic Hamiltonian (4.23), which leads to the correction

$$\Delta H^{(2)} = \sum_k \Delta \omega_k a_{\mathbf{k}}^\dagger a_{\mathbf{k}} \tag{4.57}$$

equivalent to a correction of the dispersion relation:

$$\Delta \omega_k = -\frac{2}{N} \sum_q \left(J_{\mathbf{k}} + J_{\mathbf{q}} - J_{\mathbf{k}-\mathbf{q}} - J_0 \right) n_q \tag{4.58}$$

and n_q is the Bose-Einstein distribution function. If we restrict the exchange interactions to first n.n. range, we can always write $J_{\mathbf{k}} = J\gamma_{\mathbf{k}}$ as in (4.12). If we also assume that the lattice has inversion symmetry, then it is possible to show that

$$\gamma_{\mathbf{k}+\mathbf{q}} = \gamma_{\mathbf{k}}\gamma_{\mathbf{q}} + \cdots \tag{4.59}$$

where the terms represented by $\cdots$ above are odd under inversion. Then

Exercise 4.13
Show that the correction to the dispersion relation is

$$\Delta \omega_{\mathbf{k}} = -\frac{1}{SN} \omega_{\mathbf{k}} E(T) \tag{4.60}$$

where $E(T) = $ total magnon contribution to the internal energy.

According to Eq. (4.54) we get

$$\frac{\Delta \omega_{\mathbf{k}}}{\omega_{\mathbf{k}}} \propto (T/T_c)^{5/2} \tag{4.61}$$

4.7   The Heisenberg antiferromagnet

4.7.1   Introduction

The classical equivalent of the FM Weiss state for an antiferromagnet (AFM) is the Néel [11] state. With ordinary vectors substituting the spin operators, the configuration of minimum energy for (4.1) with $J < 0$ and $B = 0$ is clearly that one in which all n.n. of a given spin are aligned exactly antiparallel to it, assuming that the lattice is bipartite. Let us consider this case, as we did in Chap. 3, in which the A and B sub-lattices are anti-parallel at $T = 0$ in the classical limit. Let us consider also a local anisotropy field B_a acting on each spin, oriented along the local magnetzation, so that it reverts as we go from an A to a B site. This is the *staggered field* we considered in Chap. 3, with the same magnitude on all sites. We add as well a uniform applied field. The resulting Hamiltonian is:

$$
\begin{aligned}
H &= J \sum_{<ij>} \mathbf{S}_{i_A} \cdot \mathbf{S}_{j_B} - \gamma B_a \left(\sum_i S_{i_A}^z - \sum_j S_{j_B}^z \right) \\
&- \gamma B \left(\sum_i S_{i_A}^z + \sum_j S_{j_B}^z \right)
\end{aligned}
\tag{4.62}
$$

We can easily extend the local boson HP transformation of the spin operators, applied before to a ferromagnet, to the case in hand. Let us choose the sublattice $A\,(B)$ with spin $\uparrow$ $(\downarrow)$. Then the natural choice for the boson operators is:

$$
S_{i_A}^+ = \sqrt{2S} \sqrt{1 - \frac{a_i^\dagger a_i}{2S}} \, a_i
$$

$$
S_{j_B}^+ = \sqrt{2S} b_j^\dagger \sqrt{1 - \frac{b_j^\dagger b_j}{2S}}
$$

$$
S_{i_A}^z = S - a_i^\dagger a_i
$$

$$
S_{j_B}^z = -S + b_j^\dagger b_j
$$

4.7.2   Antiferromagnetic spin-waves

As we did for the FM, we neglect all corrections to the linear terms in the transformation equations above, and use Bloch's theorem to Fourier transform the local boson operators to running wave operators:

$$
c_k = \frac{1}{\sqrt{N}} \sum_{i \in A} e^{i k \cdot \mathbf{R}_i} a_i
$$

$$
a_i = \frac{1}{\sqrt{N}} \sum_k e^{-i k \cdot \mathbf{R}_i} c_k
\tag{4.63}
$$

$$
d_k = \frac{1}{\sqrt{N}} \sum_{j \in B} e^{i k \cdot \mathbf{R}_j} b_i
$$

$$
b_j = \frac{1}{\sqrt{N}} \sum_k e^{-i k \cdot \mathbf{R}_j} d_k
\tag{4.64}
$$

We remark that summations in Eqs. (4.63) and (4.64) are, as indicated, only on sublattice points. By definition, the sublattice has a larger lattice spacing than the original lattice. In the special case of bipartite Bravais lattices, the volume of the basic cell of each sublattice is twice that of the original atomic lattice. Accordingly, since the k summations are performed on the first BZ of the sublattice, this zone has half the volume of the atomic BZ. Also, the number of points N of each sublattice is half the number of magnetic atoms in the crystal. We assume that the lattice has inversion symmetry, so that the structure factor is symmetric:

$$\gamma(\mathbf{k}) = \frac{1}{\nu} \sum_\delta e^{i\mathbf{k}\cdot\delta} = \gamma(-\mathbf{k}) \equiv \gamma_k$$

where δ denotes a vector joining a site $A\,(B)$ to a n.n. $B\,(A)$ site on the other sublattice, and ν is the number of n.n. in the atomic lattice. We shall neglect for the moment longer-range interactions. By following now the same procedure as for the ferromagnetic case, we obtain a quadratic Hamiltonian:

$$H_{ex}^{(2)} = -S^2 J\nu N - 2\gamma B_a SN + (JS\nu + \gamma B_a) \sum_k \left(c_k^\dagger c_k + d_k^\dagger d_k \right)$$

$$+\gamma B \sum_k \left(c_k^\dagger c_k - d_k^\dagger d_k \right) + JS\nu \sum_k \gamma_k \left(c_k^\dagger d_k^\dagger + c_k d_k \right) \qquad (4.65)$$

Exercise 4.14
Prove (4.65).

Let us simplify the notation. We define

$$\omega_e = JS\nu \;;\; \omega_a = \gamma B_a \;;\; \omega_0 = \omega_e + \omega_a$$

$$\omega_1(k) = \gamma_k \omega_e \;;\; \Delta = \gamma B \;;\; E_0 = -S^2 JN\nu - \omega_a S$$

Then (4.65) is simplified as:

$$H_{ex}^{(2)} = E_0 + \sum_k \left\{ \omega_0 \left(c_k^\dagger c_k + d_k^\dagger d_k \right) + \Delta \left(c_k^\dagger c_k - d_k^\dagger d_k \right) \right\}$$

$$+ \sum_k \omega_1(k) \left(c_k d_k + c_k^\dagger d_k^\dagger \right) \qquad (4.66)$$

This is a quadratic form, but it is not yet diagonal, since the last term in (4.66) couples spin waves of both sublattices. In their original paper Holstein and Primakoff [2] found a similar quadratic Hamiltonian upon considering simultaneously ferromagnetic exchange and dipolar interactions. The dipolar terms, as we shall see later on, generate non-diagonal terms very similar to those above.

In order to diagonalize such a quadratic form, HP introduced at this point a linear canonical transformation of the original running wave operators, which was later re-discovered by Bogoliubov in connection with the theory of superfluidity and superconductivity and is known as the *Bogoliubov transformation* [12]. Let us remark first of all that both c and $d^\dagger$ have the effect of increasing S_{tot}^z, while $c^\dagger$ and d have the opposite effect. This suggests to define normal modes created by a linear superposition of c and $d^\dagger$, since we are looking for candidates to excited states of the original exchange Hamiltonian, which must be also eigenstates of S_{tot}^z. Therefore, for each k let us perform the linear transformation

$$\alpha_k = u_k c_k - v_k d_k^\dagger$$
$$\beta_k = A_k d_k + B_k c_k^\dagger \qquad (4.67)$$

Let us for the time being suppress the index k, since we shall be working in the subspace of the operators of a given k. We can choose the coefficients u, v, A, B all real. The conditions we must impose upon the coefficients are:

1. the states created or annihilated by α and β must be linearly independent, which demands

$$[\alpha, \beta] = 0$$

This is satisfied if

$$A = u, \quad B = -v \qquad (4.68)$$

2. the new operators must satisfy the canonical commutation relations

$$[\alpha, \alpha^\dagger] = [\beta, \beta^\dagger] = 1$$

which requires

$$u^2 - v^2 = 1 \qquad (4.69)$$

Exercise 4.15
Prove (4.68) and (4.69).

Now we demand that, since α must annihilate a normal mode, it must be a solution of the Heisenberg equation of motion, with a real eigenfrequency:

$$i\hbar \frac{d\alpha(t)}{dt} = [H_{ex}, \alpha(t)] = \lambda \alpha(t) \qquad (4.70)$$

From (4.68), (4.69) and (4.70) and upon using the commutation relations for c and $c^\dagger$ we arrive at a homogeneous system of linear equations for u and v:

$$(\omega_0 + \Delta - \lambda)\, u + \omega_1(k) v = 0$$
$$\omega_1(k) u + (\omega_0 - \Delta + \lambda)\, v = 0 \qquad (4.71)$$

with eigenvalues

$$\lambda = \Delta \pm \sqrt{\omega_0^2 - \omega_1^2(k)} \qquad (4.72)$$

The equations obtained for $\beta = ud - vc^\dagger$ lead again to (4.72).

Exercise 4.16
Verify (4.72) and the assertion above.

We now call the $+(-)$ sign solution of (4.70) $\alpha(\beta)$. Introduce the notation

$$\omega(k) = \sqrt{\omega_0^2 - \omega_1^2(k)}$$

From (4.72) we get:

$$\left(\frac{u}{v}\right)_\pm = \frac{-\omega_1(k)}{\omega_0 \mp \omega(k)} \tag{4.73}$$

According to (4.69), we can introduce a parameter θ such that

$$u = \cosh\theta \ , \ v = \sinh\theta$$

Exercise 4.17
Verify that only the solution with the $+$ sign in (4.73) satisfies (4.69) and diagonalizes the Hamiltonian (4.66).

With this choice, we have:

$$\tanh 2\theta = -\frac{\omega_1(k)}{\omega_0} \tag{4.74}$$

Exercise 4.18
By susbstituting the running wave operators c_k and d_k and their hermitian conjugates in terms of α_k and β_k, obtain the diagonal Hamiltonian

$$\begin{aligned}
H_{ex}^{(2)} &= E_0 - \omega_0 N + \sum_k \left(\alpha_k^\dagger \alpha_k + 1/2\right)(\omega_k + \Delta) \\
&+ \sum_k \left(\beta_k^\dagger \beta_k + 1/2\right)(\omega_k - \Delta)
\end{aligned} \tag{4.75}$$

where

$$E_0 = -SN\omega_e - 2SN\omega_a \tag{4.76}$$

We have thus obtained two distinct AFM magnon branches, as linearly independent normal modes of the spin system in the harmonic approximation. One can accordingly call this the *free-spin-wave approximation* (FSWA) for an AFM. For each particular k the corresponding term in the hamiltonian describes a quantized harmonic oscillator, with the $1/2$ representing the contribution to the energy of the zero-point quantum fluctuations. The new ground state energy, which is obtained when the occupation numbers for all k are zero, is the eigenvalue of (4.75) for the true vacuum state, which we shall accordingly denote $|\tilde{0}>$. The expectation values of the number operators for this state vanish:

$$<\tilde{0}\,|\,\alpha_k^\dagger \alpha_k\,|\,\tilde{0}> = <\tilde{0}\,|\,\beta_k^\dagger \beta_k\,|\,\tilde{0}> = 0 \ \ .$$

The renormalized ground state energy is accordingly

$$E_0' = E_0 + \sum_k (\omega_k - \omega_0) \tag{4.77}$$

Exercise 4.19
Prove (4.76).

From (4.75) we see that the two branches have the respective eigenvalues

$$E_k^\alpha = \omega_k + \Delta \ , E_k^\beta = \omega_k - \Delta$$

and they are degenerate if the external field B is zero. If the anisotropy field is non-zero, both branches have an energy gap Δ_0 at zero external field:

$$\Delta_0 = \sqrt{\omega_a(\omega_a + 2\omega_e)} \tag{4.78}$$

As B increases, there will be a critical B for which $\Delta = \Delta_0$ at which the system becomes unstable because the lower branch energy vanishes. At this field the system goes into the *spin flop* phase, with all spins becoming approximately perpendicular to the applied field direction, which we already described in Chap. 3 in the MFA. For $B_a = B = 0$, the two branches are degenerate. In particular for a one dimensional chain

$$\omega_k = \omega_0 \mid \sin ka \mid \tag{4.79}$$

where a is the distance from an A to a B site. For small k, and for a hypercubic lattice of any dimension, we have

$$\omega_k = vk + \mathcal{O}(k^3) \tag{4.80}$$

where v is the magnon velocity.

4.7.3 Sublattice magnetization

The diagonalization process has led us to a redefinition of the operators which create or annihilate elementary excitations, as we changed from c_k, d_k to α_k, β_k (Bogoliubov transformation). We also verified that the ground state must have changed in this process, to be consistent with the assumed properties of the operators α and β. One consequence is that one finds deviations of the sublattice magnetization from the saturation value S even at $T = 0$. To verify this let us calculate the z component of the total up-sublattice spin:

$$S_A^z = \sum_i S_{i_A}^z = NS - \sum_k c_k^\dagger c_k \tag{4.81}$$

If we substitute in (4.81) $c^\dagger$ and c in terms of the new operators, we need to invert the Bogoliubov transformation.

Exercise 4.20
Prove that the inverse Bogoliubov transformation is

$$c_k = u_k \alpha_k + v_k \beta_k^\dagger$$
$$d_k = v_k \alpha_k + u_k \beta_k^\dagger \qquad (4.82)$$

Denoting the new vacuum by $| \tilde{0} \rangle$ we find that

$$< \tilde{0} | \ c_k^\dagger c_k \ | \tilde{0} >= v_k^2 < \tilde{0} | \ (1 + \beta_k^\dagger \beta_k) \ | \ \tilde{0} >= v_k^2 \qquad (4.83)$$

where we have explicitly imposed that β_k annihilate $| \tilde{0} \rangle$. We can easily calculate now the sublattice deviation. We define the relative spin deviation for one sublattice as

$$\Delta M = 1 - < S_A^z > / S$$

Then we have

$$\Delta M = \sum_k v_k^2 = 1/2 \sum_k \left(\frac{\omega_0}{\sqrt{\omega_0^2 - \omega_1^2(k)}} - 1 \right) \qquad (4.84)$$

Exercise 4.21
Verify Eq. (4.84).

Let us define the anisotropy ratio

$$\alpha = \frac{\omega_a}{\omega_e} \qquad (4.85)$$

Exercise 4.22
Calculate ΔM for a one dimensional lattice, and show that for small anisotropy ratio α

$$\Delta M = -\frac{1}{2S} \left(1 + \frac{\ln 2\alpha}{\pi} \right) \qquad (4.86)$$

Eq. (4.86) shows that in the absence of anisotropy the FSWA excludes long range AFM order in one dimension, even at zero T: one is led to conclude that, since the spin deviation at $T = 0$ diverges, the ordered Néel ground state is not the correct one. If the LRO parameter $\sigma = 0$ we must re-formulate the whole theory, since the notion of a spin-wave is based upon a translationally invariant Néel ground state. This we already mentioned in our reference to the work of Bethe [3], who proved that the ground state of the AF chain has no LRO. Numerical calculations yield for $S\Delta M$, within the FSWA, the values:

- 0.197 for the square lattice;

- 0.0078 for the $NaCl$ magnetic lattice;

- 0.0593 for the body-centered atomic lattice.

Let us remark that in a lattice with a magnetic $NaCl$ structure, the first n.n. of each spin form a simple cubic coordination, so that this structure is sometimes called simple cubic for short, in the present context.

We notice that the departure of the sublattice magnetization from the Néel saturation value at $T = 0$ is much greater in low dimensions than in 3d.

4.7.4 Ground state energy of AFM

It is interesting to compare (4.76) with the corresponding energy of the classical Néel state. As we did for the FM, in the semi-classical limit we substitute each spin operator by an ordinary vector of length

$$S_0 = \sqrt{S(S+1)}$$

Then the classical energy in the Néel configuration without external field is

$$E_0^{class} = -JN\nu S(S+1) - 2\gamma B_a \sqrt{S(S+1)}N \qquad (4.87)$$

With the help of (4.76) and (4.87) we calculate the difference between the ground state energy in the harmonic approximation and the semi-classic energy of the Néel state in the $S \to \infty$ limit:

$$E_0^{FSWA} - E_0^{class} = \sum_k \omega_k + \mathcal{O}(1/S) \qquad (4.88)$$

This result is exactly what should be expected: the quantization of the spin fluctuations in the harmonic approximation yields a ground state energy which is higher than the classical counterpart by a magnitude equal to the energy of the zero-point motion of all the harmonic modes, just as in the ferromagnetic case.

All these estimates are based on the independendent harmonic oscillator approximation or FSWA. However, more general conclusions have been arrived at regarding the ground state of the Heisenberg Hamiltonian. Under not too restrictive conditions, Lieb and Mattis [17] proved that it must be a singlet. Then it was also proved [18] that in such a case time reversal symmetry demands that the expectation value in the ground state $< S_n^z >= 0$ for all sites n, at odds with Néel's state.

In spite of this, real antiferromagnets are Néel like. One can argue, following Nagamiya et al. [14], that there is scarcely any coupling between both degenerate Néel states coupled by time reversal symmetry. An estimation by Anderson [19] of the switching time from one orientation to the time reversed one yields 3 years, in the absence of anisotropy. The non-zero anisotropy present in practically all systems would tend to make the total spin reversal even more difficult, improving the metastability of the sublattice pattern. Some variational calculations of the energy of the ground state, which assumes it to be a singlet, give similar results to those performed within the two sublattice model [20], which can be taken as an indication that the resonance energy between the two degenerate Néel states is indeed very small.

References

1. Mermin, N. D. and Wagner, H. (1966) *Phys. Rev. Lett.* **17**, 1133.

2. Holstein, T. and Primakoff, H. (1940) *Phys. Rev.* **58**, 1098.

3. Bethe, H. A. (1931) *Z. Physik* **71**, 205.

4. Mattis, D. (1963) *Phys. Rev.* **130**, 76.

5. J.C.Slater, J. C.(1930) *Phys. Rev.* **35**, 509.

6. Bloch, F. (1930) *Z. Physik* **61**, 206.

7. Möller, C. (1933) *Z. Physik* **82**, 559.

8. Keffer, F. (1966) *Handb. der Phys.* **2 Aufl., Bd.XVIII**, Springer–Verlag.

9. Klein, M. J. and Smith, R. S. (1950) *Phys. Rev.* **80**, 111.

10. Keffer, F., Kaplan, H. and Yafet, Y. (1953) *Am. J. Phys.* **21**, 250.

11. Néel, L. (1932) *Ann. Phys.* **18**, 5; (1948) **3**, 137; (1936) *Compt. rend.* **203**, 304.

12. Bogoliubov, N. N. (1947) *J. Phys. USSR* **9**, 23; (1958) *Nuovo Cimento* **7**, 794.

13. Herring, C. and Kittel, C. (1951) *Phys. Rev.* **81**, 869; (1952) *id.* **88**, 1435.

14. Nagamiya, T., Yosida, K. and Kubo, R. (1955) *"Antiferromagnetism"*, *Adv. in Phys.* **4**, 1-112.

15. Kubo, R. (1952) *Phys. Rev.* **87**, 568.

16. Anderson, P. W. (1951) *Phys. Rev.* **83**, 1260.

17. Lieb, E. and Mattis, D. (1962) *Phys. Rev.* **125**, 164;(1962) *J. Math. Phys.* **3**, 749.

18. Pratt jr., G. W. (1961) *Phys. Rev.* **122**, 489.

19. Anderson, P. W. (1952) *Phys. Rev.* **86**, 694.

20. Marshall, W. (1955) *Proc. Roy. Soc. (London)* **A 232**, 48.

Chapter 5

Magnetic anisotropy

5.1 Introduction

We have already mentioned anisotropy effects in several previous sections. Let us now study the microscopic origin of magneto-crystalline anisotropy terms in ferromagnets. First, let us determine the general structure of the anisotropic terms in the free energy. We assume for simplicity that the magnetization $\mathbf{M}$ is uniform inside the sample, thereby discarding the presence of domains. At given T and $\mathbf{B}$ we can expand Helmholtz free energy F in powers of $\mathbf{M}$[1]:

$$F = F_0 + F_a \tag{5.1}$$

where F_0 is isotropic and the rest of the terms involve anisotropic polynomials of increasing degree in the components of $\mathbf{M}$. The free energy must be invariant under time- reversal ($\mathbf{M} \rightarrow -\mathbf{M}$) so only even powers of $\mathbf{M}$ can appear in the expansion.

It is sometimes convenient to change variables to

$$u_i = M_i/ \mid \mathbf{M} \mid \equiv \alpha_i \tag{5.2}$$

where α_i are the director cosines of $\mathbf{M}$ referred to cartesian crystalline axes. Here, of course $\sum_{i=1}^{3} \alpha_i^2 = 1$.

Both F_0, and F_a must be invariant under all the operations of the symmetry group of the crystal.

As an example, in the case of uniaxial rotational symmetry it is convenient to express $\mathbf{M}$ in the spherical components $M_z, M_{\pm} = M_x \pm iM_y$, so that if the symmetry group contains rotations of the form $2\pi m/n$, where $(n, m) =$ integers and $m = 1, \cdots, n - 1$, we can include in the expansion terms like $\mathcal{R}e(M_{\pm}^{p \cdot n})$. In this case it is more convenient to express the components of the unit vector $\mathbf{u}$ in terms of the spherical angles θ (latitude) and ϕ (azimuthal angle, measured counterclockwise on the $z = 0$ plane from the semiaxis of positive x). $\theta = 0$ is the positive z semiaxis, while $\phi = 0$ is the positive x semiaxis. A real even invariant function has the form:

$$\Gamma_{2p}(\theta, \phi) = 1/2 \left[(M^+)^{2p} + (M^-)^{2p} \right] = \sin^{2p} \theta \cos 2p\phi$$

or z^{2p} or a product of both forms.

Let us review the forms of the expansion of F for different symmetry groups. We shall use Schoenflies point group notation in the following[2].

Uniaxial tetragonal symmetry : C_{4h} group.

Here $n = 4$. We assume there is a mirror symmetry plane perpendicular to the rotation symmetry axis, and this is symbolized by the subindex h in C_{4h}. The expansion of F is

$$F = F_0 + A_2 \sin^2 \theta + A_4 \sin^4 \theta + B_4 \sin^4 \theta \cos^4 \phi + \cdots \qquad (5.3)$$

where the dots contain higher order polynomials.

Uniaxial hexagonal symmetry: C_{6h} group.

For $n = 6$ every term in the free energy expansion must be invariant under rotations around the z axis by $2\pi m/6$, with $m = 1, 2, \cdots 5$, mirror reflection in the (x, y) plane and inversion at the origin (equivalent to time reversal). Therefore we have

$$F = F_0 + A_2 \sin^2 \theta + A_4 \sin^4 \theta + A_6 \sin^6 \theta \cos 6\phi + \cdots \qquad (5.4)$$

Upon expressing $\sin^{2n} \theta$ in terms of $\cos m\theta$, $m = 2, 4, \cdots, 2n$ we find :

$$\begin{aligned}
F \quad = \quad & F_0 + \frac{1}{2} K_{u1} (1 - \cos 2\theta + \frac{1}{8} K_{u2} (3 - 4\cos 2\theta + \cos 4\theta) \\
+ \quad & \frac{1}{32} K_{u3} (10 - 15 \cos 2\theta + 6 \cos 4\theta - \cos 6\theta) \\
+ \quad & \frac{1}{32} K_{u4} (10 - 15 \cos 2\theta + 6 \cos 4\theta - \cos 6\theta) \cos 6\phi + \cdots \qquad (5.5)
\end{aligned}$$

The u subindex above refers to *uniaxial* symmetry.

It is customary to quote anisotropy energy per unit volume. The coefficients in Eq. (5.5) and similar expansions contain powers of $|\mathbf{M}|$, so they depend on temperature. Numerical values for Co at $288\,K$ are[3]:

$$K_{u1} = 4.53 \times 10^6 \text{ erg cm}^{-3}$$

$$K_{u2} = 1.44 \times 10^6 \text{ erg cm}^{-3}$$

Higher order terms are smaller. Since these coefficients are positive the axis of easy magnetization for Co is the hexagonal c axis. When they are negative the z (or c) axis is a *hard axis*. In the latter case the free energy is minimized when $\theta = \pi/2$ and the magnetization lies in the c plane (orthogonal to the c axis), which is therefore an *easy plane*. If K_{u1} and K_{u2} have different signs the stable direction of $\mathbf{M}$ lies on an *easy cone*.

Cubic symmetry: O_h group

The symmetry operations were described in Sect. 1.4.1. The first anisotropic term comes in fourth order. We find, up to eighth order :

$$F = F_0 + K_1(\alpha_1^2\alpha_2^2 + \alpha_2^2\alpha_3^2 + \alpha_3^2\alpha_1^2) + K_2(\alpha_1^2\alpha_2^2\alpha_3^2)$$
$$+ \quad K_3(\alpha_1^2\alpha_2^2 + \alpha_2^2\alpha_3^2 + \alpha_3^2\alpha_1^2)^2 + \cdots \tag{5.6}$$

where K_i are called cubic anisotropic constants [3]. For Fe at $293K$,

$$K_1 = 4.72 \times 10^5 \text{ erg cm}^{-3}$$

$$K_2 = -0.075 \times 10^5 \text{ erg cm}^{-3}$$

while for Ni at $296K$,

$$K_1 = -5.7 \times 10^4 \text{ erg cm}^{-3}$$

$$K_2 = -2.3 \times 10^4 \text{ erg cm}^{-3}$$

Along the [100] and equivalent directions, the anisotropy contribution vanishes independently of the values of the coefficients. For the [111] directions we have

$$F_a = \frac{1}{3}K_1 + \frac{1}{27}K_2 + \frac{1}{9}K_3 + \cdots \tag{5.7}$$

If $K_1 > 0$ as in Fe, [100] is the easy axis, if we neglect higher order terms. In Ni, where $K_1 < 0$ the set of the [121] directions are easy. Let us consider the magnetization on the (001) plane (the $y - z$ plane). If θ=angle between the magnetization and the positive x axis, we have $\alpha_1 = \cos\theta$, $\alpha_2 = \sin\theta$, $\alpha_3 = 0$. Then Eq. (5.6) becomes:

$$F_a = \frac{1}{8}K_1(1 - \cos 4\theta) + \frac{1}{128}K_3(3 - 4\cos 4\theta + \cos 8\theta) + \cdots \tag{5.8}$$

with no contribution from K_2

Exercise 5.1
Prove Eq. (5.8).

If **M** lies on the $(1\bar{1}0)$ plane, and we call θ the angle it makes with the positive z axis, we obtain

$$\begin{aligned}
F_a &= \frac{K_1}{32}(7 - 4\cos 2\theta - 3\cos 4\theta) \\
&+ \frac{K_2}{128}(2 - 4\cos 2\theta - 2\cos 4\theta) \\
&+ \frac{1}{2048}K_3(123 - 88\cos 2\theta - 68\cos 4\theta + 24\cos 6\theta + 9\cos 8\theta) \\
&+ \cdots
\end{aligned} \tag{5.9}$$

Exercise 5.2
Prove Eq. (5.9).

Let us now calculate the anisotropy energy with $\mathbf{M}$ on the (111) plane. One can use a new reference frame with z' along the [111] axis, x' parallel to the [1$\bar{1}$0] axis and y' in the (1$\bar{1}$0) plane. Let θ be the angle between $\mathbf{M}$ and x' in the $x' - y'$ plane. Then, $\alpha_1' = \cos\theta$, $\alpha_2' = \sin\theta$, $\alpha_3' = 0$. We obtain :

$$F_a = \frac{K_1}{4} + \frac{K_2}{108}(1 - \cos 6\theta) + \frac{K_3}{16} + \cdots \qquad (5.10)$$

Exercise 5.3
Prove Eq. (5.10).

The results obtained show that in all cases considered so far F_a can be written as

$$F_a = \sum_n A_{2n}(\phi) \cos 2n\theta \qquad (5.11)$$

For example, for uniaxial symmetry we obtain, upon comparing Eq. (5.11) with Eq. (5.5):

$$A_2 = -\frac{(K_{u1} + K_{u2})}{2} - \frac{15}{32} K_{u3}(1 + \cos 6\phi)$$

Exercise 5.4
Obtain the form of the A_{2n} in terms of the coefficients defined previously, for the different cases considered above.

We have up to this point described the general form of the anisotropic terms in the series expansion of Helmholtz free energy for several symmetry types of ferromagnets. It remains now to derive the quantum mechanical origin of these terms.

5.2 Microscopic origin of anisotropy

We obtained in Sect. 1.7 the form of the single-ion anisotropy energy operator as one of the effects of the spin-orbit interaction. Therein we considered perturbation terms up to second order. We must go to forth order to find relevant single-ion anisotropy contributions to the single ion anisotropy energy if the ion lattice is cubic, since in this case the second order correction to the gyromagnetic ratio g isotropic. The fourth order single-ion terms are usually written in the form:[4]

$$H_{eff} = \frac{a}{6}\{S_x^4 + S_y^4 + S_z^4 - \frac{1}{5}S(S + 1)[3S(S(S + 1) - 1]\} \qquad (5.12)$$

These terms are only relevant if $S > 1$, since otherwise the operator above reduces to a constant.

It turns out that the spin-orbit force also generates the *anisotropic interactions* among spins which are at the origin of the anisotropy energy that we described phenomenologically in the previous section.

Let us consider a pair of magnetic ions, which can be in general of different atomic species, each of which has a non-degenerate ground state $\Gamma_0^i, i = 1, 2$. The relevant interaction terms are

$$V = \lambda(\vec{L}_1 \cdot \vec{S}_1 + \vec{L}_2 \cdot S_2) + V_{ex} \tag{5.13}$$

We need to go now up to third order in V to obtain an effective spin-spin anisotropic Hamiltonian. Consider the following processes :

- the $\vec{L}_1 \cdot \vec{S}_1$ term excites ion 1 from $\Gamma_0^{(1)}$ to an excited state n_1.

- the exchange interaction couples the excited ion 1 to ion 2 in its ground state $\Gamma_0^{(2)}$.

- the $\vec{L}_1 \cdot \vec{S}_1$ term returns ion 1 to its ground state.

These processes are contained in the third order perturbation expansion of the complete Hamiltonian for the pair of ions.

The stationary Schrödinger equation for the two spin system reads :

$$H \mid \psi \rangle = (H_0 + \epsilon V) \mid \psi \rangle = E \mid \psi \rangle \tag{5.14}$$

where the auxiliary parameter ϵ is set equal to 1 at the end of the calculation. Just as in Sect. 1.7 we perform a standard perturbation expansion of H in powers of ϵ in the orbital subspace of the ions, whereby spins are treated as c-numbers. We simplify the notation and express the formal expansion as :

$$E = E_0 + \sum_{n=1}^{\infty} \epsilon^n E^{(n)}$$

$$\mid \psi \rangle = \mid 0_1 0_2 \rangle + \sum_{n=1}^{\infty} \epsilon^n a_{(n)}^{(1)} \mid n_1 \, 0_2 \rangle + a_{(n)}^{(2)} \mid 0_1 \, n_2 \rangle \tag{5.15}$$

where indices $0_i, n_i$ refer to ground and excited states of ion i and superindices in the coefficients denote the different ions. We disregard terms where both ions are excited, since these should be smaller.

Contributions to the expansion from both ions are additive, and we also have

$$\langle n_1 0_2 \mid 0_1 0_2 \rangle = \langle 0_1 n_2 \mid 0_1 0_2 \rangle = 0$$

All states are assumed normalized to 1. The same normalization condition is imposed on $\mid \psi \rangle$.

The third order correction from ion (1) is

$$\Delta_3^{(1)} = \sum_{n_1 n_2} \frac{\langle 0_1 0_2 \mid V \mid n_1 0_2 \rangle \langle n_1 0_2 \mid V \mid n_1' 0_2 \rangle \langle n_1' 0_2 \mid V 0_1 0_2 \rangle}{(E_0^{(1)} - E_{n_1})(E_0^{(1)} - E_{n_1'})} \tag{5.16}$$

where we allow for the ions being different and having consequently a different ground state energy.

Exercise 5.5
Obtain Eq. (5.16) above.

Since we are looking for a generalization of exchange interactions we select in particular terms which are of first order in V_{ex}, with matrix elements of the form

$$\langle n_1 0_2 \mid V_{ex} \mid n_1' 0_2 \rangle = -\mathcal{J}(n_1 0_2; n_1' 0_2) \mathbf{S_1} \cdot \mathbf{S_2} \tag{5.17}$$

where the sign is chosen so as to make the interaction ferromagnetic when $\mathcal{J} > 0$. Apart from factors which depend on the overlap integral between the ions (see Sect. 2.2) the matrix element in Eq. (5.17) is proportional to

$$\int d^3\mathbf{r}_1 \, d^3\mathbf{r}_2 \, \phi_{n_1}^*(\mathbf{r}_1 - \mathbf{R}_1)\phi_0^*(\mathbf{r}_2 - \mathbf{R}_2)\frac{1}{r_{12}}\phi_0(\mathbf{r}_1 - \mathbf{R}_2)\phi_{n_1'}(\mathbf{r}_2 - \mathbf{R}_1) \tag{5.18}$$

After some algebraic manipulations, the third order effective spin exchange interaction can be expressed as (with $\mu, \nu = x, y, z$) :

$$\Delta E_3 = -\sum_{\mu\nu} \left[S_1^\mu \Gamma_{\mu\nu}^{(1)} \mathbf{S}_1 \cdot \mathbf{S}_2 S_1^\nu + S_2^\mu \Gamma_{\mu\nu}^{(2)} \mathbf{S}_1 \cdot \mathbf{S}_2 S_2^\nu \right] \tag{5.19}$$

where we have added together terms from both ions and

$$\Gamma_{\mu\nu}^{(1)} = \lambda^2 \sum_{n_1, n_1'} \frac{(0 \mid L_\mu \mid n_1)\mathcal{J}(n_1 0_2; n_1' 0_2)(n_1' \mid L_\nu \mid 0)}{\left(E_{n_1} - E_0^{(1)}\right)\left(E_{n_1'} - E_0^{(1)}\right)} \tag{5.20}$$

with the corresponding expresion $\Gamma_{\mu\nu}^{(2)}$ for ion 2.

Exercise 5.6
Obtain Eq. (5.20).

In the case $S_1 = S_2 = 1/2$ the expression (5.19) reduces to

$$H_{aniso} = -\frac{1}{4}\sum_{\mu\nu}\sum_{i=1,2}$$
$$\left[\Gamma_{\mu\nu}^{(i)} + \Gamma_{\nu\mu}^{(i)} - \delta_{\mu\nu}\left(\Gamma_{xx}^{(i)} + \Gamma_{yy}^{(i)} + \Gamma_{zz}^{(i)} \right) \right] S_1^\mu S_2^\nu \tag{5.21}$$

This is called *pseudo-dipolar* interaction because it has the same form of the dipolar one (see Chap. 7).

Exercise 5.7
Prove Eq. (5.21).

This interaction is symmetric under the interchange of spins $1 \leftrightarrow 2$, but one can select some perturbation terms of second order in V, which are bilinear in the spin-orbit and the exchange potentials, and generate the *pseudo-scalar, Dzialoshinsky-Moriya* (DM) [5, 6] anti-symmetric interaction, as shown below. Let us consider the second order contribution

$$
\Delta E_2 = -\lambda \sum_{n_1} \frac{1}{E_{n_1} - E_0^{(1)}} \cdot (\langle 0_1 0_2 \mid V_{ex} \mid n_1 0_2 \rangle \langle n_1 \mid \mathbf{L}_1 \mid 0_1 \rangle \cdot \mathbf{S_1}
$$
$$
+ \langle 0_1 \mid \mathbf{L}_1 \rangle \cdot \mathbf{S_1} \langle n_1 \mid V_{ex} \mid 0_1 0_2 \rangle) + \cdots \qquad (5.22)
$$

where dots stand for a term of the same form involving excited states of ion 2. Since $\mathbf{L}$ is pure imaginary and hermitean, we have:

$$
\langle 0_1 \mid \mathbf{L}_1 \mid n_1 \rangle = \langle n_1 \mid \mathbf{L}_1 \mid 0_1 \rangle^* = - \langle n_1 \mid \mathbf{L}_1 \mid 0_1 \rangle .
$$

Assume the matrix elements of the exchange interaction are real. Then

$$
\Delta E_2 = -\lambda \sum_{n_1} \frac{\mathcal{J}(n_1 0_2 \,;\, 0_1 0_2) \langle 0_1 \mid L_1^\mu \mid n_1 \rangle [\; S_1^\mu, (\mathbf{S}_1 \cdot \mathbf{S}_2) \;]}{\left(E_{n_1} - E_0^{(1)} \right)} + \cdots \qquad (5.23)
$$

We use now the identity

$$
[\; \mathbf{S}_1, (\mathbf{S}_1 \cdot \mathbf{S}_2) \;] = -i \mathbf{S}_1 \wedge \mathbf{S}_2
$$

and obtain the DM interaction as:

$$
H_{DM} = \lambda \mathbf{D} \cdot \mathbf{S}_1 \wedge \mathbf{S}_2 \qquad (5.24)
$$

where

$$
\mathbf{D} = -i\lambda \left[\sum_{n_1} \frac{\mathcal{J}(n_1 0_2 \,;\, 0_1 0_2) \langle 0_1 \mid \mathbf{L_1} \mid n_1 \rangle}{\left(E_{n_1} - E_0^{(1)} \right)} \right.
$$
$$
\left. - \sum_{n_2} \frac{\mathcal{J}(0_1 n_2 \,;\, 0_1 0_2) \langle 0_2 \mid \mathbf{L_2} \mid n_2 \rangle}{\left(E_{n_2} - E_0^{(2)} \right)} \right] \qquad (5.25)
$$

If the middle point of ions 1 and 2 is a center of symmetry, interchanging the ions is a symmetry operation, and $\mathbf{D} = 0$. Otherwise, in general $\mathbf{D} \neq 0$. Rules concerning the direction of $\mathbf{D}$ for different symmetries were derived by Moriya [6].

Let us return to the general form of the anisotropic exchange interaction, Eq. (5.19). The r. h. s. has the form of a fourth order polynomial in the components of neighbouring spins, of the form :

$$
S_1^\mu \Gamma_{\mu\nu} S_1^\alpha S_2^\alpha S_1^\nu
$$

In a uniformly magnetized ferromagnet, and within the mean field approximation, we can substitute the spin components by their thermodynamic averages,
which are proportional to the corresponding components of the magnetization,
and we obtain for the anisotropy energy the form

$$E_{anis} = \Gamma_{\mu\nu} M^\mu \, M^\nu \, M^\alpha M^\alpha \qquad (5.26)$$

where repeated indices are summed overi.

For cubic symmetry

$$\Gamma_{\mu\nu} = \Gamma \cdot \delta_{\mu\nu}$$

so that one gets, appart from trivial isotropic terms:

$$E_{anis} = \Gamma[M_x^2 M_y^2 + M_y^2 M_z^2 + M_z^2 M_x^2 + M_x^4 + M_y^4 + M_z^4] \qquad (5.27)$$

Allowing for other sources of fourth order contributions, the general expression
for cubic symmetry is:

$$E_{anis} = A[M_x^2 M_y^2 + M_y^2 M_z^2 + M_z^2 M_x^2] + B[M_x^4 + M_y^4 + M_z^4] \qquad (5.28)$$

The second term above can be reduced to the same form as the first by using
the identity, satisfied by the director cosines:

$$\alpha_1^4 + \alpha_2^4 + \alpha_3^4 = 1 - 2(\alpha_1^2 \alpha_2^2 + \alpha_2^2 \alpha_3^2 + \alpha_3^2 \alpha_1^2)$$

so that we obtain the form of the fourth order term in eq. (5.6).

5.3   Magneto-elastic coupling

Inside one particular domain in a FM the magnetization is uniform, except of
course at the boundaries. We assume orthogonal crystal axes.

Let us consider a given pair of spins located on opposite ends of the bond
$\mathbf{r} = n_1 \mathbf{x} + n_2 \mathbf{y} + n_3 \mathbf{z}$ in terms of the unit vectors of the three orthogonal axes.
Inside a uniformly magnetized domain, spins are parallel. Their interaction
energy, including up to pseudo- quadrupolar terms, can be written as:

$$w(\mathbf{r}) = g(r) + l(r)(\cos^2(\phi) - 1/3)$$
$$+q(r)(\cos^4(\phi) - 6/7 \cos^2(\phi) + 3/35) \qquad (5.29)$$

where ϕ is the angle between the magnetization direction and the bond $\mathbf{r}$, and the
expansion is in terms of Legendre polynomials. The range functions $g(r)$, $l(r)$
and $q(r)$ correspond to the isotropic effective exchange, dipolar and quadrupolar
interactions respectively. In the following we neglect quadrupolar terms.

The isotropic term $g(r)$ in a strained crystal may only change if the length
of the crystal bonds change, while the dipolar and quadrupolar terms will also
change upon rotational deformation of the lattice. We shall see below that the
effect of strain on the dipolar terms can induce changes in the orientation of the
magnetization.

Under strain, the unit vectors of the crystal axes are transformed into the slightly deformed vectors [7]:

$$\mathbf{x}' = (1 + \epsilon_{xx})\mathbf{x} + \epsilon_{xy}\mathbf{y} + \epsilon_{xz}\mathbf{z}$$

$$\mathbf{y}' = \epsilon_{yx}\mathbf{x} + (1 + \epsilon_{yy})\mathbf{y} + \epsilon_{yz}\mathbf{z}$$

$$\mathbf{z}' = \epsilon_{zx}\mathbf{x} + \epsilon_{zy}\mathbf{y} + (1 + \epsilon_{zz})\mathbf{z}$$

where $\epsilon_{\alpha\beta}$ are components of the symmetric second rank strain tensor. We construct now the tensor obtained by calculating all the scalar products among the strained unit vectors (the strain tensor components, are assumed small, so we neglect second order terms):

$$e_{xx} \equiv (\mathbf{x}')^2 \approx 1 + 2\epsilon_{xx}$$

$$e_{xy} \equiv \mathbf{x}' \cdot \mathbf{y}' \approx \epsilon_{xy} + \epsilon_{yx} = 2\epsilon_{xy}$$

and the corresponding expressions for the other components, which are obtained by permutation. The elastic energy of the crystal can be expressed in terms of this tensor as [8]

$$E_{el} = \frac{1}{2}c_{11}\sum_{\alpha} e_{\alpha\alpha}^2 + \frac{1}{2}c_{44}\sum_{\alpha\neq\beta} e_{\alpha\beta}^2$$

$$+ c_{12}\sum_{\alpha\neq\beta} e_{\alpha\alpha}e_{\beta\beta} \qquad (5.30)$$

We shall now calculate the magneto-elastic energy. We consider separately the contributions of pairs of spins connected by bonds along each crystal axis. If the bond is along the x axis, the director cosines of the strained unit vector $\mathbf{x}'$ referred to the unstrained crystal axes, again to first order in the strain tensor, are:

$$\beta_1 \equiv \mathbf{x}' \cdot \mathbf{x} = 1 + \epsilon_{xx}$$

$$\beta_2 \equiv \mathbf{x}' \cdot \mathbf{y} = \epsilon_{xy}$$

$$\beta_3 \equiv \mathbf{x}' \cdot \mathbf{z} = \epsilon_{xz} \qquad (5.31)$$

The interaction energy is obtained as the change of the magnetic energy due to the changes in the length and the orientation of the bond vector under strain. If the bond length along x is a_0 before straining, under strain the new bond length is

$$a = a_0\left(1 + \frac{1}{2}e_{xy}\right)$$

We can write the effective dipolar interaction energy for the pair of first nearest neighbour spins along the x axis as :

$$w_x = l(a)\left(\sum_i \alpha_i\beta_i)^2 - 1/3\right) \qquad (5.32)$$

where a and β_i under strain were given above and α_i are the director cosines of the magnetization. We can obtain the change in the magnetic interaction energy for this particular pair as

$$\delta w_x = a_0 \left(\frac{\partial l}{\partial a} \right)_{a_o} e_{xx}(\alpha_1^2 - 1/3) + l(a_0)\alpha_1\alpha_2\, e_{xy} + l(a_0)\alpha_1\alpha_3\, e_{xz} \qquad (5.33)$$

For a simple cubic magnetic lattice the total change of dipolar energy per unit volume is the sum of the three terms obtained by permuting the x axis in Eq. (5.33) with y or z, and the result is

$$E_{mag-el} \;=\; B_1 \sum_{\nu=x,y,z} e_{\nu\nu}(\alpha_\nu^2 - 1/3) + B_2 \sum_{\nu\neq\mu} e_{\nu\mu}\alpha_\nu\alpha_\mu \qquad (5.34)$$

with

$$B_1 = N a_0 \left(\frac{\partial l}{\partial a} \right)_{a_o}$$
$$B_2 = 2N\, l(a_0) \qquad (5.35)$$

where N = the number of atoms per unit volume.

Exercise 5.8
Prove Eqs. (5.35).

The same expression (5.34) is valid for the other two cubic lattices, if the coefficients are defined as

$$B_1 = \frac{N}{2} \left(6l(a_0) + a_0 \left(\frac{\partial l}{\partial a} \right)_{a_o} \right)$$
$$B_2 = N \left(2l(a_0) + a_0 \left(\frac{\partial l}{\partial a}_{a_o} \right) \right) \qquad (5.36)$$

for the f.c.c. lattice, and

$$B_1 = \frac{8N}{3} l(a_0)\,, \;\; B_2 = \frac{8N}{9} \left(1 + a_0 \left(\frac{\partial l}{\partial a} \right)_{a_o} \right) \qquad (5.37)$$

for the b.c.c. lattice.

Exercise 5.9
Prove Eqs. (5.36), (5.37).

We must add to the magneto-elastic energy in Eq. (5.34) the purely elastic energy from Eq. (5.30) in order to obtain the total energy:

$$E_{tot} = E_{el} + E_{mag-el} \qquad (5.38)$$

The strain tensor components in equilibrium are obtained by imposing that the total energy be stationary:

$$\frac{\partial E_{tot}}{\partial e_{\alpha\alpha}} = \frac{\partial E_{tot}}{\partial e_{\alpha\beta}} = 0 \ , \quad \alpha, \beta = x, y, z \tag{5.39}$$

Exercise 5.10
Show that the solution to Eqs. (5.39) is:

$$e_{xy} = -\frac{B_2 \alpha_1 \alpha_2}{c_{44}}$$

$$e_{yz} = -\frac{B_2 \alpha_3 \alpha_2}{c_{44}}$$

$$e_{zx} = -\frac{B_2 \alpha_1 \alpha_3}{c_{44}}$$

$$e_{\nu\nu} = \frac{B_1}{c_{12} - c_{11}} \left(\alpha_\nu^2 - 1/3 \right) \tag{5.40}$$

Remark that the solution above implies $\sum_\nu e_{\nu\nu} \equiv$ dilation $= \delta V/V = 0$ [7].

5.4 Magnetostriction

In the presence of magnetization the length l of an element of material experiences a relative change of the order of $\delta l/l \sim 10^{-5} - 10^{-6}$, which although small has an important influence on the domain structure and on the "technical magnetization", which is the process whereupon the magnetization $M(H)$ grows under the application of an external field H until it reaches saturation.[9] Let us consider a uniformly magnetized sample. In this case we also expect a uniform elastic deformation, as described in the previous section. The relative elongation of a segment of length l oriented in the direction of the unit vector $\mathbf{n} = (\beta_1, \beta_2, \beta_3)$ is the same as the elongation of $\mathbf{n}$, namely:

$$\frac{\delta l}{l} = \mathbf{n}' \cdot \mathbf{n} - 1 = \sum_i \beta_i^2 e_{ii} + \sum_{i \neq j} \beta_i \beta_j e_{ij} \tag{5.41}$$

which reads, upon substitution of the equilibrium values for the strain tensor components given in Eqs. (5.40):

$$\frac{\delta l}{l} = \frac{B_1}{C_{12} - C_{11}} \left(\sum_i \alpha_i^2 \beta_i^2 - 1/3 \right) - \frac{B_2}{C_{44}} \sum_{i \neq j} \alpha_i \alpha_j \beta_i \beta_j \tag{5.42}$$

If the domain magnetization direction is [100] the elongation in the same direction is :

$$\left(\frac{\delta l}{l}\right)_{100} \equiv \lambda_{100} = 2/3\,\frac{B_1}{C_{12} - C_{11}} \tag{5.43}$$

while if both are in the [111] direction we get

$$\left(\frac{\delta l}{l}\right)_{111} \equiv \lambda_{111} = -1/3\,\frac{B_2}{C_{44}} \tag{5.44}$$

If $\lambda_{100} = \lambda_{111}$ the magnetostriction is isotropic. For this case, the uniform relative elongation (5.42) for a general direction becomes :

$$\frac{\delta l}{l} = \frac{3}{2}\lambda(\cos^2\theta - 1/3) \tag{5.45}$$

where $\theta =$ angle between the strain which is being measured and the magnetization direction.

In the general case, we can substitute in Eq. (5.42) the coefficients B_1 and B_2 in terms of λ_{100} and λ_{111}:

$$\frac{\delta l}{l} = \frac{3}{2}\lambda_{100}\left(\sum_i \alpha_i^2 \beta_i^2 - 1/3\right) + 3\lambda_{111}\sum_{i\neq j}\alpha_i\alpha_j\beta_i\beta_j \tag{5.46}$$

For a polycrystalline sample, we must perform the average of Eq. (5.46) for $\alpha_i = \beta_i$ over the unit sphere. The result is

$$\overline{\lambda} = \frac{2}{5}\lambda_{100} + \frac{3}{5}\lambda_{111} \tag{5.47}$$

Exercise 5.11
Prove Eq. (5.47).

For the cubic lattices, the fractional elongation along any direction can be expressed in terms of the ones along those two particular ones. For instance, we get

$$\lambda_{110} = \frac{1}{4}\lambda_{100} + \frac{3}{4}\lambda_{111} \tag{5.48}$$

The interested reader can find in Ref. [9] detailed information and many references on the magnetostrictive properties of specific magnetic materials, as well as on several techniques for measuring λ_{100} and λ_{111}.

5.5   Inverse magnetostriction

Just as the presence of magnetization induces deformation of the crystal lattice, the inverse also occurs, namely, changes in the magnetization caused by an external stress. We have derived the expression (5.34) for the magneto-elastic energy, which couples magnetization to strains. We could obtain from it the coupling of the magnetization with the stress components, if we make recourse to the equations of elasticity that relate both tensors.

We call the stress components $X_x, X_y, ..., Y_x \cdots$, where X_α is the force per unit surface oriented perpendicular to axis x in the direction $\alpha = x, y$ or z, etc. [7]

We defined the strain tensor components in Eq.(5.30) as a scalar product, which is symmetric. Therefore it has only 6 independent components. We can take advantage of this to simplify the notation. Let us make the following substitutions for the indices:

$$xx \equiv 1,\ yy \equiv 2,\ zz \equiv 3$$

$$yz \equiv 4,\ zx \equiv 5,\ xy \equiv 6$$

The linear relation between stress and strain is

$$\mathbf{e} = \mathbf{S\,X} \tag{5.49}$$

where the elements of $\mathbf{S}$ are called the *elastic compliance constants*, or in short just the *elastic constants*. The inverse relation is written as

$$\mathbf{X} = \mathbf{C\,e} \tag{5.50}$$

and the elements of $\mathbf{C}$ are the *elastic stiffness constants*, or *moduli of elasticity*. Clearly,

$$\mathbf{S} = \mathbf{C}^{-1} \tag{5.51}$$

The elastic energy can be written as a quadratic form in the six independent components of the strain tensor as

$$E_{el} = \frac{1}{2} \sum_{\lambda,\mu} \tilde{C}_{\lambda\mu} e_\lambda e_\mu \tag{5.52}$$

Stresses are canonically conjugate to strains. For instance,

$$X_x = \frac{\partial E_{el}}{\partial e_1} = \tilde{C}_{11} e_1 + \frac{1}{2} \sum_{\mu=2}^{6} \left(\tilde{C}_{1\mu} + \tilde{C}_{\mu\,1} \right) e_\mu \tag{5.53}$$

Comparison with (5.50) yields:

$$C_{\lambda\mu} = \frac{1}{2} \left(\tilde{C}_{\lambda\mu} + \tilde{C}_{\mu\lambda} \right) = C_{\mu\lambda} \tag{5.54}$$

This symmetry reduces the number of independent elements of the 6×6 **C** matrix from 36 to 21 in the most general case. For cubic symmetry there are many more constraints, and one is left with only 3 independent stiffness constants, the same being true for the inverse matrix $\mathbf{S}$ of the elastic constants. The structure of the $\mathbf{C}$ tensor for a cubic lattice is:

$$
\begin{pmatrix}
C_{11} & C_{12} & C_{12} & 0 & 0 & 0 \\
C_{12} & C_{11} & C_{12} & 0 & 0 & 0 \\
C_{12} & C_{12} & C_{11} & 0 & 0 & 0 \\
0 & 0 & 0 & C_{44} & 0 & 0 \\
0 & 0 & 0 & 0 & C_{44} & 0 \\
0 & 0 & 0 & 0 & 0 & C_{44}
\end{pmatrix}
\tag{5.55}
$$

The inverse $\mathbf{S}$ of this matrix, namely the elastic compliance tensor for the cubic lattice is:

$$
\begin{pmatrix}
S_{11} & S_{12} & S_{12} & 0 & 0 & 0 \\
S_{12} & S_{11} & S_{12} & 0 & 0 & 0 \\
S_{12} & S_{12} & S_{11} & 0 & 0 & 0 \\
0 & 0 & 0 & S_{44} & 0 & 0 \\
0 & 0 & 0 & 0 & S_{44} & 0 \\
0 & 0 & 0 & 0 & 0 & S_{44}
\end{pmatrix}
\tag{5.56}
$$

Exercise 5.12
Prove the following relations:

$$
S_{44} = \frac{1}{C_{44}} \quad ; \quad S_{11} - S_{12} = \frac{1}{C_{11} - C_{12}}
$$

Let us suppose that a uniaxial stress of magnitude σ (in units of pressure) is applied on the ferromagnet in the direction with cosines $(\gamma_1, \gamma_2, \gamma_3)$. The stress tensor components are therefore:

$$
X_{ij} = \sigma \, \gamma_i \gamma_j = X_{ji}
\tag{5.57}
$$

The strains can be calculated in terms of the applied stresses:

$$
\begin{aligned}
e_{xx} &= \sigma \left(S_{11} \gamma_1^2 + S_{12}(\gamma_2^2 + \gamma_3^2) \right) \\
e_{xy} &= \sigma S_{44} \gamma_1 \gamma_2
\end{aligned}
\tag{5.58}
$$

We substitute now these expressions for the strains into Eq. (5.34) and we obtain the magnetoelastic energy under a uniaxial stress as:

$$
\begin{aligned}
E_{mag-el} = {} & \sigma B_1 (S_{11} - S_{12}) \left(\sum_{i=1}^{3} \gamma_i^2 \alpha_i^2 - 1/3 \right) \\
& + \sigma B_2 S_{44} \sum_{i \neq j}^{3} \gamma_i \alpha_i \gamma_j \alpha_j
\end{aligned}
\tag{5.59}
$$

Upon use of Eqs. (5.43, 5.44) to eliminate B_1 and B_2 we arrive at:

$$E_{mag-el} = -\frac{3}{2}\sigma\lambda_{100}\left(\sum_{i=1}^{3}\gamma_i^2\alpha_i^2 - 1/3\right)$$

$$-3\sigma\lambda_{111}\sum_{i\neq j}^{3}\gamma_i\alpha_i\gamma_j\alpha_j \tag{5.60}$$

In Fe, $K_1 > 0$, and the easy axis is along one of the cubic axis. For $\mathbf{M}$ aligned in the [100] direction, we have $\alpha_1 = 1$, $\alpha_2 = \alpha_3 = 0$, and

$$E_{[100]}) = -\frac{3}{2}\sigma\lambda_{100}(\gamma_1^2 - 1/3) \tag{5.61}$$

Exercise 5.13
Show that for $\mathbf{M}$ in the [111] direction we get

$$E_{[111]}(\sigma) = -\frac{3}{2}\sigma(\cos^2(\phi) - 1/3) \tag{5.62}$$

where $\phi = $ angle between the uniaxial stress and [111].

We see that in the isotropic case the same form of Eqs. (5.61, 5.62) applies with $\lambda = \lambda_{100} = \lambda_{100}$.

In conclusion, the isotropic magnetostriction energy generates in this case a uniaxial anisotropy term in the spin Hamiltonian like the one we considered several times before, with the easy axis in the direction of the applied stress.

5.6   Induced magneto-crystalline anisotropy

We have written in 5.40 the explicit expressions for the equilibrium components of the strain tensor in a uniformly magnetized material, in terms of the director cosines of $\mathbf{M}$ referred to the cubic crystal axes. If one substitutes these expressions for the strain components into (5.30) and (5.34) for the two parts of the total energy, after some straightforward algebra one finds the result:

$$E_{tot}(\alpha_i) = \Delta K_1(\alpha_1^2\alpha_2^2 + \alpha_2^2\alpha_3^2 + \alpha_3^2\alpha_1^2) \tag{5.63}$$

where we have defined

$$\Delta K_1 \equiv -\left(\frac{B_1^2}{C_{12} - C_{11}} + \frac{B_2^2}{2\,C_{44}}\right) \tag{5.64}$$

The conclusion is that the crystal deformation produced by the magnetization induces, via the magneto-elastic coupling, a magneto-crystalline anisotropy energy of the same form as the one found in Sect. **5.2**. Since both effects act together, the measured values of K_1 must be corrected by subtracting ΔK_1 if we want to obtain quantitative information on the spin-orbit coupling constant Γ defined therein [9].

References

1. Van Vleck, J. H. (1937) *Phys. Rev.* **52**, 1178.

2. Landau, L. D. and Lifshitz, E. (1966) *"Mécanique Quantique"*, MIR Editions, Moscow.

3. Chikazumi, S. (1997) "The Physics of Ferromagnetism", second edition, Chap. 12, Clarendon Press, Oxford.

4. Yosida, Kei (1998) *"Theory of Magnetism"*, Chap. 3, Springer.

5. Dzyaloshinsky, I. (1958) *J. Phys. Chem. Solids* **4**, 241.

6. Moriya, T. (1960) *Phys. Rev.* **120**, 91.

7. Kittel, Charles (1971) *"Introduction to Solid State Physics"*, fourth edition, Chap. 4, John Wiley & Sons, Inc.

8. C. Kittel, *loc. cit.* pg. 140.

9. Chikazumi, S., *loc. cit.*, Chap. 14.

Chapter 6

Green's Functions Methods

6.1   Definitions

Many physical properties of magnetic systems depend on, or can be expresed in terms of, the correlation between pairs of spin operators. In turn, these quantities can be obtained from the corresponding Green's functions. Let us first remind the definitions of both. In the Heisenberg picture, the time dependent operators which correspond to physical observables are obtained from the corresponding constant (in time) Schrödinger picture operators as

$$A(t) = e^{iHt/\hbar} A e^{-iHt/\hbar} \tag{6.1}$$

which implies

$$i\hbar \frac{dA(t)}{dt} = [\, A(t), H \,] \tag{6.2}$$

We define the correlation function of two operators A and B as:

$$F_{AB}(t, t') \equiv \langle\, A(t)B(t') \,\rangle \tag{6.3}$$

where

$$\langle\, X \,\rangle = \frac{Tr(e^{-\beta H} X)}{Tr(e^{-\beta H})} \tag{6.4}$$

Exercise 6.1
Prove that, if H has no explicit time dependence,

$$F_{AB}(t, t') = f_{AB}(t - t') \tag{6.5}$$

We can define other functions of t and t', satisfying equation (6.2) with a singular inhomogeneous right hand side term, which would correspond to Green's differential equation associated with (6.2). To this end, let us first remark that

$$i\hbar \frac{d\langle\, A(t)B(t') \,\rangle}{dt} - \langle\, [\, A(t), H \,]B(t') \,\rangle = 0 \tag{6.6}$$

117

We can now introduce some new quantities related to the ones above, which are denoted by $\langle\langle\ A; B\ \rangle\rangle$, where:

$$\langle\langle\ A; B\ \rangle\rangle \equiv -i\theta(t - t')\langle\ [A(t), B(t')]\ \rangle \tag{6.7}$$

These are functions of t and t' which satisfy:

$$i\hbar\frac{d\langle\langle\ A; B\ \rangle\rangle}{dt} - \langle\langle\ [A(t), H\ ]; B\ \rangle\rangle = \hbar\delta(t - t')\langle\langle\ A; B\ \rangle\rangle \tag{6.8}$$

It is convenient to choose A and B such that, besides their being directly conected with the observables one wants to calculate, one has a simple enough inhomogeneous term in (6.8). One very convenient choice would be any one leading to a constant factor multiplying Dirac's δ in (6.8). If A and B are one-particle boson (or fermion) operators, their conmutator (or anticonmutator respectively) has this property. So, we are led to:

$$G^r_{AB}(t, t') = -i\theta(t - t')\langle\ [A(t), B(t')]_\eta\ \rangle \tag{6.9}$$

where

$$\theta(x) = \begin{cases} 1 & x \leq 0 \\ 0 & x < 0 \end{cases}, \quad \eta = \begin{cases} -1 & \text{fermions} \\ +1 & \text{bosons} \end{cases} \tag{6.10}$$

and $[A, B]_\eta \equiv AB - \eta BA$. We also define the advanced Green's function as

$$G^a_{AB}(t, t') = i\theta(t' - t)\langle\ [A(t), B(t')]_\eta\ \rangle \tag{6.11}$$

$G^{(r,a)}_{AB}$ are bi-linear funtionals of the operators A and B.

Exercise 6.2
Prove that:

$$i\hbar\frac{\partial G^{(r,a)}_{AB}(t, t')}{\partial t} = \hbar\delta(t - t')\langle\ [A(t), B(t')]_\eta\ \rangle$$
$$+ \quad \langle\langle\ [A(t), H\ ]; B(t')\ \rangle\rangle^{(r,a)} \tag{6.12}$$

Indices "r" and "a" refer to retarded and advanced respectively. The *causal* Green's function is defined as:

$$G^{(c)}_{AB} = -i\langle\ T_\eta(A(t)B(t'))\ \rangle \tag{6.13}$$

where

$$T_\eta\left(A(t)B(t')\right) \equiv \theta(t - t')A(t)B(t') + \eta\theta(t' - t)B(t')A(t) \tag{6.14}$$

T_η is called the *"time ordering"* operator.

Exercise 6.3

Prove that:

$$i\hbar \frac{\partial G_{AB}^{(j)}(t,t')}{\partial t} =$$

$$\hbar \delta(t-t') \langle\, [A,B]_\eta\, \rangle + \langle\langle\, [A(t),H]; B(t')\, \rangle\rangle^{(j)} \qquad (6.15)$$

with $(j) = (r,a,c)$.

We define the Fourier transform:

$$G(\omega) = \frac{1}{2\pi} \int_{-\infty}^{\infty} e^{i\omega t} G(t)\,dt \qquad (6.16)$$

with

$$G(t) = \int_{-\infty}^{\infty} e^{-i\omega t} G(\omega)\,d\omega \qquad (6.17)$$

So that the Fourier transform of (6.12) is:

$$\omega G_{AB}^{(j)} = \frac{1}{2\pi} \langle\, [A,B]_\eta\, \rangle + \langle\langle\, [A,H]; B\, \rangle\rangle_\omega \qquad (6.18)$$

6.2 Spectral representation

Let us consider the eigenvalue equation of an arbitrary Hamiltonian:

$$H \mid n\,\rangle = E_n \mid n\,\rangle \qquad (6.19)$$

where n represents the set of all quantum numbers which define the eigenkets. In general, the evaluation of the previously defined Green's functions involves the calculation of averages of products of certain operators A and B evaluated in the Heisenberg picture, that is of quantities like:

$$\langle\, B(t')A(t)\, \rangle \equiv Z^{-1} \sum_n e^{-\beta E_n} < n \mid e^{i\frac{H}{\hbar}t'} B e^{-i\frac{H}{\hbar}t'} e^{i\frac{H}{\hbar}t} A e^{-i\frac{H}{\hbar}t} \mid n >=$$

$$Z^{-1} \sum_n e^{-\beta E_n} e^{i\frac{E-n}{\hbar}t'} < n \mid B \mid m > \times$$

$$e^{-i\frac{E_m}{\hbar}(t-t')} < m \mid A \mid n > e^{-i\frac{E_n}{\hbar}t} =$$

$$Z^{-1} \sum_{n,m} e^{i\frac{(E_n-E_m)}{\hbar}(t'-t)} e^{-\beta E_n} \times$$

$$< n \mid B \mid m >< m \mid A \mid n > \qquad (6.20)$$

We call $\tau \equiv t - t'$ and then we Fourier transform (6.20) with respect to τ :

$$\frac{1}{2\pi} \int_{-\infty}^{\infty} e^{i\omega\tau} \langle\, B(t-\tau)A(t)\, \rangle d\tau =$$

$$Z^{-1} \sum_{n,m} e^{-\beta E_n} < n \mid B \mid m > \times$$

$$< m \mid A \mid n > \delta(\frac{E_n - E_m}{\hbar} - \omega) \equiv I_{AB}(\omega) \qquad (6.21)$$

$I_{AB}(\omega)$ is called the *spectral density* of this particular product of operators.

Exercise 6.4
Prove that:

$$\langle\, [A(t), B(t')]_\eta\, \rangle = \int_{-\infty}^{\infty} I_{AB}(\omega)(e^{\beta\omega} - \eta)e^{-i\omega(t-t')}d\omega \qquad (6.22)$$

The Fourier representation of the Heavyside step function is

$$\theta(t) = \frac{-i}{2\pi} \int d\omega \frac{e^{-i\omega t}}{\omega + i\epsilon}, \qquad \epsilon \to 0^+ \qquad (6.23)$$

Exercise 6.5
Prove (6.23)

With the help of (6.23), we get the *Lehman spectral representation* for the Green's functions:

$$G_{AB}^{(r,a)} = \frac{1}{2\pi} \int \frac{I_{AB}(\omega)(e^{\beta\omega} - \eta)}{E - \omega \pm i\epsilon}d\omega \equiv G_{AB}^{\pm} \qquad (6.24)$$

We use now the identity

$$\frac{1}{x \pm i\epsilon} = \mathcal{P}\frac{1}{x} \mp i\pi\delta(x) \qquad (6.25)$$

valid in the sense of the equivalence of distributions, and we obtain from (6.24) a simple relationship between the imaginary part of G_{AB} and the spectral density:

$$I_{AB}(E) = \frac{-2\mathcal{I}m G_{AB}^{(r)}(E)}{e^{\beta E} - \eta} \qquad (6.26)$$

which is one example of the fluctuation-dissipation theorem. Since the integrand in (6.24) is proportional to $\mathcal{I}m\, G$, that equation is one of the Kramers-Kronig reciprocal relations between the real and the imaginary part of an analytic function. In effect, we substitute (6.26) into (6.24) and obtain

$$\mathcal{R}e\, G_{AB}^{(r)}(E) = -\frac{1}{\pi}\mathcal{P}\int_{-\infty}^{\infty} \frac{\mathcal{I}m\, G_{AB}^{(r)}(\omega)}{E - \omega}d\omega \qquad (6.27)$$

Both functions in (6.24) can be considered as the two branches (in two different Riemann sheets) of the analytic function $G(E)$ which coincides with $G^{r(a)}$ for $\mathcal{I}m\, E > (<)0$. The branch $G^{r(a)}$ is analytic in the upper (lower) half-plane. $G(E)$ has a cut on the support of $I_{AB}(E)$ along the real axis [1, 2] . We can now express the correlation function in terms of the Green's functions.

Exercise 6.6

Prove that

$$\langle\, B(t')A(t)\,\rangle = -2 \int_{-\infty}^{\infty} e^{i\omega(t-t')} \frac{\mathcal{I}m\ G_{AB}^{(r)}(\omega)\eta}{e^{\beta\omega} - \eta} d\omega \tag{6.28}$$

Let us now prove that the retarded and advanced Green's functions obey the Kramers-Kronig relations. We already showed that they are analytic in the upper(retarded) or lower(advanced) half complex plane of the energy. Consider now the functions of the complex variable E

$$F^{\pm} = \oint \phi^{\pm}(\omega; E)d\omega \tag{6.29}$$

where

$$\phi^{\pm}(\omega; E) = \frac{G^{\pm}(\omega)}{\omega - E \pm i\epsilon} \tag{6.30}$$

and the contour for $F^+(F^-)$ goes along the whole real axis, around E on an infinitesimal semicircle in the clockwise(anti-clockwise) sense and closes along a semicircle at infinity in the upper(lower) half plane. Let us now show that $\phi^{\pm}(\omega; E)$ decreases as $|\omega| \to \infty$ faster than $1/|\omega|$. In this limit we can expand ϕ in inverse powers of ω:

$$\phi^{\pm}(\omega; E) = \frac{G^{\pm}(\omega)}{\omega}\left[1 + \sum_{n=1}^{\infty}\left(\frac{E \pm i\epsilon}{\omega}\right)^n\right] \tag{6.31}$$

For finite E we can neglect all but the first term in (6.31). We show below that

$$\lim G^{\pm}(\omega)\,|_{\omega\to\infty} = \frac{c_{AB}}{\omega} + \mathcal{O}(\omega^{-2}) \tag{6.32}$$

where c_{AB} is a constant depending on the operators A, B so that the contribution to (6.29) from the infinite semicircle is zero. Then, since the contour does not encircle any singularity of ϕ we get

$$F^{\pm}(E) \equiv 0. \tag{6.33}$$

If we now separate the real and imaginary parts of (6.33) we get

$$\begin{aligned}
\mathcal{R}e\ G^{\pm}(E) &= \frac{\pm}{\pi}\int_{-\infty}^{+\infty}\mathcal{I}m\ G^{\pm}(\omega)\mathcal{P}\frac{1}{\omega - E}d\omega \\
\mathcal{I}m\ G^{\pm}(E) &= \frac{\mp}{\pi}\int_{-\infty}^{+\infty}\mathcal{R}e\ G^{\pm}(\omega)\mathcal{P}\frac{1}{\omega - E}d\omega
\end{aligned} \tag{6.34}$$

which are Kramers-Kronig relations.

It remains to verify (6.32). To this end consider the spectral representation (6.24). For large E we expand in inverse powers of E. Let us calculate the zeroth-order term:

Exercise 6.7
Show that

$$\int I_{AB}(\omega)(e^{\beta\omega} - \eta)d\omega = c_{AB} = \langle\, AB - \eta BA \,\rangle \tag{6.35}$$

which we assume to be bounded

The whole series in (6.32) can be shown to converge if the spectrum of H is bounded. We assume then that there exists a constant $0 < W < \infty$ such that

$$\mid E_n \mid < W, \quad \forall n \tag{6.36}$$

Exercise 6.8
Show that the series in (6.32 is bounded by a convergent geometric series if: a) the inequality (6.35) is satisfied, and b): the operators A, B have a bounded spectrum.

The functions mutually related by the Kramers-Kronig relations are a pair of mutual Hilbert transforms. We have proved therefore that the real and imaginary parts of an analytic function are Hilbert transforms of one another. Kramers-Kronig relations appear in the theory of linear response (dielectric constant, electric conductivity, magnetic susceptibility, complex admittance of circuits, etc.) and in dispersion relations in the theory of scattering [1].

We mention now one useful symmetry property:

Exercise 6.9
Prove that

$$G^{+}_{AB}(\omega) = \left(\, G^{+}_{A^{\dagger}B^{\dagger}}(-\omega) \,\right)^{*} \tag{6.37}$$

6.3   RPA for spin 1/2 ferromagnet

Consider the Heisenbeg exchange Hamiltonian with no external field, $S = \frac{1}{2}$ and $J_{lm} \geq 0$. Suppose we want to calculate the statistical average of $\langle\, S^z_m \,\rangle$ which, for a translationally invariant system, is site-independent, and determines the average magnetization. The identity $\mathbf{S} \cdot \mathbf{S} = S(S+1)$ yields, for spin $S = \frac{1}{2}$,

$$S^- S^+ = \frac{1}{2} - S^z \tag{6.38}$$

or

$$\langle\, S^-_n S^+_n \,\rangle = \frac{1}{2} - \langle\, S^z_n \,\rangle \tag{6.39}$$

Therefore

$$\langle\, S^z_n \,\rangle = \sigma(T) = \frac{1}{2} - \langle\, S^-_n S^+_n \,\rangle \tag{6.40}$$

Since we need the equal time correlation function of S_n^- , S_n^+, we consider the Green's function

$$G_{lm}^{(+)}(t - t_0) \equiv \langle\langle\ S_l^+(t); S_m^-(t_0)\ \rangle\rangle =$$
$$-i\theta(t - t_0)\langle\ [S_l^+(t), S_m^-(t_0)]\ \rangle \qquad (6.41)$$

where $\eta = +1$ and $(+)$ stands for retarded. Now, (6.18) reads:

$$\omega G_{lm}^{(+)}(\omega) = \frac{1}{\pi}\delta_{lm}\langle\ S_l^z\ \rangle$$
$$-\ 2\sum_{i\neq l} J_{il}(\langle\langle\ S_i^+ S_l^z; S_m^-\ \rangle\rangle_\omega - \langle\langle\ S_i^z S_l^+; S_m^-\ \rangle\rangle) \qquad (6.42)$$

Exercise 6.10
Prove equation (6.42).

The "RPA" (random phase approximation) consists of neglecting the correlations between S_i^z and S_j^+ for $i \neq j$, and substituting in (6.42) S_i^z by $\langle\ S_i^z\ \rangle$, the statistical average, whenever it appears as argument of the Green's function:

$$\langle\langle\ S_i^+ S_l^z; S_m^-\ \rangle\rangle_\omega \sim \langle\ S_l^z\ \rangle\langle\langle\ S_i^+; S_m^-\ \rangle\rangle \qquad (6.43)$$

The relation of (6.43) to a random phase argument can be established by rewriting the Heisenberg Hamiltonian and the equation of motion for the Green's functions in k-space [3]. Let us now apply (6.43) to solve (6.42) approximately, following Bogoliubov and Tyablikov [4], for all T, in the case of a spin $\frac{1}{2}$ ferromagnet. Since the system is translationaly invariant by hypothesis, that is $J_{il} = J(\mathbf{R_i} - \mathbf{R_l})$,

$$G_{lm}^{(+)}(\omega) = \langle\langle\ S_i^+; S_m^-\ \rangle\rangle_\omega^{(+)} = \frac{1}{N}\sum_k e^{i\mathbf{k}\cdot(\mathbf{R_i}-\mathbf{R_m})} G_{\mathbf{k}}^{(+)}(\omega) \qquad (6.44)$$

Therefore, substituting (6.43) and (6.44) into (6.42), we obtain the FT of G:

Exercise 6.11
Prove that

$$G_k^{(+)}(\omega) = \frac{\sigma(T)}{\pi(\omega + i\epsilon - \omega(\mathbf{k}))} \qquad (6.45)$$

where

$$\tilde{J}(\mathbf{k}) \equiv \sum_{\mathbf{R}} J(\mathbf{R})e^{i\mathbf{k}\cdot\mathbf{R}} \qquad (6.46)$$

and

$$\omega(\mathbf{k}) = 2\sigma(T)[\tilde{J}(0) - \tilde{J}(\mathbf{k})] \qquad (6.47)$$

We can now apply the identity (6.40) to calculate $\sigma(T)$. First, we need the equal time correlation function $\langle\, S_n^-(t)S_m^+(t)\,\rangle$ (which is time independent). We have (6.28) at our disposal, which reduces, for $t = t'$ to:

$$\langle\, BA\,\rangle = -2\int d\omega \frac{\mathcal{I}mG_{AB}^{(+)}(\omega)}{e^{\beta\omega}-1} \tag{6.48}$$

For the special choice $B = S_i^-$, $A = S_i^+$,

$$\langle\, S_i^- S_i^+\,\rangle = -\frac{2}{N}\sum_k \int d\omega \frac{\mathcal{I}mG_k^{(+)}(\omega)}{e^{\beta\omega}-1} \tag{6.49}$$

Now,

$$\mathcal{I}mG_k^{(+)}(\omega) \equiv \mathcal{I}mG_k(\omega+i\epsilon) =$$
$$-\pi\delta\left(\omega - 2\sigma(T)(\tilde{J}(0) - \tilde{J}(k))\right)\frac{\sigma(T)}{\pi} \tag{6.50}$$

then

$$\langle\, S_i^- S_i^+\,\rangle = 2\sigma(T)\Psi(T) \tag{6.51}$$

where

$$\Psi(T) = \frac{1}{N}\sum_k \frac{1}{e^{\beta\omega_k}-1} \tag{6.52}$$

and

$$\omega_k = 2\sigma(T)(\tilde{J}(0) - \tilde{J}(k)) \geq 0 \tag{6.53}$$

Substituting $\langle\, S_i^- S_i^+\,\rangle$ from (6.51) into identity (6.40) we have

$$\sigma(T) = \frac{\frac{1}{2}}{1 + 2\Psi(T)} \tag{6.54}$$

The quantity $\Psi(T)$ was introduced in chapter 4, as the statistical average of the total number of spin deviations in a FM within the FSWA. In the present formalism, however, it is clear that we cannot maintain that interpretation, unless $\Psi \to 0$. We are led to the conclusion that as regards the spin deviation, the FSWA expression is obtained upon expanding the RPA formula in a series in Ψ and retaining only terms up to first order. We shall see later on that the spin deviation in the FSWA is obtained from the RPA formula for general S, in the limit $S \to \infty$.

Exercise 6.12
Prove that

$$\lim_{T\to 0} \Psi(T) = 0 \tag{6.55}$$

which implies

$$\sigma(T)\,|_{T\to 0} \to \frac{1}{2} \tag{6.56}$$

If we assume that there is a finite temperature T_c, such that

$$\Psi(T) \longrightarrow \infty \ , \quad T \nearrow T_c \tag{6.57}$$

then we can prove that this singularity gives rise to the phase transition from the low T ordered FM phase to the high T disordered PM phase. Let us assume then that as $T \nearrow T_c$, $\sigma \to 0$. From (6.52) we have in 3d,

$$\Psi(T) = \frac{v_0}{8\pi^3} \int_{1^{st}BZ} d^3\mathbf{k} \frac{1}{e^{2\beta\sigma\epsilon_k} - 1} \tag{6.58}$$

where $\omega_k = 2\sigma \left(\tilde{J}(0) - \tilde{J}(\mathbf{k}) \right)$. Since σ is small by assumption near T_c we expand the exponential in the denominator of (6.58) in a Taylor series. The dominant term in the expansion of Ψ is

$$\Psi = \frac{v_0}{16\pi^3\sigma\beta} \int \frac{d^3\mathbf{k}}{\epsilon_k} + \mathcal{O}(1) \tag{6.59}$$

On the other hand, from (6.54) we get:

$$2\sigma\Psi \big|_{T\nearrow T_c} \to \frac{1}{2} \tag{6.60}$$

$$\langle \frac{1}{f(k)} \rangle = F^{-1} = \frac{V}{8\pi^3} \int \frac{d^3\mathbf{k}}{1 - \gamma_k} \tag{6.61}$$

For any f, $\langle f(\mathbf{k}) \rangle$ denotes the average of $f(\mathbf{k})$ over the first Brillouin zone. In the literature [5] the moments of

$$f(k) \equiv 1 - \gamma_k$$

are denoted as

$$F^{(n)} \equiv \langle (f(k))^n \rangle^M \tag{6.62}$$

so that in particular

$${}^M\langle \frac{1}{f(k)} \rangle = F^{-1} = \frac{v_0}{8\pi^3} \int \frac{d^3\mathbf{k}}{1 - \gamma_k}^M \tag{6.63}$$

Comparing (6.60) and (6.58) we obtain the critical temperature:

$$\beta_c \equiv \frac{1}{k_B T_c} = \frac{2V}{8\pi^3} \int \frac{d^3\mathbf{k}}{\tilde{J}(0) - \tilde{J}(\mathbf{k})} = \frac{2}{\tilde{J}(0)} \langle \frac{1}{1 - \gamma_\mathbf{k}} \rangle \tag{6.64}$$

where $\gamma_k \equiv \tilde{J}(k)/\tilde{J}(0)$. For first n. n. range exchange we get

$$\beta_c J \nu = 2F^{(-1)} \tag{6.65}$$

ν being the number of first n. n. in the lattice.

6.4   Comparison of RPA and MFA

We get the mean field approximation if we neglect completely the non diagonal terms in the matrix equation (6.42) for the Green-function, which means that (6.42) reduces to:

$$\omega G_{lm}^{(+)MFA}(\omega) = \frac{\delta_{lm}\langle S_l^z \rangle}{\pi} + 2\sum_{i\neq l} J_{il}\langle S_i^z \rangle \langle\langle S_l^+; S_m^- \rangle\rangle \tag{6.66}$$

or

$$G_{lm}^{(+)MFA}(\omega) = \frac{\delta_{lm}\sigma}{\pi(\omega - 2\sigma\tilde{J}(0) + i\epsilon)} \tag{6.67}$$

Exercise 6.13
Prove that for first n. n. exchange the form (6.67) for G and Eq. (6.54) yield

$$\sigma(T) = \frac{1}{2}\tanh{(\beta\sigma\nu J)} \tag{6.68}$$

which is the MFA result.

For $\sigma \to 0$, (6.68) leads to:

$$\beta_c \nu J = 2 \tag{6.69}$$

Comparing (6.69) with (6.65),

$$\frac{1}{k_B T_c^{RPA}} = \frac{1}{k_B T_c^{MFA}} F^{(-1)} \tag{6.70}$$

Since $F^{(-1)} \geq 1$, $T_c^{RPA} \leq T_c^{MFA}$.
  By expanding Ψ near $T = T_c$ in powers of σ one finds:

$$\Psi = \frac{T}{4T_c\sigma} - \frac{1}{2} + \alpha\sigma\frac{T_c}{T} + \ldots \tag{6.71}$$

which allows to obtain the behaviour of σ as $(T \to T_c)$:

$$\sigma \sim \frac{1}{\sqrt{\alpha}}\sqrt{1 - \frac{T}{T_c}} \tag{6.72}$$

This implies that the RPA yields the same critical exponent $\beta = 1/2$ for σ as the MFA.

Exercise 6.14
Verify equations (6.71) and (6.72), and check that

$$\alpha = \frac{1}{3}F^{(1)}F^{(-1)}$$

6.5   RPA for arbitrary spin

This was first derived by R. Tahir-Kheli and D. Ter-Haar [6] and later simplified and extended by H. Callen [7], whose treatment we shall now follow. First, we define a new Green's function which depends on a parameter a :

$$G^{(r)}(l - m, t; a) = -i\theta(t)\langle\, [\, S_l^+(t), B_m\,]\, \rangle \tag{6.73}$$

where we define the operator B_m as

$$B_m = e^{aS_m^z} S_m^- \tag{6.74}$$

and we apply again the RPA to the Green's function which results upon calculating the time derivative of (6.73) (we take $\hbar = 1$):

$$i\frac{\partial G^{(r)}(l - m, t; a)}{\partial t} =$$
$$\delta(t)\delta_{lm}\langle\, [\, S_l^+(t), B_m\,]\, \rangle +$$
$$\sum_{j\neq l} J_{(lj)}\, [\, \langle\langle\, S_j^z S_l^+; B_m\, \rangle\rangle - \langle\langle\, S_l^z S_j^+; B_m\, \rangle\rangle\,] \tag{6.75}$$

We obtain, for the ($\mathbf{k}$, ω)-Fourier transform of (6.75) in RPA,

$$G^{(r)}(k\omega; a) = \frac{\langle\, [S_l^+, B_m]\, \rangle}{\omega - \omega(k) + i\epsilon} \tag{6.76}$$

where as in (6.45),

$$\omega(k) = 2\sigma(T)[\tilde{J}(0) - \tilde{J}(k)] \tag{6.77}$$

For a uniform (infinite) system, σ is site independent, and so is the statistical average on the right hand side of equation (6.76). We apply the general relation (6.28) between the correlation function and the retarded Green's function, and find:

$$\langle\, e^{aS_l^z} S_l^- S_l^+\, \rangle = \frac{1}{N}\sum_k \frac{\langle\, [\, S_l^+, e^{aS_l^z} S_l^-\,]\, \rangle}{e^{\beta\omega(k)} - 1} \tag{6.78}$$

Both averages appearing in (6.78) can be obtained if one calculates the function (l-independent)

$$\Phi(a) = \langle\, e^{aS_l^z}\, \rangle \tag{6.79}$$

Let us first look at the l. h. s. of (6.78). We use the identity

$$S^- S^+ = S(S + 1) - (S^z)^2 + S^z$$

to write the average as

$$\langle\, e^{aS_l^z} S_l^- S_l^+\, \rangle =$$
$$S(S + 1)\Phi(a) - \Phi''(a) + \Phi'(a) \tag{6.80}$$

Exercise 6.15
Prove Eq. (6.80), where $\Phi'(a)$, $\Phi''(a)$ are the first and second derivatives of $\Phi(a)$.

The average on the r.h.s. of (6.78) requires the explicit calculation of

$$\Theta(a) \equiv \langle\, [\, S^+, e^{aS^z} S^- \,]\, \rangle \tag{6.81}$$

To this end, we use the identity

$$[\, S^+, (S^z)^n \,] \equiv X_n \equiv [\, (S^z - 1)^n - (S^z)^n \,] S^+ \tag{6.82}$$

In order to obtain (6.82), first

Exercise 6.16
Prove the recursion relation:

$$X_n = X_{(n-1)} S^z - (S^z)^{(n-1)} S^+ \tag{6.83}$$

and then prove (6.82) by applying Péano complete induction principle of algebra.

We can now calculate $\Theta(a)$:

Exercise 6.17
Prove that:

$$\begin{aligned}
\Theta(a) = {}& S(S+1)(e^{-a} - 1)\Phi(a) \\
& + (e^{-a} + 1)\Phi'(a) - (e^{-a} - 1)\Phi''(a)
\end{aligned} \tag{6.84}$$

We use now the function $\Psi(T)$ defined in (6.52), and substitute (6.80), and (6.84) into (6.78), obtaining a differential equation for $\Phi(a)$:

$$\Phi''(a) - \Phi'(a)\frac{1 + \Psi + e^a\Psi}{1 + \Psi - e^a\Psi} - S(S+1)\Phi(a) = 0 \tag{6.85}$$

whose complete solution requires two boundary conditions. The first one is:

$$\Phi(0) = 1 \tag{6.86}$$

The second boundary condition is obtained from the identity (spectral decomposition of S^z)

$$\prod_{p=-S}^{S} (\, S^z - p \,) \equiv 0 \tag{6.87}$$

whose statistical average can be expressed in terms of derivatives of $\Phi(a)$ at $a = 0$:

$$\left(\prod_{p=-S}^{S}(\frac{d}{da} - p)\right) \Phi(a)|_{a=0} = 0 \tag{6.88}$$

Callen [7] obtained the solution of (6.85) with the boundary conditions (6.86) and (6.88):

$$\Phi(a) = \frac{\Psi^{2S+1}e^{-Sa} - (1+\Psi)^{2S+1}e^{(S+1)a}}{[\Psi^{2S+1} - (1+\Psi)^{2S+1}][(1+\Psi)e^a - \Psi]} \tag{6.89}$$

Now we can calculate $\langle S_z \rangle$:

$$\langle S_z \rangle = \Phi'(0) = $$
$$\frac{(S-\Psi)(1+\Psi)^{2S+1} + (S+1+\Psi)\Psi^{2S+1}}{(1+\Psi)^{2S+1} - \Psi^{2S+1}} \tag{6.90}$$

One verifies that for $S = 1/2$ this equation reduces to the result of last section.

6.6 RPA for ferromagnets

6.6.1 Paramagnetic phase

Let us now consider a uniform static external field B applied along the z axis. The initial $(B \to 0)$ longitudinal susceptibility of a FM, χ_{zz}, is defined as

$$V\chi_{zz} = \lim_{B \to 0} \frac{\gamma N \sigma}{B} \quad , \quad \sigma = \langle S_n^z \rangle \tag{6.91}$$

where V is the total volume of the system. In order to calculate the longitudinal response function one cannot use G_{zz}, since it vanishes identically, because $[S_R^z, S_{R'}^z] \equiv 0$, with R, R' arbitrary sites on the lattice. We can use instead G_{xx} or G_{yy} to this end, where

$$G_{xx} = \langle\langle S_l^x; S_m^x \rangle\rangle$$

and the corresponding definition for G_{yy}. If we write the perturbed Hamiltonian in the presence of the field :

$$H = H_0 - \gamma B \sum_R S_R^z \tag{6.92}$$

where H_0 is the Heisenberg Hamiltonian, we verify that the only change in the expression (6.76) for the Green's function is to add the term γB to the frequency $\omega(\mathbf{k})$. We call G^{+-} the Green's function (6.76) for $a = 0$. Upon substituting $S^{x,y}$ in terms of $S^{\pm}$ we find

$$\begin{aligned} G_{xx} &= 1/4(G^{++} + G^{+-} + G^{-+} + G^{--}) \\ G_{yy} &= 1/4(G^{++} - G^{+-} - G^{-+} + G^{--}) \end{aligned} \tag{6.93}$$

with a simplified notation referring to the spin operators involved. Due to total spin conservation, $G^{++} = G^{--} = 0$.

If we only consider first n. n. exchange,

$$G_k^{+-}(\omega) = \frac{\sigma}{\pi} \cdot \frac{1}{\omega + i\epsilon - 2\sigma J\nu(1 - \gamma_k) - h} \tag{6.94}$$

where $h = \gamma B$.

For $k = 0$, the case of a uniform field, the dispersion relation $\omega(k) = 0$ and we have

$$G^{xx}(k = 0, \omega) = G^{yy}(k = 0, \omega) =$$

$$\frac{1}{4N} \sum_R \left[G^{+-}(R, \omega) + G^{-+}(R, \omega) \right] \qquad (6.95)$$

Appealing again to total spin conservation,

$$\left[\sum_R S^+(R), H_0 \right] = 0 \qquad (6.96)$$

and therefore we have exactly

$$G^{+-}(k = 0, \omega) = \frac{\sigma}{\pi} \frac{1}{\omega - \gamma B + i\epsilon}$$

$$G^{-+}(k = 0, \omega) = -\frac{\sigma}{\pi} \frac{1}{\omega + \gamma B + i\epsilon} \qquad (6.97)$$

We find

$$G^{xx}(k = 0, \omega) = \frac{\sigma}{4\pi} \left(\frac{1}{\omega - \gamma B} - \frac{1}{\omega + \gamma B} \right) \qquad (6.98)$$

And finally

$$V\chi_{zz} = -\lim_{B \to 0} 2\pi N \gamma^2 G^{xx}(k = 0, \omega = 0) \qquad (6.99)$$

The self-consistency condition (6.90) involves the function Ψ defined in (6.52) We have seen that in 1 or 2 dimensions Ψ diverges at any finite T, and consequently $\sigma = 0$ for $T > 0$. In other words, a Heisenberg FM in low dimensions at $T > 0$ is in the paramagnetic state.

At $T = 0$ however, if we let $B \to 0$ *after* $T \to 0$ we get $\Psi = 0$, and so for a Heisenberg FM in any dimension there is long range order at $T = 0$.

In the paramagnetic state, as $h \to 0$, $\sigma \to 0$ as well, so we define the ratio

$$\lambda = \lim_{h \to 0} \left(\frac{h}{2\sigma\nu J} \right) = \frac{\gamma^2 \chi_{zz}^{-1}}{2J\nu} \qquad (6.100)$$

Then $\omega(\mathbf{k}) = 2\sigma J\nu(1 + \lambda - \gamma_{\mathbf{k}}) \to 0$ as $h \to 0$. Now we can take the limit $h \to 0$ of Ψ [8] in the PM phase, and obtain:

$$\lim_{h \to 0} \sigma\Psi = S(S + 1)/3 \qquad (6.101)$$

which leads to an integral equation for λ:

$$\frac{1}{N} \sum_k \frac{1}{1 - \gamma_k + \lambda(T)} = \frac{T_c^{MFA}}{T} \qquad (6.102)$$

Solving (6.102) for each T gives us χ_{zz} , according to (6.100). In less than $3d$ we follow this program for any $T > 0$. In $3d$ the spontaneous magnetization $\sigma \neq 0$ for $T < T_c$, so that $\lambda \equiv 0$ in the ordered FM phase. Precisely at the critical temperature, $\sigma = 0$, but we know this is the temperature at which the susceptibility diverges, so that $\lambda(T_c^{RPA}) = 0$, and imposing $\lambda = 0$ we obtain the equation for T_c^{RPA}

$$\frac{1}{N} \sum_k \frac{1}{1 - \gamma_k} = \frac{T_c^{MFA}}{T_c^{RPA}} \tag{6.103}$$

For $T > T_c^{RPA}$ (paramagnetic phase) $\lambda \neq 0$ is the solution of the integral equation (6.102), valid for any dimension and any spin S within RPA.

At very large $T \gg T_c^{MFA}$, the r. h .s. of (6.102) is small, so that λ must be large. We call

$$\eta_k \equiv \frac{1 - \gamma_k}{\lambda} \quad ,$$

which is small if $\lambda \gg 1$, and we expand (6.102) in a geometric series in η_k. We define the small parameter $\alpha \equiv T_c^{MFA}/T$. Up to terms of order λ^{-2} we find

$$\frac{\lambda - 1}{\lambda^2} = \alpha \tag{6.104}$$

The large solution of (6.104) for λ is

$$\lambda \approx \frac{1 - \alpha}{\alpha} = \frac{T - T_c^{MFA}}{T_c^{MFA}} \tag{6.105}$$

which is Curie-Weiss law. *Notice that extrapolating this line, λ vanishes at T_c^{MFA}.*

We have seen that $T_c^{RPA} < T_c^{MFA}$. The fact that the extrapolation of the Curie-Weiss law to low T gives an intersection *above* T_c^{RPA} agrees with experiments.

We can also calculate the equal-time correlation function

$$\langle \, \mathbf{S_0} \cdot \mathbf{S_R} \, \rangle$$

in the paramagnetic phase. Since in this phase the system is isotropic in the absence of a magnetic field , we have

$$\langle \, S_0^x S_R^x \, \rangle = \langle \, S_0^y S_R^y \, \rangle = \langle \, S_0^z S_R^z \, \rangle \tag{6.106}$$

which implies

$$\langle \, \mathbf{S_0} \cdot \mathbf{S_R} \, \rangle = 3/2 \langle \, S_0^- S_R^+ \, \rangle \tag{6.107}$$

The transverse correlation function on the r.h.s. above can be obtained as before in terms of the Green's function, by application of the fluctuation-dissipation theorem. We obtain

$$\langle \, \mathbf{S_0} \cdot \mathbf{S_R} \, \rangle = \frac{3T k_B}{2 \nu J} \frac{1}{N} \sum_k \frac{e^{-i\mathbf{k} \cdot \mathbf{R}}}{1 - \gamma_k + \lambda} \tag{6.108}$$

Exercise 6.18
Prove Eqs. (6.107) and (6.108).

The high T limit (6.105) is valid at any dimension, but in the low T limit for one and two space dimensions one needs to work explicitly each case to obtain λ and χ. Let us then look at the linear chain first.

6.6.2   Linear FM chain

We write the summation in (6.102) as an integration inside the first BZ, as usual, and call $T_c^{MFA} = T_1$ for $d = 1$. The integral is standard and we get

$$\frac{1}{\sqrt{\lambda(\lambda + 2)}} = \frac{T_1}{T} \tag{6.109}$$

which is valid at any $T > 0$. From (6.100) we have

$$\chi_{zz} = \frac{\gamma^2}{2\nu J \lambda} \tag{6.110}$$

which for $d = 1$, with $\nu = 2$, leads to

$$\chi_{zz}^{1d} = \frac{\gamma^2}{4J(\sqrt{1 + T^2/T_1^2} - 1)} \tag{6.111}$$

Exercise 6.19
Prove Eqs. (6.109) to (6.111).

For $T \gg T_1$,

$$\chi_{zz}^{1d} \approx \frac{\gamma^2 T_1}{4J(T - T_1)} \tag{6.112}$$

while for $T \to 0$ we get the divergent form

$$\chi_{zz}^{1d} \approx \frac{\gamma^2 T_1}{T^2} \tag{6.113}$$

Let us now calculate the transverse equal-time correlation function, from which we can extract the correlation length in the PM phase:

$$\langle\, S_0^- S_R^+ \,\rangle = \frac{1}{\beta J}\frac{1}{N}\sum_k \frac{e^{ikR}}{1 - \cos ka + \lambda} \tag{6.114}$$

The integral is again standard, and the result is

$$\langle\, S_0^- S_R^+ \,\rangle = \frac{1}{\beta J \sqrt{\lambda(\lambda + 2)}}\, e^{-R/\xi} \tag{6.115}$$

where the correlation length for the FM chain is

$$a\xi^{-1} = -\log\left[\,1 + \lambda - \sqrt{\lambda(\lambda+2)}\,\right] \tag{6.116}$$

Exercise 6.20
Verify Eqs. (6.115) and (6.116).

We obtain the limits

$$\xi_{T\to 0} = aT_1/T \tag{6.117}$$

and

$$\xi_{T\to\infty} = \frac{a}{\log(2T/T_1)} \tag{6.118}$$

6.6.3 Square FM lattice

In order to obtain an explicit expression for λ from Eq. (6.102) we need to calculate the integral

$$I(\lambda) = \frac{1}{4\pi^2}\int_{-\pi}^{\pi} dX \int_{-\pi}^{\pi} dY \frac{1}{1 + \lambda - 1/2(\cos X + \cos Y)} \tag{6.119}$$

where we use the notation $ak_x = X$, $ak_y = Y$. It is very convenient to make a change of variables, defining

$$u = 1/2(X + Y)\,,\ v = 1/2(X - Y)\ .$$

The integration region in the (u, v) plane has twice the area of the BZ. The integral in the new variables reads, with due account of the factor $1/2$ from the Jacobian,

$$I(\lambda) = \frac{1}{\pi^2}\int_{0}^{\pi} du \int_{0}^{\pi} dv \frac{1}{1 + \lambda - \cos u \cos v} \tag{6.120}$$

The double integral can be reduced to a single one by use of the formula:

$$\frac{1}{\pi}\int_{0}^{\pi} \frac{d\alpha}{A - B\cos\alpha} = \frac{1}{\sqrt{A^2 - B^2}} \tag{6.121}$$

Let us define the parameter

$$m \equiv \frac{1}{(1 + \lambda)^2}\quad.$$

Then

$$I(\lambda) = \frac{2}{\pi(1 + \lambda)}K(m) \tag{6.122}$$

where $K(m)$ = complete elliptic integral of the first kind [9]. Finally, the equation λ satisfies for the square lattice is

$$\frac{1}{(1 + \lambda)\pi}K\left(\frac{1}{(1 + \lambda)^2}\right) = \frac{T_2}{T} \tag{6.123}$$

where $T_2 = T_c^{MFA}$ for $2d$ (square lattice). As $T \to 0$ the r. h. s. of (6.123) diverges. The K function has a logarithmic singulariry near $m = 1$:

$$\lim_{m \to 1} K(m) = \frac{1}{2} \log \left(\frac{16}{1 - m} \right) \tag{6.124}$$

so that K has a logarithmic divergence at $\lambda = 0$. The singularities of both sides of Eq. (6.123) must coincide, and we obtain for $T \to 0$

$$\lambda_{T \to 0} \to 8 e^{-(\pi T_2/T)} \tag{6.125}$$

The susceptibility at low T can be obtained now from (6.110):

$$\chi_{zz} = \frac{\gamma^2}{64J} e^{\pi T_2/T} \tag{6.126}$$

and we verify that it diverges exponentially as $T \to 0$.

On the other hand, when $T \gg T_2$, $\lambda \to \infty$ and $m \to 0$. In this limit,

$$\lim_{m \to 0} K(m) = \frac{\pi}{2} \tag{6.127}$$

so that for large T

$$\lambda = \frac{T - T_2}{T_2} \tag{6.128}$$

and we recover Curie-Weiss law.

Let us now calculate the correlation function in the paramagnetic phase, which we know that for $d = 2$ implies $T > 0$. In this case we have no way to calculate the integral in a closed form, so we look for the asymptotic limit $R \to \infty$. We want to calculate

$$\langle\, S_0^- S_R^+ \,\rangle = \frac{1}{\beta J \nu} \frac{1}{N} \sum_k \frac{e^{i\mathbf{k} \cdot \mathbf{R}}}{1 + \lambda - \gamma_k} \tag{6.129}$$

For very large R the exponential oscillates very rapidly, leading to a cancellation of the terms for large k. Then we can limit the integral to small values of k, and expand the structure factor γ_k around $k = 0$, in a power series in the two components of $\mathbf{k}$,

$$\gamma_k = 1 - q^2/4 + \mathcal{O}(q^4) \quad , \quad q = ak \ .$$

Although the integral over $\mathbf{k}$ is limited to the first BZ, we can safely extend it to ∞, since the dominant contribution comes anyway from the region around the origin. Then the integral in (6.129) becomes:

$$\frac{1}{N} \sum_k (\cdots) = \frac{1}{4\pi^2} \int_0^\infty q \, dq \int_0^{2\pi} d\phi \, \frac{e^{iq(R/a)\cos\phi}}{\lambda + q^2/4} \tag{6.130}$$

The exponential in (6.130) is the generating function of the Bessel functions of the first kind :

$$e^{im\cos\phi} = J_0(m) + 2\sum_{n=1}^{\infty} J_n(m)i^n \cos n\phi \qquad (6.131)$$

Therefore,

$$\frac{1}{N}\sum_k (\cdots) = \frac{1}{2\pi}\int_0^\infty \frac{J_0(qR/a)q\,dq}{\lambda + q^2/4} \qquad (6.132)$$

The integral can be found in the tables [10]:

$$\int_0^\infty \frac{xJ_0(ax)\,dx}{x^2 + b^2} = K_0(ab) \qquad (6.133)$$

where $K_0(x)$ = modified Bessel function. For large argument the asymptotic form is [9]

$$K_0(x)_{x\to\infty} \longrightarrow \sqrt{\frac{\pi}{2x}}e^{-x}\left(1 + \mathcal{O}(\frac{1}{x})\right) \qquad (6.134)$$

Finally, we can write

$$\frac{1}{N}\sum_k (\cdots) = \frac{2}{\pi}\sqrt{\frac{\pi\xi}{2R}}e^{-R/\xi} \qquad (6.135)$$

with $(a/\xi) = 2\sqrt{\lambda}$. Therefore at low T

$$\xi/a = \frac{1}{2\sqrt{2}}\,e^{\pi T_2/2T} \qquad (6.136)$$

which is very similar to results obtained with other methods, in particular in a reformulation of spin wave theory with a fixed number of bosons [11] and in a *"large-N"* expansion calculation [12]. One must keep in mind that (6.136) is an asymptotic approximation so that the smaller T becomes, the larger is the distance at which (6.135) and (6.136) are valid, since we must have $R/\xi \gg 1$ for the approximations performed to be valid.

6.7 FM with a finite applied field

If the applied field h is large, that is if $\beta h \gg 1$, the expression (6.52) for Ψ simplifies:

$$\Psi \approx \frac{1}{N}\sum_k e^{-\beta\omega(\mathbf{k})} \qquad (6.137)$$

where

$$\omega(\mathbf{k}) = h + Jv\sigma(1 - \gamma_k) \qquad (6.138)$$

Assume that we are dealing with a simple hypercubic lattice in d dimensions. Then the structure factor is

$$\gamma_k = \frac{1}{d} \sum_{m=1}^{d} \cos a k_m \qquad (6.139)$$

The integrand in (6.137) can be decomposed as a product of identical functions of the d independent coordinates. Each integral is of the form

$$\int_0^\pi dt \ e^{-A\cos t} = I_0(A) \qquad (6.140)$$

where I_0 is a modified Bessel function [9] and $A = 2\sigma J/k_BT$. Therefore,

$$\Psi \approx e^{-(h+J\nu\sigma)/k_BT} \left(I_0(\frac{2\sigma J}{k_BT}) \right)^d$$
$$= e^{-h/k_BT} \left(e^{-2\sigma J/k_BT} I_0(\frac{2\sigma J}{k_BT}) \right)^d \qquad (6.141)$$

since $\nu = 2d$ in a hypercubic lattice. If $k_BT \ll J$, we can use the asymptotic form [9]

$$e^{-z} I_0(z) \to \frac{1}{\sqrt{2\pi z}} \qquad (6.142)$$

Therefore, in the case $J \gg k_BT$ and $h \gg k_BT$ we obtain

$$\Psi \sim e^{-h/k_BT} \left(\frac{k_BT}{4\pi\sigma J} \right)^{d/2} \qquad (6.143)$$

Since by assumption the field is large and the temperature low, σ will be almost saturated. This means that $\Psi \ll 1$, and we can expand σ in (6.90) for small values of Ψ, that is [7]

$$\sigma = S - \Psi + (2S+1)\Psi^{2S+1} - \cdots \qquad (6.144)$$

and we obtain finally

$$\sigma \approx S - e^{-h/k_BT} \left(\frac{k_BT}{4\pi\sigma J} \right)^{d/2} \qquad (6.145)$$

6.8   RPA for antiferromagnet

Let us now turn to a two-sublattice antiferromagnet with Hamiltonian

$$H = \sum_{<ab>} J_{ab} \mathbf{S_a} \cdot \mathbf{S_b} - h_a \sum_a S_a^z + h_a \sum_b S_b^z \qquad (6.146)$$

where the negative sign of the exchange interactions has been taken into account. We include, as in Chap. 3, a staggered anisotropy field B_a and we call $h_a = \gamma B_a$. Points in the $\uparrow$ ($\downarrow$) sublattice are called a (b).

6.8.1   Spin 1/2 AFM

We define now retarded Green's functions as for the FM

$$G_{11}(\mathbf{R}_a - \mathbf{R}_0, t) = \langle\langle\, S_a^+\,;\, S_0^-\,\rangle\rangle$$
$$G_{21}(\mathbf{R}_b - \mathbf{R}_0, t) = \langle\langle\, S_b^+\,;\, S_0^-\,\rangle\rangle \qquad (6.147)$$

where R_0 belongs to the up sublattice. We call $\sigma = \langle\, S_a^z\,\rangle = -\langle\, S_b^z\,\rangle$. Now the Fourier transformed equations of motion for the pair of Green's functions are

$$(\omega - h - 2\sigma J_0)G_{11} + \sigma J_k G_{21} = \frac{\sigma}{\pi}$$
$$(\omega + h + 2\sigma J_0)G_{21} - \sigma J_k G_{11} = 0 \qquad (6.148)$$

where we dropped the subindex a in h and the tilde on J_k to simplify the notation. Let us now define the quantities

$$\epsilon_{1k} = h + 2\sigma(J_0 - J_k)$$
$$\epsilon_{2k} = h + 2\sigma(J_0 + J_k)$$
$$\epsilon_k^2 = \epsilon_{1k}\epsilon_{2k} \qquad (6.149)$$

Then with a little algebra one finds

$$G_{11} = \frac{\sigma}{4\pi\epsilon_k}\left(\frac{2\epsilon_k + \epsilon_{1k} + \epsilon_{2k}}{\omega - \epsilon_k} - \frac{\epsilon_{1k} + \epsilon_{2k} - 2\epsilon_k}{\omega + \epsilon_k}\right)$$
$$G_{21} = -\frac{\sigma^2 J_k}{2\pi\epsilon_k}\left(\frac{1}{\omega - \epsilon_k} - \frac{1}{\omega + \epsilon_k}\right) \qquad (6.150)$$

The poles of the Green's functions are $\omega = \pm\epsilon_k$. To see what spectrum this yields, let us specialize to the case with only first n. n. interactions, where $J_k = J\nu\gamma_k$. Then,

$$\epsilon_k = \sqrt{(h + 2J\nu\sigma)^2 - (2J\nu\sigma\gamma_k)^2} \qquad (6.151)$$

In the absence of anisotropy we find for small k a linear dispersion relation

$$\epsilon_k = sk \qquad (6.152)$$

Exercise 6.21
Show that for the square lattice in 2d we find

$$s = 8J\sigma(T)a/\sqrt{2}$$

where a = atomic lattice constant.

This is a dispersion relation typical of a massless boson excitation, like photons or acoustic phonons. We explicitly indicate the T dependence of the AF magnon velocity through σ. As T increases, the velocity decreases, and it vanishes with σ at T_N. Although some renormalization of s with temperature is to be expected, its scaling with the magnetization, as found above, is a direct consequence of the RPA renormalization of the magnon excitation energy in (6.151). If there is an anisotropy staggered field h the dispersion relation has a gap at $k = 0$. For small k we obtain from (6.151) the expansion

$$\epsilon_k = \Delta + \frac{\omega_e^2}{2\Delta} A(ka)^2 \qquad (6.153)$$

where $\Delta^2 = \omega_a^2 + 2\omega_a \omega_e$ and we are using the notation of Chap. 3: $\omega_a = h$, $\omega_e = Jv\sigma$. The geometric factor A in (6.153) depends on the lattice, and for the square lattice it is $1/2$.

6.8.2   Arbitrary spin AFM

Let us now extend the RPA for the AFM to arbitrary values of S. To this end we define as for the FM a new Green's function, which depends on a parameter v as:

$$G_{11}(\mathbf{R}_n, \omega; v) = \langle\langle\, S_n^+; e^{vS_0^z} S_0^- \,\rangle\rangle_\omega \qquad (6.154)$$

and the quantity

$$\Theta(v) = \langle\, [S_n^+, e^{vS_0^z} S_0^-]\, \rangle \qquad (6.155)$$

In particular, $\Theta(0) = 2\langle\, S_0^z\, \rangle$. Let us also define

$$U(v) = \langle\, e^{vS_0^z} S_0^- S_0^+\, \rangle \qquad (6.156)$$

In the RPA we obtain for the imaginary part of the retarded Fourier transformed Green's function :

$$\mathcal{I}mG_{11}(\mathbf{k}, \omega + i\epsilon; v) \;=\; \frac{\Theta(v)}{4\epsilon_k}[-(2\epsilon_k + \epsilon_{1k} + \epsilon_{2k})\delta(\omega - \epsilon_k) +$$
$$(\epsilon_{1k} + \epsilon_{2k} - 2\epsilon_k)\delta(\omega + \epsilon_k)] \qquad (6.157)$$

The fluctuation-dissipation theorem relates U and Θ through

$$U(v) = \frac{1}{4N}\sum_k \frac{\Theta(v)}{4\epsilon_k}\left[\frac{2\epsilon_k + \epsilon_{1k} + \epsilon_{2k}}{e^{\beta\epsilon_k} - 1} + \frac{2\epsilon_k - \epsilon_{1k} - \epsilon_{2k}}{e^{-\beta\epsilon_k} - 1}\right] \qquad (6.158)$$

which can also be written as

$$U(v) = \left(\frac{1}{2N}\sum_k \frac{\epsilon_{1k} + \epsilon_{2k}}{2\epsilon_k}\coth(\beta\epsilon_k/2) - 1/2\right)\Theta(v) \qquad (6.159)$$

In the sums above N is the number of sites in each magnetic sublattice, namely half the total number of spins in the system. Let us call

$$\Phi \equiv \frac{1}{2N}\sum_k \frac{\epsilon_{1k} + \epsilon_{2k}}{2\epsilon_k}\coth(\beta\epsilon_k/2) \qquad (6.160)$$

In order to make contact with Callen's formalism for the RPA we write

$$U(v) = \Psi\Theta(v) \tag{6.161}$$

and define $\Phi \equiv \Psi + 1/2$. Since we only need local averages on a given sublattice, we recognize that the problem is formally identical to the one for the FM, so that the general solution for the local magnetization is

$$\langle\, S_0^z \,\rangle = (S + 1/2)\frac{(\Phi + 1/2)^{2S+1} + (\Phi - 1/2)^{2S+1}}{(\Phi + 1/2)^{2S+1} - (\Phi - 1/2)^{2S+1})} - \Phi \tag{6.162}$$

which is the form (6.90) takes in terms of Φ. Let us now quote the expansions of (6.162) in the limits of Φ very small or very large. For $\Phi \to 0$ we get [7]

$$\sigma = S - \Phi + (2S + 1)\Phi^{2S+1} - (2S + 1)^2\Phi^{2S+2} + \mathcal{O}(\Phi^{2S+3}) \tag{6.163}$$

while for $\Phi \to \infty$ ($\Psi \to \infty$)

$$\Phi\sigma \to S(S + 1)/3 \tag{6.164}$$

6.8.3   Zero-point spin deviation

In $3d$ an AFM at $T = 0$ has a finite sublattice spin deviation, due to the zero-point fluctuations of the sublattice magnetization which we already described in Chap. 4 within the FSWA. From (6.160) we find that as $\beta \to \infty$, $\Phi \to$ constant, and this constant depends on the dimensionality and on the parameters of the system. If we restrict ourselves to first n. n. interactions, we find for a square lattice [8] (see section 4.3.3):

$$\lim_{T\to 0} \Phi = 0.697 \tag{6.165}$$

For $S = 1/2$ this yields $\sigma = 0.358$. For $S \to \infty$ instead

$$\sigma \to S - \Psi = S - 0.197 \quad, \tag{6.166}$$

which agrees with the spin wave theory results [15], Eqs.(4.51) and (4.83) In fact expression (6.166) is obtained within the FSWA for any spin, which shows that in this respect the FSWA is the $S \to \infty$ limit of the RPA.

In $1d$, $\Psi \to \infty$ in the absence of anisotropy, which in the FSWA implies that σ diverges. In the RPA this implies on the contrary that $\sigma \to 0$ when $h \to 0$. Both results agree in the sense that there is no LRO in $1d$ in the absence of anisotropy. We remind that in chapter 4 we defined the anisotropy parameter $\alpha = h/(JS\nu)$, in terms of which

$$\Psi^{(1d)} = -\frac{1}{2S}\left(1 + \log\frac{2\alpha}{\pi}\right) \tag{6.167}$$

The weak divergence of Ψ implies that in RPA even a very small anisotropy leads to a finite σ at $T = 0$. The absence of LRO in the isotropic $1d$ AFM was rigourously proven in 1931 by Hans Bethe [13], who obtained a complete analytic solution for the ground state of the isotropic Heisenberg AFM chain with first n. n., known as the *Bethe Ansatz* theory.

In $3d$, Ψ is finite. Results for some lattices were quoted in chapter 4.

6.8.4   Correlation length

By applying the fluctuation-dissipation theorem, we have

$$\langle\, S^-(0)S^+(\mathbf{R})\,\rangle = -2\int d\omega \mathcal{I}m\langle\langle\, S^+(\mathbf{R}); S^-(0)\,\rangle\rangle_{\omega+i\epsilon} N(\omega) \qquad (6.168)$$

where

$$N(x) \equiv \frac{1}{e^{\beta x} - 1}$$

and

$$\langle\langle\, S^+(\mathbf{R}_\alpha); S^-(0)\,\rangle\rangle_{\omega+i\epsilon} = \frac{1}{N}\sum_k G_{\alpha 1}(\mathbf{k}, \omega + i\epsilon)e^{-i\mathbf{k}\cdot\mathbf{R}_\alpha}$$

Here $\alpha = 1(2)$ if $\mathbf{R}_\alpha$ is the position vector of a spin in sublattice ↑ (↓).

If the lattice has inversion symmetry then $G(-\mathbf{k}, \omega) = G(\mathbf{k}, \omega)$, so that we can write (6.168) as

$$\langle\, S^-(0)S^+(\mathbf{R}_\alpha)\,\rangle = -2\int e^{-i\mathbf{k}\cdot\mathbf{R}_\alpha} N(\omega)\mathcal{I}m G_{\alpha 1}(\mathbf{k}, \omega + i\epsilon)d\omega \qquad (6.169)$$

Now we take the imaginary part of (6.150) and substitute it in the last equation.

For points on the same (up) sublattice we get

$$\langle\, S^-(0)S^+(\mathbf{R}_a)\,\rangle = \sigma\Delta(\mathbf{R}) + \frac{2\sigma}{N}\sum_k e^{-i\mathbf{k}\cdot\mathbf{R}_a}\frac{\epsilon_{1k} + \epsilon_{2k}}{4\epsilon_k}\coth\beta\epsilon_k/2 \quad (6.170)$$

with $\Delta_{\mathbf{R}} = 0$, $\mathbf{R} \neq 0$; $\Delta(\mathbf{0}) = 1$.

For points on different sublattices,

$$\langle\, S^-(0)S^+(\mathbf{R_b})\,\rangle = -\frac{1}{N}\sum\frac{\sigma^2 J\nu\gamma_k}{4\epsilon_k}\coth(\beta\epsilon_k/2)\,e^{-i\mathbf{k}\cdot\mathbf{R_b}} \qquad (6.171)$$

We shall now look at these formulae in the paramagnetic (PM) phase, so we take the limit $\sigma \to 0$. We remark that since we maintain an anisotropy staggered applied field h, we can still consider the system as infinitesimally polarized with a two-sublattice magnetic structure, and get a finite λ in the limit $h \to 0$. We find:

$$\langle\, S^-(0)S^+(\mathbf{R}_a)\,\rangle = \frac{1}{2N\beta J\nu}\sum_k e^{-i\mathbf{k}\cdot\mathbf{R}_a}\left[\frac{1}{1 + \lambda - \gamma_k} + \frac{1}{1 + \lambda + \gamma_k}\right] \qquad (6.172)$$

and

$$\langle\, S^-(0)S^+(\mathbf{R_b})\,\rangle = \frac{-1}{2N\beta J\nu}\sum_k e^{-i\mathbf{k}\cdot\mathbf{R_b}}\left[\frac{1}{1 + \lambda - \gamma_k} - \frac{1}{1 + \lambda + \gamma_k}\right] \qquad (6.173)$$

One important simplification arises from a property of the structure factor γ_k in the case of first n. n. interactions. In this case it is possible to find a vector $\mathbf{Q}$ with the property

$$\mathbf{Q}\cdot(\mathbf{R}_\alpha - \mathbf{R}_\beta) = 2m\pi$$

if α, β belong to the same sublattice, and

$$\mathbf{Q} \cdot (\mathbf{R}_\alpha - \mathbf{R}_\beta) = (2m+1)\pi$$

if they belong to different sublattices. We then have

$$\gamma_{\mathbf{k}+\mathbf{Q}} = -\gamma_{\mathbf{k}} \tag{6.174}$$

For the vector $\mathbf{Q}$ in the square lattice we have the choices

$$\mathbf{Q} = (\pi/a)\,(\pm 1, \pm 1) \quad .$$

Any vector in this set carries the first magnetic BZ into the first atomic one, and for each of these translations there is a change of sign of γ_k. Taking advantage of this symmetry of the structure factor we can write the sum over the magnetic BZ as a sum over the atomic one, and for $0 < T \ll T_2$ we find for the correlator exactly the same expression as in the case of a FM, except that the sign changes as we change from the $\uparrow$ to the $\downarrow$ sublattice:

$$\langle\, S^-(0)S^+(\mathbf{R}_a)\,\rangle = \frac{1}{2N\beta J\nu} \sum_{k\in\, BZ_{at}} e^{-i\mathbf{k}\cdot\mathbf{R_a}} \frac{1}{1+\lambda-\gamma_k} \tag{6.175}$$

and

$$\langle\, S^-(0)S^+(\mathbf{R_b})\,\rangle = -\frac{1}{2N\beta J\nu} \sum_{k\in\, BZ_{at}} e^{-i\mathbf{k}\cdot\mathbf{R_b}} \frac{1}{1+\lambda-\gamma_k} \tag{6.176}$$

Exercise 6.22
Prove Eqs. (6.175) and (6.176).

The calculation of the Néel temperature T_N can be done now by simply taking formula (6.103) for the Curie temperature T_c of a FM and remembering that the integral in k space must be extended over the atomic BZ, which we indicate by a prime on the summation simbol. In the absence of anisotropy, we have:

$$\frac{1}{N} \sum_{k}{}' \frac{1}{1-\gamma_k} = \frac{S(S+1)\nu J}{3k_B T_N} \tag{6.177}$$

Regarding the correlation length, if $T > 0$ in $1d$ we have the same expresion as for the FM chain, except that we keep track of the oscillating sign:

$$\langle\, \mathbf{S_0} \cdot \mathbf{S_R}\,\rangle = (-1)^\nu S(S+1)e^{-R/\xi} \tag{6.178}$$

where $\nu = 1(0)$ for different (the same) sublattice points at distance R, and where ξ is the same as in (6.116). The corresponding result for the square lattice is

$$\langle\, \mathbf{S_0} \cdot \mathbf{S_R}\,\rangle = (-1)^\nu \frac{3k_B T}{2J} \left(\frac{\xi}{2\pi R}\right)^{1/2} e^{-R/\xi} \quad, \tag{6.179}$$

with ξ from (6.136). A similar result is obtained in the two-loop approximation of Chakravarty, Halperin and Nelson [14]. Let us look now at the $T \to 0$ limit for the AFM chain. In this limit, $\coth \beta \epsilon_k / 2 \to 1$, so that Φ is

$$\Phi = \frac{1}{N} \sum_k \frac{1}{\sqrt{1 - m^2 \gamma_k^2}} \tag{6.180}$$

with $m \equiv (1 + \lambda)^{-1}$. The resulting integral is proportional to the complete elliptic integral of the first kind [10] and for $\lambda \to 0$, which is the limit for $T \to 0$, we find

$$\Phi \to \frac{2}{\pi} \log \frac{4}{\lambda} \tag{6.181}$$

On the other hand we know that as Φ diverges we have $\Phi\sigma \to S(S+1)/3$. Then

$$\lambda = 4 \exp - \left(\frac{S(S+1)\pi}{6\sigma} \right) \tag{6.182}$$

From the definition of λ we have $h = \lambda \sigma \nu J$, so as the staggered field $h \to 0$ we also have $\sigma \to 0$ and then $\lambda \to 0$ as $T \to 0$. Finally,

$$\chi = \frac{\gamma^2}{2\nu J \lambda} \to \infty$$

that is, the initial staggered susceptibility diverges as $T \to 0$.

According to (6.117) ξ diverges in $1d$ for the FM chain as $T \to 0$, so in the AFM we have

$$\langle \mathbf{S}_0 \cdot \mathbf{S}_{na} \rangle = (-1)^n S(S+1) \tag{6.183}$$

where a is the atomic lattice constant. This result is clearly wrong at long distances, since it implies the existence of long range order. A calculation for the $S = 1/2$ AFM chain within the Green's function formalism was published by Kondo and Jamaji [16] in which a RPA-type decoupling is performed in the next order of the hierarchy of equations, thus allowing to incorporate correctly the vanishing of σ, and to obtain self-consistently the correlators $c_n \equiv \langle S_1^z S_n^z \rangle$. The results for the correlators c_n for $n = 1, 2$ agree reasonably well both with the exact result [18], $c_1 = -0.59086$ and with the numerical calculation on a chain of 11 spins performed by Bonner and Fisher [19], which yields $c_2 = 0.25407$. The agreement is much better for the first nearest neighbours. One suspects that in order to obtain reasonable results for c_n with the decoupling approach , one must go down to the n-th level of the hierarchy of equations. At any rate, c_1 is just what is needed for calculating the specific heat, as we already mentioned in Chap. 3. The agreement of the specific heat calculated with the Green's function decoupling in the second level with the numerical estimates based on finite size scaling is fairly good.

In $3d$ there is a finite transition temperature T_N, so that the correlation length is finite for $T > T_N$. In this phase λ is finite and we can use (6.108), with the sum over k extended inside the atomic BZ. If we are interested in the long distance behaviour of the correlator it is enough to keep up to the quadratic terms in the expansion of γ_k for small k in the denominator of (6.108). Then, transforming the sum into an integral we have

$$\langle\, \mathbf{S_0} \cdot \mathbf{S_R}\, \rangle = \frac{3k_B T}{(2\pi)^3 \nu J} \int_{BZ} a^3 \mathrm{d}^3 \mathbf{k} \frac{e^{-i\mathbf{k}\cdot\mathbf{R}}}{Aa^2 k^2 + \lambda} \tag{6.184}$$

where a =lattice constant and A is a numerical coefficient which depends on the lattice. For large R we can extend the integral to the whole volume, which eventually tends to infinite, and then (6.184) is just the Yukawa potential:

$$\langle\, \mathbf{S_0} \cdot \mathbf{S_R}\, \rangle = \frac{\xi e^{-R/\xi}}{4\pi A R} \tag{6.185}$$

where

$$\xi = a\sqrt{A/\lambda} \tag{6.186}$$

Exercise 6.23 *Prove (6.185) and (6.186).*

Observe that ξ is inversely proportional to $\sqrt{\chi}$. At high T then

$$\xi \sim \frac{1}{\sqrt{T - T_c^{MFA}}}$$

while as $T \searrow T_N$, ξ diverges as $(T - T_N)^{-1/2}$, and it is infinite all the way down to $T = 0$ in the ordered phase. In fact in the AFM phase the correlator (6.184) $\sim R^{-2}$.

6.9   RPA susceptibility of AFM

We study now the response of an AFM to a uniform static external field. Is is convenient to include some anisotropy in the model from the start, and we choose now an exchange anisotropy term, which originates from the combined effect of the spin-orbit interaction of the un-paired electrons of the ions and the crystal-field. The resulting effect in the effective spin Hamiltonian is an anisotropic exchange interaction. We assume that the exchange zz term is different from the others, and restrict ourselves to first n. n. exchange interactions among a ($\uparrow$) and b ($\downarrow$) spins:

$$\begin{aligned} H &= \sum_{a,b}(J_{ab}\mathbf{S_a} \cdot \mathbf{S_b} + K_{ab}S_a^z S_b^z) \\ &\quad -\gamma B(\sum_a S_a^z + S_b^z) \end{aligned} \tag{6.187}$$

We shall calculate the generalized retarded Green's functions defined in Eq. (6.154), where we introduced the operator

$$B_0(v) \equiv e^{vS_0^z}$$

where as before the origin is taken at an up spin site (sublattice a) and we also define

$$B_{b_0} \equiv e^{vS_{b_0}^z}$$

where b_0 denotes the origin of the b (down) sublattice. As before, indices $1(2)$ refer to $\uparrow$ ($\downarrow$) sublattices. The retarded Green's functions satisfy the RPA equations:

$$
\begin{aligned}
(E - \gamma B)G_{11}(a, E; v) &= \frac{\Theta_a(v)}{2\pi}\delta_{\mathbf{R}_a, 0} + 2\sum_b J_{ab}\sigma_a G_{21}(b, E; v) \\
&\quad - 2(J_{ab} + K_{ab})\,\sigma_b G_{11}(a, E; v)
\end{aligned}
\tag{6.188}
$$

and

$$
\begin{aligned}
(E - \gamma B)G_{21}(b, E; v) &= 2\sum_a J_{ab}\sigma_b G_{11}(a, E; v) \\
&\quad - 2(J_{ab} + K_{ab})\,\sigma_a G_{21}(b, E; v)
\end{aligned}
\tag{6.189}
$$

with a simplified notation in which $a(b)$ substitutes $\mathbf{R}_{(\mathbf{a}, \mathbf{b})}$, and all sites have coordinates referred to the origin in the a sublattice.

Let us remark that in the presence of a uniform external field we do not have any more the time-reversal symmetry $\sigma_b = -\sigma_a$. We notice that this symmetry is apparent in the self-consistency equation (6.162), since one verifies that there $\sigma(\Phi) = -\sigma(-\Phi)$. We must now retain explicitly the reference to the sublattice. In particular, we generalize the definition of $\Theta(v)$ in Eq. (6.155) and consider two different functions $\Theta_{(a,b)}(v)$. Fourier transforming to k space we get:

$$
\begin{aligned}
(E - \gamma B)G_{11}(k, E; v) &= \frac{\Theta_a(v)}{2\pi} + 2J_k\sigma_a G_{21}(k, E; v) \\
&\quad - 2(J_0 + K_0)\sigma_b G_{11}(k, E; v)
\end{aligned}
\tag{6.190}
$$

and

$$
\begin{aligned}
(E - \gamma B)G_{21}(k, E; v) &= 2J_k\sigma_b G_{11}(k, E; v) \\
&\quad - 2(J_0 + K_0)\sigma_a G_{21}(k, E; v)
\end{aligned}
\tag{6.191}
$$

For first nearest neighbours $J_k = J\nu\gamma_k$ and the roots of the secular determinant of the system of equations above are

$$E_k^{\pm} - \omega_0 = (J + K)\nu(\sigma_a + \sigma_b) \pm \sqrt{\Delta_k} \tag{6.192}$$

where $\omega_0 = \gamma B$ and the discriminant is

$$\Delta_k = (J + K)^2\nu^2(\sigma_a + \sigma_b)^2 - 4\nu^2\sigma_a\sigma_b\left[(J + K)^2 - J^2\gamma_k^2\right] \tag{6.193}$$

We see that in the absence of a magnetic field we have

$$E_k^{\pm} = \pm E_0 = \pm 2\nu J\sigma\sqrt{(1 + \alpha)^2 - \gamma_k^2} \tag{6.194}$$

and we recover exactly the dispersion relation (6.151).

Notice that the anisotropy parameter α plays exactly the same role as the parameter λ did for the FM.

If we follow now the same procedure as before, we find

$$
\begin{aligned}
G_{11}(k, E) &= \frac{\sigma_a}{\pi}\left[\frac{1}{E - E_+} + \frac{1}{E - E_-}\right] \\
&+ \frac{2(J + K)\nu\sigma_a(\sigma_a - \sigma_b)}{\pi(E_+ - E_-)}\left[\frac{1}{E - E_+} - \frac{1}{E - E_-}\right]
\end{aligned}
\tag{6.195}
$$

and

$$
G_{21}(k, E) = \frac{2(J + K)\nu\sigma_a^2}{\pi(E_+ - E_-)}\left[\frac{1}{E - E_+} - \frac{1}{E - E_-}\right]
\tag{6.196}
$$

The self-consistency equation (6.158) relates $\Theta(v)$ with the function $U(v)$ defined in (6.156). The only change we need to introduce now is to identify the sublattice, so we write

$$
\Theta_a(v)\Psi_a = -2\frac{1}{N}\sum_k \int \mathcal{I}m G_{11}(k, E + i\epsilon; v))N(E)dE
\tag{6.197}
$$

Then

$$
\begin{aligned}
\Psi_a &= \frac{1}{2N}\sum_k\left[\frac{1}{e^{\beta E_+} - 1} + \frac{1}{e^{\beta E_-} - 1}\right] \\
&+ \frac{1}{2N}\sum_k \frac{2(J + K)\nu(\sigma_a - \sigma_b)}{\pi(E_+ - E_-)}\left[\frac{1}{e^{\beta E_+} - 1} - \frac{1}{e^{\beta E_-} - 1}\right]
\end{aligned}
\tag{6.198}
$$

while for the b sublattice one should simply interchange a and b above. The poles $E_\pm$ are invariant under that interchange, while the sign of the second term changes. Therefore, with the convenient change of notation $a(b) \to +(-)$, we have

$$
\begin{aligned}
\Psi_\pm &= \frac{1}{2N}\sum_k\left[\frac{1}{e^{\beta E_+} - 1} + \frac{1}{e^{\beta E_-} - 1}\right] \\
&\pm \frac{1}{2N}\sum_k \frac{2(J + K)\nu(\sigma_+ - \sigma_-)}{\pi(E_+ - E_-)}\left[\frac{1}{e^{\beta E_+} - 1} - \frac{1}{e^{\beta E_-} - 1})\right]
\end{aligned}
\tag{6.199}
$$

We make the assumption that under a small field the linear response is the same for both sublattices, so we write

$$
\sigma_\pm = \langle\, S_\pm^z \,\rangle = \pm\sigma_0 + \alpha B
\tag{6.200}
$$

where σ_0 is the spontaneous sublattice spin polarization.

The coefficient α is proportional to the susceptibility:

$$
\chi_\| = \gamma\rho\alpha
\tag{6.201}
$$

where $\rho = N/V$ is the volume concentration of the magnetic spins in the sample. We now recall that σ can be expressed in terms of Ψ through Eq. (6.162), where $\Phi = \Psi + 1/2$, and this is true for both sublattices, so that we express α, using (6.200), as

$$\alpha = \frac{1}{2} \sum_{m=\pm} \frac{\partial \langle S_m^z \rangle}{\partial \Psi_m} \left(\frac{\partial \Psi_m}{\partial B} \right)_{B=0} \tag{6.202}$$

Exercise 6.24
Prove that

$$\alpha = \lim_{B \to 0} \frac{1}{2} \sum_{m=\pm} \left[\left(\frac{(2S+1)\Psi_m(1+\Psi_m)^S}{(1+\Psi_m)^{2S+1} - \Psi_m^{2S+1}} \right)^2 - 1 \right] \left(\frac{\partial \Psi_m}{\partial B} \right) \tag{6.203}$$

In the PM phase $(T > T_N)$ Ψ diverges as $B \to 0$, but (6.203) has a finite limit:

Exercise 6.25
Prove that the limit (6.203) yields for α^{-1} the expression

$$\alpha^{-1} = \frac{3k_B T}{S(S+1)N} \sum_k \frac{\gamma - J_0 \alpha}{(\gamma - J_0 \alpha)^2 - (J_0 \alpha \gamma_k)^2} \tag{6.204}$$

where $J_0 = 2\nu J$.

If we choose $T = T_N$ the last equation must be identical to (6.177), which requires that $\alpha(T_N) = \gamma/2J_0$, and also implies that

$$\gamma - J_0 \alpha(T_N) > 0$$

for consistency. We found in the MFA that $\chi_\|(T_N)$ is the absolute maximum of $\chi_\|(T)$, so that the inequality above is satisfied for all T in the MFA. If this be also true in the RPA, (6.201) and (6.204) would determine $\chi_\|$ in the PM phase within this approximation. We shall assume the inequality above to be valid. We ought now to obtain $\chi_\|$ in the RPA for $T < T_N$. Before that we must make a digression, since in the absence of anisotropy the AFM is unstable upon application of a longitudinal uniform field.

6.10   Spin-flop transition

We remind that this transition is the result of the instability of the AFM phase under the action of a static uniform field parallel to the sublattice magnetization, a phenomenon we studied within the MFA in Chap. 3. From the dynamical point of view one expects that a soft magnon be responsible for this instability.

Therefore let us study the dispersion relation when both anisotropy and external field are present. The frequencies of both branches are

$$E_{\pm} = \gamma B - (J + K)\nu(\sigma_a + \sigma_b) \pm \sqrt{\Delta_k} \qquad (6.205)$$

where

$$\Delta_k = [(J + K)\nu(\sigma_a + \sigma_b)]^2 - (2J\nu\gamma_k)^2 \sigma_a \sigma_b \qquad (6.206)$$

The vanishing of the lower branch will entail the divergence of Φ and in consequence an instability of the AFM phase, since then the LRO parameter σ would vanish. From (6.200) we get

$$\begin{aligned}
\sigma_a + \sigma_b &= 2\alpha B \\
\sigma_a \sigma_b &= B^2 \alpha^2 - \sigma_0^2
\end{aligned} \qquad (6.207)$$

Then

$$\Delta_k = 4(J + K)^2 \nu^2 \alpha^2 B^2 - 4(J\nu\gamma_k)^2(\alpha^2 B^2 - \sigma_0^2) \qquad (6.208)$$

Since B is a small perturbation, $\alpha^2 B^2 - \sigma_0^2 \leq 0$ and we have

$$\Delta_k = \Delta_0 - (2J\nu)^2(1 - \gamma_k^2)(\sigma_0^2 - \alpha^2 B^2) \leq \Delta_0 \qquad (6.209)$$

Therefore the critical condition for the vanishing of the lower $(-)$ branch is

$$B_{SF}[\,\gamma - 2(J + K)\nu\alpha\,] = \sqrt{\Delta_0} \qquad (6.210)$$

If the anisotropy is small ($K \ll J$) we can expand (6.210) and retain only the first order term in K/J. One finally obtains for the critical field for spin-flop:

$$B_{SF}^2 \simeq \frac{8K\nu^2 J\sigma_0^2}{\gamma^2 - 4\gamma\alpha J\nu} \qquad (6.211)$$

Now, $\gamma\alpha = \chi_{\parallel}/\rho$. We recall the MFA result

$$\chi_{\perp}^{MFA} = \frac{\gamma^2}{4J\nu\rho}$$

Then we find

$$B_{SF}^2 = \frac{2K\nu\sigma_0^2\rho}{\chi_{\perp}^{MFA} - \chi_{\parallel}} \qquad (6.212)$$

We see that in the absence of anisotropy ($K = 0$) an infinitesimal field destabilizes the AFM phase. We have seen in Chap. 3 that the MFA predicts that in the new stable configuration the sublattices align approximately perpendicular to the field. When $K = 0$ both magnon branches are degenerate when $B = 0$, a result we already obtained with the FSWA in Chap. 4. The effect of the field is to break this degeneracy, which is the result of the time reversal symmetry of the Hamiltonian. The energy of the lower branch decreases as B increases, and eventually becomes negative. A negative excitation energy is naturally a symptom of instability of the assumed ground state configuration and as a result the system goes into a new ground state and the excitations must be redefined accordingly. In the spin-flop phase we still have two ordered sublattices, and the excitations are spin waves, which were obtained by Wang and Callaway [20].

6.11 $\chi_\parallel$ at low T

If the Hamiltonian has no anisotropy terms the system becomes completely isotropic at T_N, so that $\chi_\parallel(T_N) = \chi_\perp(T_N)$. If $T < T_N$ we have an AFM phase with broken symmetry, and $\chi_\parallel(T) \neq \chi_\perp(T)$. We show now that

$$\lim_{T \to 0} \chi_\parallel(T) = 0$$

Let us consider a small applied field along the spin quantization axis at low T. Then the poles of the Green's function are

$$E_\pm(k) = \pm E_k + \delta \qquad (6.213)$$

where E_k are the unperturbed poles in the absence of field and

$$\delta = B[\gamma - 2(J + K)\nu\alpha] \qquad (6.214)$$

According to Eq. (6.201), $\alpha\gamma = \chi_\parallel/\rho$. We are interested in the zero-field susceptibility, so that eventually $\delta \to 0$. The system has an exchange anisotropy $K > 0$, so that it will not undergo the transition to the spin-flop phase with an infinitesimal field. If $\beta J \gg 1$ the spontaneous sublattice spin polarization σ_0 approaches the zero T limit, so $\Psi \approx 0$ and we have from (6.163)

$$\left(\frac{\partial \sigma_\pm}{\partial \Psi_\pm}\right)_{\Psi \to 0} \to -1 \qquad (6.215)$$

Then

$$\chi_\parallel(T) = -\frac{\gamma}{4}\left(\frac{\partial(\Psi_+ + \Psi_-)}{\partial B}\right)_{B=0} \qquad (6.216)$$

Exercise 6.26
Prove Eq. (6.216) and show that from (6.199) we get:

$$\chi_\parallel(T) = \frac{\beta}{4}\left[\gamma^2 - 4(J + K)\nu\frac{\chi_\parallel(T)}{\rho}\right]A(\beta) \qquad (6.217)$$

where

$$A(\beta) = \frac{1}{N}\sum_k \frac{e^{\beta E_k}}{(e^{\beta E_k} - 1)^2} \qquad (6.218)$$

The second term in the square brackets on the r. h. s. of Eq. (6.217) can be neglected since $\chi_\parallel(T)$ is very small. The sum in Eq. (6.218) can be simplified because $\beta J \gg 1$ and we can use the small k approximation $E_k \approx sk$ because contributions from large values of k will be exponentially small in (6.218). After $B \to 0$ we can let the anisotropy vanish, and this is why we have used the linear dispersion relation which is obtained for $K = 0$. Then

$$\chi_\parallel(T) = \frac{\rho\gamma^2 a^3 k_B^2 T^2}{24s^3} \qquad (6.219)$$

The magnon velocity depends on J and on the geometry of the lattice. In particular for the b. c. c. atomic lattice,

$$\chi_{\parallel}(T) = \frac{\gamma^2 \rho}{3S^3 J} \left(\frac{k_B T}{16 S J} \right)^2 \tag{6.220}$$

With anisotropy the dispersion relation has a gap at $k = 0$ and $(A(\beta), \chi_{\parallel}(T))$ vanish exponentially for $T \to 0$.

6.12 Transverse susceptibility

We assume now that the field is perpendicular to the sublattice magnetization, say in the x direction. One fairly direct way to obtain the response of the system is through the free energy. The Hamiltonian in the presence of the field is

$$H = H_0 - \mathbf{M} \cdot \mathbf{B} \tag{6.221}$$

We can consider the Helmholtz free energy as a function of the total magnetic moment, the temperature and the magnetic field. Then the zero-field magnetic moment in the x direction is

$$M_x = -\left(\frac{\partial F}{\partial B_x} \right)_{B=0} \tag{6.222}$$

and for the susceptibility we have

$$\chi^{xx} = \frac{1}{V} \left(\frac{\partial M_x}{\partial B_x} \right)_{B=0} = -\frac{1}{V} \left(\frac{\partial^2 F}{\partial B_x^2} \right)_{B=0} \tag{6.223}$$

From the definition of F we find

$$\chi^{xx} = \left(\frac{\beta}{V} \right) [\langle M_x^2 \rangle - (\langle M_x \rangle)^2] \tag{6.224}$$

where the statistical averages are evaluated in the absence of the external field. Because of this, M_x averages to zero, since we assumed that the spins were aligned in the $\pm z$ directions, and we are left with

$$\chi_{\perp} = \beta \gamma^2 \rho \frac{1}{2N} \sum_{l,m} \langle S_l^- S_m^+ \rangle \tag{6.225}$$

The expression above is valid both in the AF and the PM phases. In the PM phase we must show that $\chi_{\perp}$ coincides with $\chi_{\parallel}$ as calculated before, due to isotropy. Eq. (6.224) is again the fluctuation-dissipation theorem, in this case in the static limit.

We remind that for any pair of operators A, B,

$$\sum_R \langle A_R B_0 \rangle = -\frac{2}{N} \sum_{R, k} e^{-i k \cdot R} \int \mathcal{I}m \langle \langle B_0; A_R \rangle \rangle_{k, E + i\epsilon} N(E) dE \tag{6.226}$$

and we recall that

$$\sum_R e^{-i\mathbf{k}\cdot\mathbf{R}} = N\Delta_{k,0} \tag{6.227}$$

where $\Delta_{k,0}$ is Kronecker's delta. We calculate now the sums in (6.225) directly by summing (6.170) and (6.171):

$$\sum_{R_\alpha} \langle\, S_{R_\alpha}^- S_0^+ \,\rangle = -2\sum_k \Delta_{k,0} \int \mathcal{I}m G_{\alpha 1}(k, E + i\epsilon) N(E)\mathrm{d}E \tag{6.228}$$

Let us consider the AFM phase. For points on the same sublattice

$$\sum_{R_+} \langle\, S_{R_+}^- S_0^+ \,\rangle = \sum_k \frac{\Delta_{k,0}}{2\nu\beta} \frac{(J + K)}{(J + K)^2 - J^2\gamma_k^2} \tag{6.229}$$

and for points on different sublattices

$$\sum_{R_-} \langle\, S_{R_-}^- S_0^+ \,\rangle = -\sum_k \frac{\Delta_{k,0}}{2\nu\beta} \frac{J\gamma_k}{(J + K)^2 - J^2\gamma_k^2} \tag{6.230}$$

Adding up (6.229) and (6.230) and substituting in (6.225) we find:

$$\chi_\perp = \frac{\gamma^2\rho}{2\nu(2J + K)} \tag{6.231}$$

which coincides with the MFA result and is temperature independent.

6.13   Single-site anisotropy

We already found this anisotropy in Chap. 1, as one of the effects of the spin-orbit interaction in the presence of a crystal field. Let us now study the thermodynamic and the dynamic effects of a term in the Hamiltonian of the uniaxial anisotropy form

$$H_a = -\frac{1}{2}D\sum_i (S_i^z)^2 \tag{6.232}$$

We assume $D > 0$ (otherwise spins will align predominantly perpendicular to the z axis, in the x, y plane, which is why the case $D < 0$ is called the XY model). Let us consider the terms that result upon the commutation of S_n^+ with the anisotropy term in the Hamiltonian. We call the resulting operator $-A_n^{(2)}$:

$$-A_n^{(2)} = [\, S_n^+, (S_n^z)^2 \,] = -(S_n^z S_n^+ + S_n^+ S_n^z) \tag{6.233}$$

We cannot apply the argument that led to the RPA, since both operators on the r. h. s. of (6.233) refer to the same site and are strongly correlated. One can instead obtain the equation of motion of the new Green's function which contains the operator $A_n^{(2)}$. Devlin [21] defines three recurrent sets of operators:

$$A_n^{(1)} = S_n^+$$
$$A_n^{(\nu)} = -[A_n^{(\nu-1)}, (S_n^z)^2] \ , \quad 2S \geq \nu \geq 2 \qquad (6.234)$$

$$B_n^{(\nu)} = [A_n^{(\nu)}, S_n^+]$$
$$C_n^{(\nu)} = [A_n^{(\nu)}, S_n^-] \qquad (6.235)$$

Exercise 6.27
Prove that $C_n^{(\nu)}$=polynomial in S_n^z of order ν, which contains only even (odd) powers if ν is even (odd).

Exercise 6.28
Prove that $\forall \nu$,

$$[A_n^{(\nu)}, S_n^z] = -A_n^{(\nu)} \qquad (6.236)$$

The last step in this preparatory background is to show that there are only $2S - 1$ linearly independent operators $A_n^{(\nu)}$. This comes about because the matrix S^+ connects only a state with azimuthal quantum number m with that with $m + 1$. In the manifold of spin S, where the spin matrices are of order $2S + 1$, this operator contains $2S$ non-zero parameters and it is therefore a linear combination of $2S$ linearly independent matrices out of the $(2S + 1)^2$ dimensional basis set that spans the whole manifold of matrices. On the other hand, one can prove that

$$A_n^{(\nu)} = (2S_n^z - 1)A_n^{(\nu-1)} \qquad (6.237)$$

Exercise 6.29
Prove (6.237).

Since all powers of S^z are diagonal, all matrices $A_n^{(\nu)}$ have the same structure and they are therefore different linear combinations of the same $2S$ basis matrices. This implies that there cannot be more than $2S$ linearly independent matrices of this particular structure. As a consequence we can write, with a choice of sign convenient for what follows,

$$A_n^{2S+1} = -\sum_{h=1}^{2S} a_h A_n^h \qquad (6.238)$$

Exercise 6.30
Show that for $S = 1$, $A^3 = A$.

We consider now $2S$ Green's functions (we include an external field):

$$\langle\langle\ A_l^{(\nu)}; S_m^-\ \rangle\rangle, \quad \nu = (1, \cdots, 2S)$$

and find the Fourier transformed equations of motion for $\nu \leq 2S - 1$:

$$
\begin{aligned}
(E - \gamma B)\langle\langle\ A_l^{(\nu)}; S_m^-\ \rangle\rangle_E &= \frac{\delta_{lm}}{2\pi}\langle\ C_l^{(\nu)}\ \rangle + \frac{D}{2}\langle\langle\ A_l^{(\nu+1)}; S_m^-\ \rangle\rangle_E \\
&\quad + \sum_j \Big(\ 2J\langle\langle\ A_l^{(\nu)} S_j^z; S_m^-\ \rangle\rangle_E \\
&\quad - J\langle\langle\ B_l^{(\nu)} S_j^-; S_m^-\ \rangle\rangle_E\ \Big) \\
&\quad - J\sum_j \langle\langle\ C_l^{(\nu)} S_j^+; S_m^-\ \rangle\rangle_E \quad\quad (6.239)
\end{aligned}
$$

The equation for $\nu = 2S$ becomes:

$$
\begin{aligned}
(E - \gamma B)\langle\langle\ A_l^{(2S)}; S_m^-\ \rangle\rangle_E &= \frac{\delta_{lm}}{2\pi}\langle\ C_l^{(2S)}\ \rangle - \frac{D}{2}\sum_{h=1}^{2S}\langle\langle A_l^{(h)}; S_m^-\ \rangle\rangle_E \\
&\quad + \sum_j \big(\ 2J\langle\langle\ A_l^{(2S)} S_j^z; S_m^-\ \rangle\rangle_E \\
&\quad - J\langle\langle\ B_l^{(2S)} S_j^-; S_m^-\ \rangle\rangle_E \\
&\quad - J\langle\langle\ C_l^{(2S)} S_j^+; S_m^-\ \rangle\rangle_E\ ) \quad\quad (6.240)
\end{aligned}
$$

Exercise 6.31
Verify (6.239) and (6.240).

We can now perform the approximations analogous to the usual RPA:

$$
\begin{aligned}
\langle\langle\ A_l^{(\nu)} S_j^z; S_m^-\ \rangle\rangle_E &\approx\ \langle\ S_j^z\ \rangle\langle\langle\ A_l^{(\nu)}; S_m^-\ \rangle\rangle_E \\
\langle\langle\ B_l^{(\nu)} S_j^-; S_m^-\ \rangle\rangle_E &\approx\ 0 \\
\langle\langle\ C_l^{(\nu)} S_j^+; S_m^-\ \rangle\rangle_E &\approx\ \langle\ C_l^{(\nu)}\ \rangle\langle\langle\ S_j^+; S_m^-\ \rangle\rangle_E \quad\quad (6.241)
\end{aligned}
$$

Let us simplify the notation, and call $\langle\ C_l^{(\nu)}\ \rangle = Z_\nu$, which is site independent. For spin S Devlin's equations are [21]:
For $\nu \leq 2S - 1$:

$$
\begin{aligned}
\frac{Z_\nu}{2\pi} &= (E - \gamma B - 2\sigma(J_0)G^{(\nu)}(k, E) + Z_\nu J_k G^{(1)}(k, E) \\
&\quad - \frac{D}{2}G^{(2\nu+1)}(k, E) \quad\quad (6.242)
\end{aligned}
$$

For $\nu = 2S$:

$$\frac{Z_{2S}}{2\pi} = (E - \gamma B - 2\sigma J_0)G^{(2S)}(k, E) + Z_{2S}J_k G^{(1)}(k, E)$$

$$- \frac{D}{2} \sum_{h=1}^{2S} a_h G^{(h)}(k, E) \tag{6.243}$$

For a given S we obtain a magnon spectrum with $2S$ branches. $D/J \to \infty$ is the MFA limit. For the opposite limit $D/J \to 0$ Devlin's results are similar to those of Lines [22]. We refer the reader to Devlin's papers for details of calculations for FM and AFM systems and for comparison with other decoupling schemes and the MFA [22, 23, 24, 25].

6.14   Dynamic linear response

We shall now consider the effect of a time dependent external perturbation on a spin system, and show that the linear dynamic susceptibility can be expressed in terms of the retarded Green's functions. The central fact to stress is that in the linear approximation the Green's functions involved are those of the system in thermodynamic equilibrium.

To the same order of approximation, the power transferred from the system to the external probe, or in other words, the rate of energy dissipated by our system, turns out to be proportional to the imaginary part of the dynamic susceptibility. This is the reason for the expresssion *fluctuation-dissipation theorem* as the general relation of proportionality between the power absorbed from an external probe and the fluctuations of the relevant variables in the unperturbed system. A relation of this form is a particular example of the general Kubo formula [26].

First of all, let us briefly revise the basic ideas of linear response theory. The complete Hamiltonian of a perturbed system is

$$H(t) = H_0 + V(t) \tag{6.244}$$

H_0 is the Hamiltonian of the unperturbed system, which is not time dependent. The external probe couples to this system through the time-dependent potential $V(t)$.

The density matrix, in the Schrödinger picture, satisfies the linear differencial equation:

$$i\hbar \frac{\partial \rho^s}{\partial t} = [H(t), \rho^s] \tag{6.245}$$

Transforming (6.245) to the interaction picture, each operator is transformed as:

$$A(t) = U_0(t)A^s(t)U_0^\dagger(t) \tag{6.246}$$

where $U_0(t) \equiv e^{iH_0 t/\hbar}$. In Schrödinger's picture (SP) dynamic operators (corresponding to observables of the unperturbed system) are not time-dependent.

In the interaction picture, ρ satisfies:

$$i\hbar\frac{\partial\rho(t)}{\partial t} = [\,V(t),\rho(t)\,] \tag{6.247}$$

where $V(t) = U_0(t)V^s(t)U_0^\dagger(t)$

Exercise 6.32
Prove Eqs. (6.247).

In the case of a magnetic system driven by an external field the interaction term in the Hamiltonian is:

$$V(t) = -\gamma\sum_l \mathbf{B}_0(R_l,t)\cdot\mathbf{S}_l(t) \tag{6.248}$$

with $\mathbf{S}_l(t)$ in the interation picture. We assume that $V(t)$ is "small", which implies that in Eq. (6.247) we can decompose the density matrix as a sum of an unperturbed part and a perturbation which can be expanded in a series of powers of $V(t)$:

$$\rho = \rho_0 + \triangle\rho(t) \tag{6.249}$$

ρ_0 is the density matrix of the system in thermodynamic equilibrium at the given temperature. We sustitute now Eq. (6.249) in Eq. (6.247):

$$i\hbar\frac{\partial(\rho_0 + \triangle\rho)}{\partial t} = [\,V,\rho_0 + \triangle\rho\,] = [\,V,\rho_0\,] + \mathcal{O}(V^2) \tag{6.250}$$

We assume that we only need to consider changes in the density matrix to first order. As a consequence, physical quantities of interest change only to this order.

It is customary at this point to introduce a mathematical procedure to ensure convergence of the integrals over time involving the perturbation, known as *adiabatic switching* [26]: it is assumed that $V^s(-\infty) = V^s(+\infty) = 0$, $\rho^s(-\infty) = \rho^s(+\infty) = \rho_0$, which can be obtained if $V^s(t)$ contains an exponential factor $e^{-\epsilon|t|}$, $\epsilon = 0^+$.Let us integrate (6.250)

$$i\hbar[\,\rho(t) - \rho(-\infty)\,] = \int_{-\infty}^t [\,V(t'),\rho_0\,]dt' \tag{6.251}$$

that is

$$i\hbar\triangle\rho(t) = \int_{-\infty}^t [\,V(t'),\rho_0\,]dt' \tag{6.252}$$

Let us now calculate the statistical average of the spin operator S_l^α (in the interation picture)

$$\begin{aligned}
\langle\,S_l^\alpha(t)\,\rangle &= Tr\,(S_l^\alpha(t)\rho(t)) = Tr\,(S_l^\alpha(t)\rho_0)\\
&\quad + Tr\,(S_l^\alpha(t)\triangle\rho(t))
\end{aligned} \tag{6.253}$$

Since $Tr\,(S_l^\alpha(t)\rho_0) = Tr\,(S_l^\alpha\rho_0) = \langle\,S_l^\alpha\,\rangle_0$,

$$\langle\,S_l^\alpha(t)\,\rangle = \langle\,S_l^\alpha\,\rangle_0 + Tr\,(S_l^\alpha(t)\Delta\rho(t)) \qquad (6.254)$$

Traces are invariant under unitary transformations, so that all the traces above can be calculated in the SP where operators are time independent. We get finally:

$$\langle\,\Delta S_l^\alpha(t)\,\rangle = -\frac{\gamma}{\hbar}\int_{-\infty}^{\infty} e^{-\epsilon|t'|}\theta(t-t')dt'\sum_{\beta,m} B_0^\beta(m,t')G_{\alpha\beta}^{(r)}(l,m;t-t') \qquad (6.255)$$

The Green's function appears in (6.255) because one can exploit the cyclic invariance of the trace to obtain the commutator of the spin operators.

Exercise 6.33
Prove Eq. (6.255).

We define now the average local magnetization at time t as:

$$\langle\,\mathbf{M}(\mathbf{R}_l,t)\,\rangle = \frac{g\mu_B}{v_0}\langle\,\mathbf{S}_l(t)\,\rangle \qquad (6.256)$$

where $v_0 = V/N$ is the atomic volume. Let us separate the induced part $\mathbf{m}$ in Eq. (6.256):

$$\mathbf{m}(\mathbf{R}_l,t) = \langle\,\mathbf{M}(\mathbf{R}_l,t)\,\rangle - \langle\,\mathbf{M}_0(\mathbf{R}_l)\,\rangle \qquad (6.257)$$

where

$$\langle\,\mathbf{M}_0(\mathbf{R}_l)\,\rangle = \langle\,\mathbf{M}_0(\mathbf{R}_l,-\infty)\,\rangle$$

is the equilibrium magnetization in the absence of the external field. Then:

$$m^\alpha(\mathbf{R}_l,t) = -\frac{\gamma^2}{\hbar}\sum_l\int_{-\infty}^{\infty} dt'e^{-\epsilon|t'|}\sum_m B_0^\beta(m,t')G_{\alpha\beta}^{(r)}(l-m;t-t') \qquad (6.258)$$

Let us Fourier transform Eq. (6.258) with respect to t and $\mathbf{R}_l$:

$$
\begin{aligned}
m^\alpha(\mathbf{q},\omega) &= \frac{1}{2\pi}\int dt\sum_l e^{-i(\mathbf{q}\cdot\mathbf{R}_l-\omega t)}m(l,t)\\[6pt]
&= -\frac{\gamma^2}{\hbar v_0}\sum_l\int dt e^{-i\mathbf{q}\cdot\mathbf{R}_l}\int_{-\infty}^{\infty} e^{-\epsilon|t'|}dt'e^{i\omega(t-t')}e^{i\omega t'}\\[6pt]
&\quad\times\sum_{h,\beta} B_0^\beta(h,t')G_{\alpha\beta}^{(r)}(l-m,t-t')\\[6pt]
&= -\frac{\gamma^2}{\hbar}\int_{-\infty}^{\infty} dt'e^{-\epsilon|t'|}e^{i\omega t'}\\[6pt]
&\quad\times\sum_{h,l,\beta} G_{\alpha\beta}^{(r)}(l-h,\omega)B_0^\beta(h,t')e^{-i\mathbf{q}\cdot\mathbf{R}_l} \qquad (6.259)
\end{aligned}
$$

In equation (6.259) we have incorporated the adiabatic factor $e^{-\epsilon|t'|}$ which "switches" on and off the perturbation $V(t)$. We now substitute the inverse space Fourier transforms of G and B_0 into (6.259):

$$B_0^\beta(h, t') = \frac{1}{N} \sum_q e^{-i\mathbf{q}\cdot\mathbf{R}_h} B_0^\beta(\mathbf{R}_h, t') \tag{6.260}$$

and

$$G_{\alpha\beta}^{(r)}(l - h, \omega) = \frac{1}{N} \sum_q e^{-i\mathbf{q}\cdot(\mathbf{R}_l - \mathbf{R}_h)} G_{\alpha\beta}^{(r)}(\mathbf{q}, \omega) \tag{6.261}$$

and we get

$$m^\alpha(q, \omega) = -\frac{\gamma^2}{\hbar v_0} \sum_K G_{\alpha\beta}^{(r)}(\mathbf{q}, \omega) B_0^\beta(\mathbf{q} + \mathbf{K}, \omega) \tag{6.262}$$

where the sum over $\mathbf{K}$ = reciprocal lattice vectors, takes account of the fact that $\mathbf{q}$ must belong to the first Brillouin zone, since m and G, being defined only on lattice sites, do not admit an expansion in arbitrarily small wavelengths. This restriction of course does not apply to the external field, in principle, although in most cases the field wavelength will probably be much longer than a lattice constant.

Exercise 6.34
Prove Eqs. (6.262).

We define now the dynamic susceptibility tensor χ' as:

$$\mathbf{m}_\alpha(\mathbf{q}, \omega) = \int d^3\mathbf{q}' \chi'_{\alpha\beta}(\mathbf{q}, \mathbf{q}'; \omega) \mathbf{B}_0^\beta(\mathbf{q}', \omega) \tag{6.263}$$

So that, comparing (6.263) with (6.262), we have:

$$\chi'_{\alpha\beta}(\mathbf{k}, \mathbf{k}'; \omega) = \chi_{\alpha\beta}(\mathbf{k}; \omega) \sum_K \delta^3(\mathbf{k}' - \mathbf{k} - \mathbf{K}) \tag{6.264}$$

where

$$\chi_{\alpha\beta}(k; \omega) = -\frac{\gamma^2}{\hbar v_0} G_{\alpha\beta}^{(r)}(k, \omega) \tag{6.265}$$

which is the main result of this section and, as anticipated, establishes the proportionality between the linear dynamic response function and the retarded Green's function.

6.15   Energy absorbed from external field

By definition, the power delivered by the radiation field to the spin system is

$$\langle \dot{Q}(t) \rangle = \frac{d}{dt}(Tr\rho(H_0 + V(t))) = Tr\frac{d\rho}{dt}H(t) + Tr\rho\frac{dV}{dt} \tag{6.266}$$

Exercise 6.35

Prove that

$$\langle \dot{Q}(t) \rangle = -\gamma \sum_l \frac{\partial \mathbf{B}_0(l,t)}{\partial t} \cdot \langle \mathbf{S}_l(t) \rangle \tag{6.267}$$

Assume that either $\langle \mathbf{S}_l \rangle_0 = 0$ or $\langle \mathbf{S}_l \rangle_0 \perp d\mathbf{B}_0/dt$. Then,

$$\langle \dot{Q}(t) \rangle = -\gamma \sum_l^l \frac{d\mathbf{B}_0(l,t)}{dt} \cdot \mathbf{m}(l,t) \tag{6.268}$$

In terms of the Fourier tramsforms of $\mathbf{B}_0$ and $\mathbf{m}$, we can calculate the total energy absorbed by the spin system over the whole infinite interval $(-\infty, \infty)$, by assuming the adiabatic switching on and off of the external field. Then

$$
\begin{aligned}
\Delta Q &= Q(\infty) - Q(-\infty) = \int \langle \dot{Q}(t) \rangle dt \\
&= -\frac{\gamma}{v_0} \int \mathbf{m}(k,t) i\omega \mathbf{B}_0(-k,-\omega) d^3k d\omega
\end{aligned}
\tag{6.269}
$$

Exercise 6.36

Prove Eqs. (6.269).

We can express (6.269) in terms of the susceptibility:

$$m^\alpha(k,\omega) = \int d^3k \chi'(k,k';\omega)_{\alpha\beta} B_0^\beta(k',\omega) \tag{6.270}$$

Then

$$\Delta Q = -i\frac{\gamma}{v_0} \int d^3k' \int d^3k \int d\omega \, \omega \chi'(k,k';\omega)_{\alpha\beta} B_0^\beta(k',\omega) B_0^\alpha(-k,-\omega) \tag{6.271}$$

Since $\mathbf{B}_0(-k,-\omega) = \mathbf{B}_0^*(k,\omega)$,

$$
\begin{aligned}
\Delta Q &= i \int d^3k \int d^3k' \left[\int_{-\infty}^0 d\omega + \int_0^\infty d\omega \right] \\
&\quad \times \omega B_0^{\alpha*}(k,\omega) \chi'_{\alpha\beta}(k,k';\omega) B_0^\beta(k',\omega)
\end{aligned}
\tag{6.272}
$$

Suppose that the external radiation field $\mathbf{B}_0$ has no large k components, and that it is polarized along α. Then, $q = q'$ in equation (6.263) (or $K = 0$ in (6.264)). We can use Eq. (6.265) and the symmetry property of the Green's function quoted in Eq. (6.37) to obtain:

$$\Delta Q = \frac{2}{(2\pi)^4} \int d^3k \int_0^\infty \omega d\omega \mid B_0^\alpha(k,\omega) \mid^2 \mathcal{I}m\chi(k,\omega)_{\alpha\alpha} \tag{6.273}$$

This is a special case of the fluctuation-dissipation theorem.

6.16   Susceptibility of FM

As an example, let us obtain the transverse susceptibility of the Heisenberg FM in the RPA. Eq. (6.265) expresses the susceptibility tensor components in terms of retarded Green's functions. In particular we have

$$\chi_{xx}(k,\omega) = -\frac{\gamma^2 \rho}{\hbar} G_{xx}^+(k,\omega) \tag{6.274}$$

where $G_{xx}^+(k,\omega)$ is the space-time Fourier transform of

$$G_{xx}^+(R,t) = \langle\langle\, S^x(R,t); S^x(0,0) \,\rangle\rangle^+ \tag{6.275}$$

We already found in Eq. (6.93) that

$$G_{xx} = 1/4 \left(\, G^{+-} + G^{-+} \,\right)$$

Upon use of Eq. (6.37) we find the relation

$$G_{k\omega}^{-+(+)} = \left(\, G_{k,-\omega}^{+-(+)} \,\right)^* \tag{6.276}$$

so that for G_{xx} we get the expression

$$G_{xx}(k,\omega) = -\frac{\sigma}{\pi}\, \frac{E_k}{(\omega+i\epsilon)^2 - E_k^2} \tag{6.277}$$

where $E_k = 2\nu J\sigma(1-\gamma_k) + \gamma B$. In the absence of an external longitudinal field ($B=0$) we effectively find the soft mode associated with a global rotation of spins, a $k=0$ magnon with vanishing excitation energy, so that the static transverse susceptibility $\chi_{xx} \propto k^{-2}$, as mentioned in Chap. 3 in connection with Goldstone's theorem.

6.17   Corrections to RPA

The low temperature expansion of the magnetization as obtained from RPA yields for FM a spurious term in $(T/T_c)^3$, which is not present in the correct perturbation expansion obtained by Dyson [27]. Several attempts have been made to correct this flaw of the approximation. Callen's alternative method of decoupling [7] leads to agreement with the dominant terms of both Dyson's low temperature expansion for the magnetization and Opechowski's [28] high T expansion for the magnetic susceptibility. For more details we refer the reader to the literature on this subject [3, 29].

References

1. Bogolyubov, N. N. and Parasyuk B. (1956) *Dokl. Akad. Nauk. USSR* **109**, 717.

2. Tyablikov, S. V., (1967) *Methods in the Quantum Theory of Magnetism*, Plenum Press, New York, Sect. 26.

3. Keffer, F. (1966) *Handbuch der Physik, 2 Aufl.* **Bd. XVIII**, Springer–Verlag, Berlin-Heidelberg.

4. Bogoliubov, N. N. and Tyablikov, S. V. (1959) *Sov. Phys.- Doklady* **4**, 604.

5. Hewson, A. C. and Ter-Haar, D. (1964), *Physica* **30**, 890.

6. Tahir-Kheli, R. and Ter-Haar, D. (1962), *Phys. Rev.* **127**, 38; 95.

7. Callen, H. (1963) *Phys. Rev.* **130**, 890.

8. Yablonskiy, D. A. (1991) *Phys. Rev* **B 44**, 4467.

9. Abramowitz, Milton and Stegun, Irene (1965) *Handbook of Mathematical Functions*, Dover Publications Inc., New York.

10. Gradshteyn, I. S. and Ryzhik, I. M. (1994) *Table of Integrals, Series and Products*, Editor: Alan Jeffrey, Academic Press, San Diego.

11. Takahashi, M. (1990) *Phys. Rev.* **B 42**, 766.

12. Arovas, D.P. and Auerbach, A. (1988) *Phys. Rev.* **B 38** , 316.

13. Bethe, H. A. (1931) *Z. Phys.* **71**, 205.

14. Chakravarty, S., Halperin, B. I. and Nelson, D. R. (1987) *Phys. Rev.* **B 39**, 2344.

15. Takahashi, M. (1989) *Phys. Rev.* **B 40**, 2494.

16. Kondo, J. and Yamaji, K. (1972) *Prog. Theoret. Phys. (Japan)* **47**, 807.

17. L.Hulthén, L. (1938) *Arkiv Mat. Astron. Fysik*

18. Orbach, R. (1958) *Phys. Rev.* **112**, 309.

19. Bonner, J. C. and Fisher, M. E. (1964) *Phys. Rev.* **A 135**, 640.

20. Wang, Y.-L. C. and Callen, H. B. (1964) *J. Phys. Chem. Solids* **25**, 1459.

21. Devlin, John F. (1971) *Phys. Rev.* **B4**, 136.

22. Lines, M. E. (1967) *Phys. Rev.* **156**, 534.

23. Narath, A. (1965) *Phys. Rev.* **140**, A584.

24. Anderson, F. B. and Callen, H. B. (1964) *Phys. Rev.* **136**, A1068.

25. Murau, T. and Matsubara, T. (1968) *J. Phys. Soc. Japan* **25**, 352.

26. Kubo, R. (1957) *J. Phys. Soc. Japan* **12**, 570.

27. Dyson, F. J. (1956) *Phys. Rev.* **102**, 1217; *idem*, 1230.

28. Opechowski, W. (1959) *Physica* **25**, 476.

29. Haas, C. W. and Jarrett, H. S. (1964) *Phys. Rev.* **135**, A 1089.

Chapter 7

Dipole-Dipole Interactions

7.1 Dipolar Hamiltonian

We have been studying until now the exchange Hamiltonian

$$H_e = - \sum_{n \neq m} J_{nm} \mathbf{S_n} \cdot \mathbf{S_m} \qquad (7.1)$$

In dielectric magnetic materials, for which we expect (7.1) to be a good approximation, spins are well localized around each magnetic ion in the system. Then each corresponding magnetic dipole is concentrated in a very small region of space, of a typical size smaller than any distance between the different dipoles in the system. We are then justified in using the point dipole approximation to describe the magnetic interactions. Usually the contribution to the magnetic interaction energy of higher order magnetic multipoles of the ions is negligible in comparison with the dipolar one. The dipole-dipole interaction Hamiltonian is obtained by simply considering a magnetic dipole moment $\gamma \mathbf{S}_n = \mathbf{m}_n$ associated with each atom with spin $\mathbf{S}_n$, and substituting it into the expression for the mutual potential energy of a system of classical magnetic dipoles:

$$H^{dip} = \frac{\gamma^2}{2} \sum_{n \neq m} \frac{1}{R_{nm}^3} \left[\mathbf{S}_n \cdot \mathbf{S}_m - 3 \frac{(\mathbf{S}_n \cdot \mathbf{R}_{nm})(\mathbf{S}_m \cdot \mathbf{R}_{nm})}{R_{nm}^2} \right] \qquad (7.2)$$

where $\gamma = g\mu_B$ and $\mathbf{R}_{nm} = \mathbf{R}_n - \mathbf{R}_m$. We can, on the other hand, consider (7.2) as a specific example of a bilinear form in the spin operators which obeys definite symmetry requirements. Consider the most general bilinear interation:

$$W = \sum_{l \neq m} \mathcal{J}_{lm}^{\alpha\beta} S_l^\alpha S_m^\beta \quad , \ (\alpha, \beta) = (x, y, z) \qquad (7.3)$$

If the diadic $\mathcal{J}$ has cilindrical symmetry around the vector $\mathbf{r}_l - \mathbf{r}_m \equiv \mathbf{R}_{lm}$ it can be written as:

$$\mathcal{J}^{\alpha\beta} = A^{\alpha\beta}(| R_{lm} |) \frac{\mathbf{e}^\alpha R_{lm}^\alpha \mathbf{e}^\beta R_{lm}^\beta}{R_{lm}^2} \qquad (7.4)$$

where $\{\mathbf{e}^\mu\}$ are the versors of thge coordinate axes and we have taken the coefficient A symmetric under interchange of sites l and m. We can separate an isotropic part:

$$\mathcal{J} = A\delta_{\alpha\beta}\mathbf{e}^\alpha\mathbf{e}^\beta + B\frac{\mathbf{R}_{lm}\mathbf{R}_{lm}}{R_{lm}^2} \tag{7.5}$$

with A,B scalar functions of R_{lm}. If $\mathcal{J}$ is traceless,

$$3A + B = 0 \tag{7.6}$$

Then:

$$\mathcal{J}^{\alpha\beta} = J(R_{lm})\mathbf{e}^\alpha(\delta_{\alpha\beta} - \frac{3R_{lm}^\alpha R_{lm}^\beta}{R_{lm}^2})\mathbf{e}^\beta \tag{7.7}$$

which has the form of the dipolar interaction. The traceless condition is equivalent to the statement that the dipolar potential satisfies Laplace's differential equation for $R_{lm} \neq 0$. The symmetry requirement imposed to get (7.7) is however too restrictive for a lattice, where only some discrete rotations around $\mathbf{R}_{lm}$ are symmetry operations. For example, for cubic crystals we might add to (7.6) a term [7]

$$b\frac{\mathbf{e}^\alpha(R_{lm}^\alpha)^2\mathbf{e}^\alpha}{R_{lm}^2} \tag{7.8}$$

If the coeficient b depends on α, this term has the form of the anisotropic exchange already considered. The forms (7.7) and (7.8) are symmetric under exchange of l and m, but we could also construct quadratic forms which are anti-symmetric under permutation of two spins. For instance,

$$\mathcal{J}_A = \sum_{\alpha\beta\gamma}\mathbf{e}^\alpha\epsilon_{\alpha\beta\gamma}D^\gamma_{(lm)}\mathbf{e}^\beta \tag{7.9}$$

is such a form if D is a pseudo-vector.

The double contraction of (7.9) with spins $\mathbf{S}_l$ and $\mathbf{S}_m$ gives

$$H_m = \sum_{l\neq m}\mathbf{D}_{lm} \cdot \mathbf{S}_l \wedge \mathbf{S}_m \tag{7.10}$$

which is the Moriya-Dzialoshinsky interaction obtained in Sect. 5.2.

Let us return to Eq. (7.2) and write $\mathbf{R}_{nm}$ in terms of circular components, so that $R_{nm}^\pm = x_{nm} \pm i\, y_{nm}$. Then the expression under the sum in (7.2) becomes

$$\frac{1}{R_{nm}^3}\left(\frac{S_n^+ S_m^- + S_n^- S_m^+}{2} + S_n^z S_m^z\right)$$
$$-\frac{3}{R_{nm}^5}\left[(\frac{1}{2}S_n^+ R_{nm}^- + \frac{1}{2}S_n^- R_{nm}^+ + S_n^z z_{nm})\right.$$
$$\left.\times(\frac{1}{2}S_m^+ R_{nm}^- + \frac{1}{2}S_m^- R_{nm}^+ + S_m^z z_{nm})\right]$$

We expand the expression in square brackets above:

$$[\cdots] = \frac{1}{4}S_n^+ S_m^+ (R_{nm}^-)^2 \tag{7.11}$$

$$+\frac{1}{4}S_n^- S_m^- (R_{nm}^+)^2 + \frac{1}{4}S_n^+ S_m^- R_{nm}^- R_{nm}^+ \tag{7.12}$$

$$+\frac{1}{4}S_n^- S_m^+ R_{nm}^- R_{nm}^+ + \tag{7.13}$$

$$+\frac{1}{2}S_n^+ R_{nm}^- z_{nm} S_m^z + \frac{1}{2}S_m^+ z_{nm} R_{nm}^- S_n^z \tag{7.14}$$

$$+\frac{1}{2}S_n^- R_{nm}^+ z_{nm} S_m^z + \frac{1}{2}S_m^- R_{nm}^+ z_{nm} S_n^z + S_n^z S_m^z z_{nm}^2$$

Exercise 7.1
Show that one can rewrite the dipolar Hamiltonian (7.2) as:

$$H_d = \frac{1}{2}\gamma^2 \sum_{n \neq m} \left\{ \frac{S_n^z S_m^z}{R_{nm}^3}\left(1 - \frac{3z_{nm}^2}{R_{nm}^2}\right) - \frac{S_n^+ S_m^-}{4R_{nm}^3}\left(1 - \frac{3z_{nm}^2}{R_{nm}^2}\right) \right.$$
$$\left. - \frac{3}{4}\left(S_n^+ S_m^+ B_{nm} + h.c. \right) - \frac{3}{2}\left(S_n^+ S_m^z F_{nm} + h.c. \right) \right\} \tag{7.15}$$

The coefficients in Eq. (7.15) are defined as:

$$B_{am} = \frac{(R_{am}^-)^2}{R_{am}^5}$$

$$F_{am} = \frac{R_{am}^- z_{am}}{R_{am}^5} \tag{7.16}$$

Let us now obtain the RPA equations for the exchange-dipolar Hamiltonian, including the Zeeman interaction with an external field B. We must evaluate the Green's function $G_{a,b}^{(r)}(t,t')$:

$$G_{ab}^r(t,t') = -i\theta(t-t')\langle [S_a^+(t); S_b^-(t')]\rangle = \langle\langle S_a^+; S_b^-\rangle\rangle \tag{7.17}$$

As in Chap. 6 we obtain the equation of motion:

$$i\hbar\frac{dG_{ab}^r(t,t')}{dt} = \langle\langle [S_a^+, H]; S_b^-\rangle\rangle + \hbar\delta(t-t')\langle [S_a^+, S_b^-]\rangle \tag{7.18}$$

Since $H = H_e + H_d$, we must now evaluate the commutator of S_a^+ with H_d:

$$
\begin{aligned}
[S_a^+, H_d] \;=\; & \gamma^2 \sum_{m \neq a} \frac{1}{R_{am}^3}\Big(1 - \frac{3z_{am}^2}{R_{am}^2}\Big)[S_a^+, S_a^z]S_m^z \\[2mm]
& - \frac{\gamma^2}{8} \sum_{m \neq a} \frac{1}{R_{am}^3}\Big(1 - \frac{3z_{am}^2}{R_{am}^2}\Big)[S_a^+, S_a^-]S_m^+ \\[2mm]
& - \frac{3\gamma^2}{8} \sum_{m \neq a} B_{am}^*[S_a^+, S_a^-]S_m^- - \frac{3\gamma^2}{4} \sum_{m \neq a}[S_a^+, S_a^z]S_m^+ F_{am} \\[2mm]
& - \frac{3\gamma^2}{4} \sum_{m \neq a}\{[S_a^+, S_a^-]S_m^z + [S_a^+, S_a^z]S_m^-\}F_{am}^* \qquad (7.19)
\end{aligned}
$$

The terms with coefficients F, F^* contribute higher order terms and we shall neglect them.

To simplify the notation we drop the index r on the retarded Green's functions.

Exercise 7.2
Show that $G_{ab}^{+-}(\omega)$ satisfies in the RPA, and with the same approximations as above, the equation:

$$
\begin{aligned}
\hbar\omega G_{ab}^{+-}(\omega) \;=\; & \langle S_a^z\rangle\frac{\delta_{ab}}{\pi} - 2\sum_{m \neq a} J_{ma}\big(\langle S_a^z\rangle G_{mb}^{+-}(\omega) - \langle S_m^z\rangle G_{ab}^{+-}(\omega)\big) \\[2mm]
& + \gamma B G_{ab}^{+-}(\omega) - \gamma^2 \sum_{m \neq a}\Big(1 - \frac{3z_{ma}^2}{R_{ma}^2}\Big)\frac{1}{R_{ma}^3}\langle S_m^z\rangle G_{ab}^{+-}(\omega) \\[2mm]
& - \frac{\gamma^2}{4}\sum_{m \neq a} 2\Big(1 - \frac{3z_{ma}^2}{R_{ma}^2}\Big)\frac{1}{R_{ma}^3}\langle S_a^z\rangle G_{mb}^{+-}(\omega) \\[2mm]
& - \frac{3\gamma^2}{4}2\langle S_m^z\rangle \sum_{m \neq a} B_{am}^* G_{mb}^{--}(\omega) \qquad (7.20)
\end{aligned}
$$

where

$$
G_{ab}^{--}(\omega) = \langle\langle S_a^-; S_b^-\rangle\rangle^{(r)} \qquad (7.21)
$$

The appearance of this term reflects the fact that H_d *does not conserve* $\sum_m S_m^z$. The presence of G^{--} requires obtaining its equation of motion. It is more convenient to Fourier transform now all Green's functions to k space. We write:

$$
G_{ab}(\omega) = \frac{1}{N}\sum_k G_k(\omega)e^{i\mathbf{k}\cdot(\mathbf{R}_a - \mathbf{R}_b)} \qquad (7.22)
$$

Exercise 7.3
Obtain the system of two coupled equations for the Green's functions $G_k^{+-}(\omega)$, $G_k^{--}(\omega)$:

$$
\begin{aligned}
\omega G_k^{+-}(\omega) \;=\; & \frac{\sigma}{\pi} + \epsilon_k G_k^{+-}(\omega) + \gamma B G_k^{+-}(\omega) \\[2mm]
& - [\,\gamma^2 d_0 + \frac{\gamma^2}{2}d_k\,]\,\sigma G_k^{+-}(\omega) - \frac{3\gamma^2}{2}\sigma B_k^* G_k^{--}(\omega) \qquad (7.23)
\end{aligned}
$$

and

$$\omega G_k^{--}(\omega) = \left(-\epsilon_k + \gamma B - \gamma^2 \sigma \left(d_0 + \frac{\gamma^2}{2} \sigma d_k \right) \right) G_k^{--}(\omega)$$
$$- \frac{3\gamma^2}{2} \sigma B_k G_k^{+-}(\omega) \tag{7.24}$$

where

$$\sigma = \langle S_m^z \rangle$$

$$d_k = \sum_{m \neq a} \frac{1}{R_{am}^3} \left(1 - \frac{3z_{am}^2}{R_{am}^2} \right) e^{i\mathbf{k}\cdot\mathbf{R}_{am}}$$

$$B_k = \sum_{m \neq a} \frac{(R_{am}^-)^2}{R_{am}^5} e^{i\mathbf{k}\cdot\mathbf{R}_{am}}$$

$$\epsilon_k = 2\sigma(J(0) - J(k)) \tag{7.25}$$

Calling

$$R_k = \gamma B + \epsilon_k - \gamma^2 \sigma \left(d_0 + \frac{d_k}{2} \right) \tag{7.26}$$

we write the secular equation of the linear system (7.23) and (7.24) as:

$$[(E - R_k)(E + R_k)] + \frac{9}{4}\gamma^4 \mid B_k \mid^2 \sigma^2 = 0 \tag{7.27}$$

The roots of (7.27) are

$$\omega_k^2 = R_k^2 - \frac{9}{4}\gamma^4 |B_k|^2 \sigma^2 \tag{7.28}$$

We shall see below that for the ferromagnetic, uniform, ground state to be stable against the excitation of these modes, we must require that:

$$R_k \geq 0 \quad , \quad \forall k \tag{7.29}$$

From the expression (7.28) for the dispersion relstion we see that we must have

$$\mid R_k \mid > \frac{3}{2}\gamma^2 \mid B_k \mid \sigma, \; \forall k \tag{7.30}$$

for stability of the FM phase. If (7.29) and (7.30) are not satisfied we expect the appearence of a different equilibrium phase. Cohen and Keffer [1] studied this problem within the FSWA and concluded that a s.c. FM lattice is unstable, while the FM b.c.c. and f.c.c. ones are at least metastable.

The Fourier transforms of the dipolar tensor elements were calculated for an infinite lattice. In this case, translation invariance makes all functions involved independent of the individual sites of each coupled pair. This simplification is clearly not possible if one or both sites are near an inhomogeneity like a surface

or in general any deffect which breaks that symmetry.
We eliminate G^{--} in (7.23) and (7.24) to obtain $G_k^{+-}(\omega)$, and find:

$$G_k^{+-}(E) = \frac{\sigma}{2\pi\omega_k} \left\{ \frac{E + R_k}{E - \omega_k} - \frac{E + R_k}{E + \omega_k} \right\} \tag{7.31}$$

One can now repeat the process of calculation we followed for the pure exchange
AFM and obtain the corresponding expressions for the zero-point spin deviation
and T_c in the RPA. It is not possible however in this case to obtain analytic
forms for R_k and B_k, except in the small k limit. Before evaluating $R(k)$ and
$B(k)$ let us get a further physical insight of the exchange-dipolar magnons by
appealing to the Holstein-Primakoff (HP) method of bosonization of the spin
operators which was applied in Chap. 4 to the pure exchange case.

7.2   Dipole-exchange spin-waves

Let us remind ourselves of the HP transformation equations:

$$\begin{aligned}
S_i^- &= \sqrt{2S} a_i^\dagger \hat{f}_i \\
S_i^+ &= \sqrt{2S} \hat{f}_i a_i \\
S_i^z &= S - \hat{n}_i
\end{aligned} \tag{7.32}$$

and

$$\hat{f}_i = \sqrt{1 - \frac{\hat{n}_i}{2S}}$$

$$\hat{n}_i = a_i^\dagger a_i \quad .$$

The complete Hamiltonian is

$$H = H_e + H_Z + H_d \tag{7.33}$$

We express now the Zeeman, exchange and dipolar contributions to the Hamiltonian in terms of the local boson operators. The Zeeman and exchange parts
we have obtained already:

$$\begin{aligned}
H_Z &= \gamma B N S + \gamma B \sum_n a_n^\dagger a_n \\
H_e &= -S^2 \sum_{l \neq m} J_{lm} - \sum_{l \neq m} J_{lm} \left\{ 2S(a_l^\dagger \hat{f}_l \hat{f}_m a_m - a_l^\dagger a_l) + a_l^\dagger a_l a_m^\dagger a_m \right\}
\end{aligned} \tag{7.34}$$

Let us define the dipolar tensor for each pair of sites separated by $\mathbf{x}$ as

$$\mathcal{D}^{\alpha\beta}(\mathbf{x}) = -\frac{\partial^2}{\partial x_\alpha \partial x_\beta} \frac{1}{|\mathbf{x}|} \tag{7.35}$$

We write the dipolar part of H as

$$H_d = \sum_{i=1}^{3} H_d^{(i)}$$

where

$$
\begin{aligned}
H_d^{(1)} &= \frac{S^2\gamma^2}{2} \sum_R{}' \mathcal{D}^{zz}(R) - \frac{\gamma^2}{4} \sum_R{}' \mathcal{D}^{zz}(R_{lm})\{\, 2S(\, a_l^\dagger \hat{f}_l \hat{f}_m a_m \\
&\quad + 2a_l^\dagger a_l\,) - 2a_l^\dagger a_l a_m^\dagger a_m \,\} \tag{7.36} \\
H_d^{(2)} &= -\frac{3\gamma^2}{4}\sqrt{2S} \sum_R{}' F(R)\hat{f}_l a_l (S - a_m^\dagger a_m) + \text{h.c.} \tag{7.37} \\
H_d^{(3)} &= -\frac{3\gamma^2 S}{4}\sqrt{2S} \sum_R{}' B(R)\hat{f}_l a_l \hat{f}_m a_m + \text{h.c.} \tag{7.38}
\end{aligned}
$$

where a prime on a summation means that we exclude $R = 0$. The program now consists of extracting from Eqs. (7.34) to (7.38) a quadratic form in the boson operators and then diagonalizing it to obtain the new normal modes. The quadratic form in the local basis which approximates to second order the complete Hamiltonian of Eq. (7.33) is

$$
\begin{aligned}
H_2 &= -2S \sum_{l\neq m} J_{lm}(a_l^\dagger a_m - a_l^\dagger a_l) + \gamma B \sum_l a_l^\dagger a_l \\
&\quad - \frac{\gamma^2}{4} \sum_{l\neq m} \mathcal{D}^{zz}(lm)(a_l^\dagger a_m + 2a_l^\dagger a_l) \\
&\quad - \frac{3\gamma^2}{4} \sum_{l\neq m} (\, B_{lm} a_l a_m + \text{c.c.} \,) + C \tag{7.39}
\end{aligned}
$$

where the constant C is

$$C = -\gamma BNS - S^2 \sum_{l\neq m} \frac{J_{lm}}{2} + \frac{S^2\gamma^2}{2} \sum_{l\neq m} \mathcal{D}^{zz}(lm) \tag{7.40}$$

From $H_d^{(2)}$ we get a linear term in the boson local operators but it vanishes in an infinite crystal if the lattice is symmetric under one of the reflections $x \to -x$, $y \to -y$ or $z \to -z$ since the sum

$$\sum_{l\neq m} F_{lm} = 0 \tag{7.41}$$

by symmetry in this case. We proceed now to diagonalize the quadratic Hamiltonian (7.39). This is done in three stages:

1) *Transformation to spin-waves.*
We go over to the Bloch (plane wave) representation

$$a_l = \frac{1}{\sqrt{N}} \sum_k e^{-i\mathbf{k}\cdot\mathbf{r}_1} b_k \tag{7.42}$$

The transformed hamiltonian in k space is:

$$\begin{aligned} H_2 &= \sum_k \{ R_k b_k^\dagger b_k + \frac{1}{2}\beta_k b_k b_{-k} \\ &\quad + \frac{1}{2}\beta_k^* b_k^\dagger b_{-k}^\dagger \} \end{aligned} \tag{7.43}$$

where we simplified somehow the notation by calling

$$\beta_k = -\frac{3}{2}\gamma^2 S B_k \tag{7.44}$$

2) *Separate the k space into the subspaces $k^z > 0$, $k^z < 0$.*
Then, we call $\sum^{(+)}$ = summation over $k^z > 0$.
Then

$$H_2 = C + \sum_k^{(+)} H_k \tag{7.45}$$

where

$$H_k = R_k(\, b_k^\dagger b_k + b_{-k}^\dagger b_{-k} \,) + \beta_k b_k b_{-k} + \beta_k^* b_{-k}^\dagger b_k^\dagger \tag{7.46}$$

and R_k was defined in Eq. (7.26). We are explicitly using the symmetry $R_k = R_{-k}$, $B_k = B_{-k}$.

3) *Diagonalize H_k:*
Now we perform the linear Bogoliubov [2] transformation (section 3.3.2):

$$\begin{aligned} b_k &= u_k c_k + v_k c_{-k}^\dagger \\ b_{-k} &= u_{-k} c_{-k} - v_{-k} c_k^\dagger \end{aligned} \tag{7.47}$$

We shall choose $u_k = u_{-k}$ = real and

$$v_k =| \, v_k \, | \, e^{i\alpha_k} \tag{7.48}$$

One verifies that conmutators are preserved if

$$| \, u_k \, |^2 - | \, v_k \, |^2 = 1 \tag{7.49}$$

Substituting Eq. (7.47) in (7.46), one verifies that the new off-diagonal terms vanish if:

$$\beta_k u_k^2 - 2R_k u_k v_k^* + \beta_k^* (v_k^*)^2 = 0 \tag{7.50}$$

We shall show below that

$$\beta_k =| \, \beta_k \, | \, e^{-i2\phi_k}$$

where ϕ_k is the azimuthal angle of $\mathbf{k}$ in the coordinate system in which z is taken along the magnetization. If we substitute (7.48) into (7.50) and choose $\alpha_k = 2\phi_k$, we can rewrite (7.50) as a quadratic equation for the variable

$$z = |u/v| \tag{7.51}$$

which has the solutions

$$z = R_k \pm \sqrt{R_k^2 - \beta_k^2} \tag{7.52}$$

Exercise 7.4
Show that (7.50) is compatible with (7.48) and (7.49) only if $R_k > 0$, $\forall k$ and $|\beta_k| < |R_k|$.

This proves the assertion we made before Eq. (7.29).
    According to the condition (7.49) we can parametrize $|u|$ and $|v|$ as

$$|u_k| = \cosh\mu_k$$
$$|v_k| = \sinh\mu_k \tag{7.53}$$

To gain some insight into the physics of the excitations associated with the operators $c_k, c_k^\dagger$ let us go back to the Hamiltonian in the form of Eq. (7.45). The Heisenberg equations of motion of operators b_k and $b_{-k}^\dagger$ are coupled. What we are assuming implicitly upon making Bogoliubov transformation (7.47) is that the linear combination of magnons (7.47) is a normal mode, which can only happen if b_k and $b_{-k}^\dagger$ have the same time dependence e^{-iEt}. Substituting explicitly this time dependence we obtain for the eigen-modes the following system of two equations:

$$(E - R_k)b_k + \beta_k^* b_{-k}^\dagger = 0$$
$$\beta_k b_k + (E + R(k))b_{-k}^\dagger = 0 \tag{7.54}$$

Exercise 7.5
Obtain Eq. (7.54).

This system is identical with (7.23) and (7.24), if we disregard the inhomogeneity on the right hand member of (7.23), so that we obtain the same eingenvalue equation (for low T). From (7.54) we get for the eigenvalue

$$\omega_k = \sqrt{R_k^2 - \frac{9}{4}\gamma^2 S^2 |B_k|^2} \tag{7.55}$$

which coincides with Eq. (7.28) with the only difference that in RPA S is replaced by the self-consistent average σ.

One verifies that Bogoliubov transformation restores the canonical harmonic form of the Hamiltonian:

$$H = \sum_k^+ \omega_k \ (\ c_k^\dagger c_k + c_{-k}^\dagger c_{-k} \ )$$
(7.56)

Exercise 7.6
Verify Eq. (7.56)

One finds with a little algebra that

$$v_k = v_{-k} = \sqrt{\frac{R_k - \omega_k}{2\omega_k}} e^{-2i\phi_k}$$

$$u_k = u_{-k} = \frac{|\ \beta_k\ |}{\sqrt{2\omega_k(R_k - \omega_k)}}$$
(7.57)

Let us quote a useful relation:

$$\frac{R_k \pm \beta_k \cos 2\phi_k}{\omega_k} = \cosh^2 \mu_k + \sinh^2 \mu_k$$
$$\pm \sinh \mu_k \cosh \mu_k \cos 2\phi_k$$
(7.58)

Exercise 7.7
Verify Eq. (7.58)

We now proceed to calculate R_k and B_k and to evaluate the eigenmode energies. This must be done in a special way for the uniform precession mode with $k = 0$, which is a *magnetostatic mode*. As we shall see shortly, the effect of the surface of the sample cannot be eliminated in this case, and the corresponding demagnetization field, due to the divergence of **M** at the surface, must be incorporated into the local field.

7.3 Uniform precession ($k = 0$) mode

Let us consider the sum:

$$d_0 = \sum_{m \neq 0} (1 - \frac{3z_m^2}{R_m^2}) \frac{1}{R_m^3}$$
(7.59)

where we can choose the origin at the position of an arbitrary spin. This freedom, as we already noticed, is due to the assumption that all spins are sufficiently far from the surface of the system. In fact, for a truly infinite crystal with inversion symmetry, the sum in Eq. (7.59) is exactly zero, so that its contribution in a finite sample is entirely due to the surface. We use now Lorentz method to calculate d_0. Let us define a separation parameter a as the radius of the Lorentz sphere. We require that the volume of the sphere be macroscopic, that is $a \gg a_0 =$ lattice constant. We divide the summation (7.59) into $|R| \le a$ and $|R| > a$. For finite k we require aswell that $ka \ll 1$, or $\lambda \gg a$, where λ is the magnon wavelength. This means that within the Lorentz sphere we can consider the magnetization as exactly ($k = 0$) or approximately ($k \ne 0$) uniform. Then, if the symmetry is cubic we shall neglect the contribution from points inside the Lorentz sphere. We are then left with the sum (7.59) for $|R| > a$. Then $|R_m| > a \gg$ lattice constant, and we can substitute the sum by an integral:

$$d_0 \approx \frac{N}{V} \int_{|r|>a} d^3r (1 - \frac{3z^2}{r^2}) \frac{1}{r^3} = \frac{N}{V} \int_{|r|>a} d^3r \nabla \cdot \left(\frac{\mathbf{z}}{r^3}\right) \qquad (7.60)$$

with $\mathbf{z} \equiv z\mathbf{e}_z$ and $\mathbf{e}_z=$versor of the z axis. Application of Gauss' theorem yields:

$$d_0 = \frac{N}{V} \oint_{surface} \frac{\mathbf{z}}{r^3} \cdot d\mathbf{S} \qquad (7.61)$$

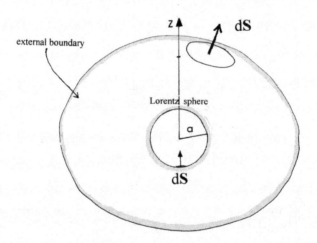

Figure 7.1: *Boundary surfaces for the integration in Eq. (7.61).*

Since the spherical region $r < a$ is excluded, we have one internal and one external boundary, the elements of area on each of them being oriented as indicated in Fig. 7.1. Then the surface integral splits into two parts:

$$d_0 = \frac{N}{V} \oint_{ext.\ bound.} \mathbf{z} \cdot \frac{d\mathbf{S}}{r^3} + \frac{N}{V} \oint_{r=a} \mathbf{z} \cdot \frac{d\mathbf{S}}{r^3} \qquad (7.62)$$

The integral over the external boundary depends on the shape of the sample, and it is defined as

$$\oint_{ext.\ bound.} \mathbf{z} \cdot \frac{d\mathbf{S}}{r^3} \equiv 4\pi N_z \qquad (7.63)$$

which is the α component of the vector $\mathbf{N}$ defined as

$$4\pi(\mathbf{N})_\alpha = \oint_{ext.\ bound.} (\mathbf{r})_\alpha \frac{(d\mathbf{S})_\alpha}{r^3} \tag{7.64}$$

It is easy to show that the sum rule

$$N_x + N_y + N_z = 1 \tag{7.65}$$

follows from the fact that the solid angle subtended by a closed surface from an interior point is 4π.

The coefficients N_α are called *demagnetization factors*.

In an uniformly magnetized ferromagnetic sample which contains localized magnetic moments $\mathbf{m}_i$, $\ i = 1, \cdots N$, we write the dipolar energy in the mean field approximation as

$$E_{dip} = \frac{1}{2} \sum_{i \neq j}^{N} \langle m_i^\lambda \rangle \mathcal{D}^{\lambda\mu}(\mathbf{R}_{ij} \langle m_j^\mu \rangle \tag{7.66}$$

Since

$$\langle \mathbf{m}_i \rangle = \mathbf{m} \ , \forall \ i$$

we have

$$E_{dip} = \frac{1}{2} \sum_{i \neq j}^{N} \mathcal{D}^{\lambda\mu}(\mathbf{R}_{ij}) \langle m^\lambda \rangle \langle m^\mu \rangle \tag{7.67}$$

We recognize the sum in Eq. (7.67) as the Fourier transform for $k = 0$ of the dipolar tensor. The eigenvalues of this tensor are

$$4\pi(N/V) \ N_\eta$$

where $\eta = X, Y, Z$ are the indices of the principal axes of the Fourier transformed dipolar tensor. Then the final form of E_{dip} is:

$$\frac{E_{dip}}{V} = \frac{4\pi}{2} M^2 \left(N_X \cos^2 \alpha + N_Y \cos^2 \beta + N_Z \cos^2 \gamma \right) \tag{7.68}$$

where $\cos\alpha$, etc., are the director cosines of the magnetization $\mathbf{M}$ referred to the principal axes. One verifies that this expression can be written, in terms of the internal field defined in Eq. (7.67) for $B = 0$, as

$$\frac{E_{dip}}{V} = -\frac{1}{2} \mathbf{B}_i \cdot \mathbf{M} \tag{7.69}$$

where the component of $\mathbf{B}_i$ along each principal axis $\lambda = X, Y, Z$ is

$$\mathbf{B}_i^X = -4\pi N_X \ M_X \ , \quad etc.$$

E_{dip} depends on the shape of the sample, as is clear from the definition of the demagnetization factors, and it is called the *"shape-anisotropy energy"*.

Exercise 7.8
Obtain (7.65) and show that the internal surface integral is $-4\pi/3$.

Then

$$d_0 = \frac{N}{V}\left(4\pi N_z - \frac{4\pi}{3} \right) \tag{7.70}$$

Now we turn to B_k. The field inside the Lorentz sphere vanishes for cubic symmetry. Then

$$B_k = \frac{N}{V} \int_{r>a} d^3r \frac{(r^-)^2}{r^5} e^{i\mathbf{k}\cdot\mathbf{r}}$$

For $k = 0$ this simplifies if the sample has reflection symmetry ($x \to -x$, $y \to -y$, in which case

$$B_0 \simeq \frac{N}{V}\int_{r>a} d^3r \frac{x^2 - y^2}{r^5} =$$

$$\frac{1}{3}\frac{N}{V}\int_{r>a} d^3r \left[(1 - \frac{3x^2}{r^2}) - (1 - \frac{3y^2}{r^2}) \right] \frac{1}{r^3} =$$

$$\frac{4\pi N}{3V}(N_x - N_y) \tag{7.71}$$

according to definition (7.64). The contributions from the surface integrals on the Lorentz sphere cancel each other.

On the other hand, for $k = 0$, (7.26) yields:

$$R_0 = \gamma B - \frac{3}{2}\gamma^2\sigma d_0 \tag{7.72}$$

We call the frequency of the uniform precession mode ω_u. Then from Eq. (7.28) we obtain

$$\omega_u^2 = (\gamma B - \frac{3}{2}\gamma^2 d_0\sigma)^2 - \frac{9}{4}\gamma^4\sigma^2 B_0^2 \tag{7.73}$$

We can now subtitute (7.71) and (7.72) into (7.73), and obtain

$$\omega_u = \sqrt{(\gamma B_i + 4\pi\gamma N_x M)(\gamma B_i + 4\pi\gamma N_y M)} \tag{7.74}$$

where the internal field B_i is defined as

$$B_i = B - 4\pi N_z \tag{7.75}$$

The magnetization M is

$$M = \gamma\sigma\frac{N}{V} \tag{7.76}$$

While σ is replaced by S in the FSWA, we can calculate it self-consistently within the RPA.

Exercise 7.9
Obtain (7.75).

7.4   Eigenmodes for k $\neq$ 0

In this case, we can expand the plane wave in both integrals of (7.25) in products of spherical harmonics

$$e^{i\mathbf{k}\cdot\mathbf{r}} = \sum_{l,|m|<l} -4\pi(+i)^l j_l(kr) Y_l^{m*}(\Omega) Y_l^m(\Omega') \tag{7.77}$$

where $\Omega \equiv (\theta, \phi)$ are the spherical angles of $\mathbf{r}$ and $\Omega' \equiv (\theta_k, \phi_k)$ those of $\mathbf{k}$, in the same reference system in which the spin quantization axis z is parallel to the magnetization. Substituting (7.77) into (7.25), and noticing that

$$1 - \frac{3z^2}{r^2} \equiv C Y_2^0(\Omega)$$

we find:

$$d_k = -\frac{CN}{V} \int_{r>a} r^2 \mathrm{d}r j_l(kr) \times$$

$$\int \mathrm{d}\Omega \sum_{lm} Y_2^0(\Omega) Y_l^{m*}(\Omega)(i)^l 4\pi \frac{1}{r^3} Y_l^m(\Omega') \tag{7.78}$$

Due to the orthonormality relations of the spherical harmonics

$$\langle Y_l^{m*}|Y_{l'}^{m'}\rangle = \delta_{ll'}\delta_{mm'}$$

we obtain

$$d_k = -4\pi C(i)^2 \frac{N}{V} \int_{r>a} \mathrm{d}r \frac{j_2(kr)}{r} Y_2^0(\Omega') =$$

$$4\pi \frac{N}{V}(1 - 3cos^2\theta_k) \int_{r>a} \mathrm{d}r \frac{j_2(kr)}{r} \tag{7.79}$$

Now

$$\int^x \mathrm{d}r \frac{j_2(kr)}{r} = \frac{j_1(kx)}{kx}$$

In the case

$$kR \gg 1$$

where R is the size of the sample, $j_1(kR)/kR$ is negligible at the surface, whatever the shape, and we drop the surface contribution. We remark that the results that follow cannot be taken over to the case $kR \leq 1$, which is the region of the *magnetostatic* modes.
We find

$$d_k = 4\pi \frac{N}{V}(3\cos^2\theta_k - 1)\frac{j_1(ka)}{ka} \tag{7.80}$$

We notice that a was chosen with the condition $ka \ll 1$, and

$$\lim_{x\to 0}\left(\frac{j_1(x)}{x}\right) = \frac{1}{3}$$

so that

$$d_k = \frac{4\pi}{3} \frac{N}{V} (3\cos^2\theta_k - 1) = \frac{N}{V} \left(\frac{8\pi}{3} - 4\pi\sin^2\theta_k \right) \tag{7.81}$$

Repeating the whole procedure for B_k we obtain

$$B_k = -\frac{N}{V} \frac{4\pi}{3} \sin^2\theta_k e^{-2i\phi_k} \tag{7.82}$$

Upon substituting (7.65) and (7.70) into (7.28) we arrive at

$$\begin{aligned}
\omega_k^2 &= (\gamma B - 4\pi\gamma M N_z + \epsilon_k) \times \\
&\quad (\gamma B - 4\pi\gamma M N_z + \epsilon_k + 4\pi\gamma M \sin^2\theta_k)
\end{aligned} \tag{7.83}$$

We conclude that the dominant contribution of dipolar interactions to the magnon energy is independent on $|\mathbf{k}|$ when $ka \ll 1$ and $kR \gg 1$. The last condition implies that the results obtained are valid away from the magnetostatic region where $kR \leq 1$. Notice that for any finite sample we cannot extend these results to the neighbourhood of $k = 0$.

We found a dispersion relation which depends on the orientation of $\mathbf{k}$ relative to the magnetization. On the other hand, the exchange energy for small k is isotropic in a cubic crystal, and approximately quadratic. Then if we keep the orientation of $\mathbf{k}$ fixed and vary the magnitude of k we shall obtain a curve, where the dependence on k comes through the exchange part. The series of curves thus obtained as we vary the orientation of $\mathbf{k}$ is called the *magnon manifold*. The interesting fact is that in many important cases the uniform precession mode is degenerate with states in this manifold. There is then the possibility of relaxing the uniform mode through processes in which a $\mathbf{k} = 0$ magnon scatters against some static imperfection of the crystal and decays into one of these degenerate modes with $\mathbf{k} \neq \mathbf{0}$. The extra linear momentum is transferred to the static imperfection and the energy is conserved. This is the interpretation of experiments measuring the relaxation rate of the uniform mode. These are scattering events involving two magnons (two-magnon processes). In order that these processes conserve energy the uniform mode must be contained within the magnon manifold. Since the results we just obtained are shape dependent, we analyze this for a spherical sample. In this case Eq. (7.74) yields the energy

$$\omega_u = \gamma B.$$

This is positive for any external field, but one should keep in mind that the local field must point in the direction of the applied field if the system is to keep magnetized in that direction, so that the minimum applied field which ensures saturation of the sample is $4\pi M/3$. Otherwise the sample will be subdivided into magnetic domains, and the total net magnetization will be smaller than the saturation one at that temperature. Provided $B > 4\pi M N_z$ we verify that if the sample is spherical ω_u is contained within the magnon manifold band for the same B. The sphere is the most favourable shape of the sample for this relaxation process. The degeneracy of the uniform mode with some of the modes within the manifold is verified for all ellipsoidal shapes of the sample, with the only exception of the disk nornal to the field.

7.4.1   Demagnetization factors

We list now the demagnetization factors and the values of ω_u for various shapes.

- **Sphere:** $N_x = N_y = N_z = 1/3$; $\omega_u = \gamma B$

- **Infinite disk, $B \perp$ disk :** $N_x = N_y = 0$, $N_z = 1$; $\omega_u = \gamma(B - 4\pi M)$

- **Infinite disk, $B \parallel$ disk:** $N_x = 1$, $N_y = N_z = 0$; $\omega_u = \gamma\sqrt{B(B + 4\pi M)}$

- **Infinite cylinder, $B \parallel$ cylinder:** $N_x = N_y = 1/2$, $N_z = 0$; $\omega_u = \gamma(B + 2\pi M)$

- **Infinite cylinder, $B \perp$ cylinder:** $N_x = N_z = 1/2$, $N_y = 0$; $\omega_u = \gamma\sqrt{B(B - 2\pi M)}$

7.5   Ellipticity of spin precession

Let us look at the anisotropy of the spin precession around the quantization z axis in a magnon mode. We must calculate the ratio of the variances of the x and y transverse spin components, that is the quantity

$$e = \frac{\langle(S_i^x)^2\rangle}{\langle(S_i^y)^2\rangle} = -\frac{\langle(S_i^+ + S_i^-)^2\rangle}{\langle(S_i^+ - S_i^-)^2\rangle} \tag{7.84}$$

If we expand the local ascending and lowering operators in terms of spin wave operators we have

$$S_i^+ \pm S_i^- = \sqrt{\frac{2S}{N}} \sum_k e^{-i\mathbf{k}\cdot\mathbf{R_i}}(b_{-k}^\dagger \pm b_k) \tag{7.85}$$

We want to evaluate the expectation values in Eq. (7.84) in a particular magnon eigenstate

$$| \tilde{k}\rangle \equiv c_k^\dagger| \tilde{0}\rangle$$

where $| \tilde{0}\rangle$ is the ground state of the complete exchange-dipolar Hamiltonian. Substituting the Bogoliubov transformation from the b_k to the c_k one gets

$$
\begin{aligned}
e_k &= \frac{\langle(S_i^x)^2\rangle_k}{\langle(S_i^y)^2\rangle_k} \\
&= \frac{\cosh^2\mu_k + \sinh^2\mu_k - \sinh\mu_k\cosh\mu_k\cos 2\phi_k}{\cosh^2\mu_k + \sinh^2\mu_k + \sinh\mu_k\cosh\mu_k\cos 2\phi_k} \\
&= \frac{R_k - \beta_k\cos 2\phi_k}{R_k + \beta_k\cos 2\phi_k}
\end{aligned}
\tag{7.86}
$$

where use was made of Eq. (7.58). In a semi-classical image, the local spin is a vector. During its precessional motion around the z axis the end point of this vector describes a periodic curve that we can project on the x, y plane. The anisotropy of this curve is measured by the quantity e, or e_k for a particular magnon state as considered above. The expression (7.86) is maximum for $\phi = \pi/2$ and minimum for $\phi = 0$, so that the short axis of the curve is contained in the azimuthal plane containing $\mathbf{k}$. We see that if the dipolar anisotropy parameter, defined as

$$E_d = \frac{\gamma^2}{Ja^3} \tag{7.87}$$

is very small, so is the ellipticity

$$\epsilon \equiv \mid e - 1 \mid \tag{7.88}$$

since in that case $\beta_k \ll R_k$.

7.6   Effect of magnons on total spin

In the analysis of relaxation processes it is important to distinguish between those which change only the transverse or the longitudinal component of the total spin or both. Let us first calculate the effect of magnons on the square of the total spin vector operator,

$$\mathcal{S} = \sum_i^N \mathbf{S}_i$$

In terms of the local boson HP operators the z component is

$$\mathcal{S}_z = NS - \sum_i^N a_i^\dagger a_i \tag{7.89}$$

and the transverse components can be approximated at low T by the lowest HP approximation:

$$\mathcal{S}^+ \approx \sqrt{2S} \sum_i^N a_i = \sqrt{2SN}b_0 \tag{7.90}$$

while the h. c. of this expression gives $\mathcal{S}^-$. For the z component

$$\mathcal{S}_z = NS - \sum_k b_k^\dagger b_k \tag{7.91}$$

if we make use of the canonical transformation to spin waves (7.42) in (7.90) and (7.91).

It will prove convenient to separate in (7.91) the term $k = 0$:

$$\mathcal{S}_z = NS - \sum_{k \neq 0} b_k^\dagger b_k - b_0^\dagger b_0 \tag{7.92}$$

If we substitute in (7.91) the spin-wave operators $\{b_k\}$ in terms of the true magnon operators $\{c_k\}$ we obtain for $\mathcal{S}_z$ the expression

$$\mathcal{S}_z = NS - \sum_k a_i^\dagger a_i = NS - \sum_k (\mid u_k \mid^2 + \mid v_k \mid^2)c_k^\dagger c_k + \sum_k \mid v_k \mid^2) \qquad (7.93)$$

where the last sum is the zero-point spin-deviation due to the dipolar interaction.

We turn now to the other terms. If the temperature is not too high, or in general if just a few magnons are excited in the system, the first term in (7.91) is much larger than the magnon contribution, so the average of the square is approximately

$$\langle(\mathcal{S}_z)^2\rangle = (NS)^2 - 2NS\sum_{k \neq 0}\langle b_k^\dagger b_k\rangle - 2NS\langle b_0^\dagger b_0\rangle \qquad (7.94)$$

while

$$\frac{1}{2}\langle \mathcal{S}^+\mathcal{S}^- + \mathcal{S}^-\mathcal{S}^+\rangle = 2SN\left(\langle b_0^\dagger b_0\rangle + 1/2\right) \qquad (7.95)$$

Therefore

$$\langle(\mathcal{S})^2\rangle = (NS)(NS+1) - 2NS\sum_{k \neq 0}\langle b_k^\dagger b_k\rangle \qquad (7.96)$$

We perform now the Bogoliubov transformation from the $\{ b_k \}$ to the $\{ c_k \}$ operators and obtain

$$\langle(\mathcal{S})^2\rangle - (NS)(NS+1) = -2NS\sum_{k \neq 0}\{(\mid u_k \mid^2 + \mid v_k \mid^2)n_k + \mid v_k \mid^2\} \qquad (7.97)$$

where $n_k = \langle c_k^\dagger c_k\rangle$ and the indicated average has been performed over an eigenstate of the exchange-dipolar Hamiltonian, or over the corresponding canonical statistical ensemble at temperature T. The last term above is the zero-point contribution to the spin-deviation, which describes the decrease of the magnetization of the ground state so this is a renormalization of $\mathcal{S}^2$ independent of the excitation state. Therefore, the change of $\mathcal{S}^2$ from its ground state value due to the presence of magnons is

$$\Delta\mathcal{S}^2 = -2NS\sum_{k \neq 0}\{(\mid u_k \mid^2 + \mid v_k \mid^2)n_k \qquad (7.98)$$

We see that the effect of a single magnon with $k \neq 0$ is to decrease the length of the total spin of the system, as well as decreasing the z component in one unit. This conclusion has important consequences regarding the physics of the different processes which are effective in relaxing the magnetization to its equilibrium value.

7.7 Magnetostatic modes

We consider now the case in which the magnetization has a spatial variation in a scale comparable with the sample size R, that is $kR \sim 1$. We are interested in macroscopic samples and in the long-wave limit, so that we can still take $ka \ll 1$, with a = lattice constant. Then the exchange terms in the energy, proportional to $(ka)^2$, are negligible: we are still in the long wave limit as regards exchange. Let us also take $kc \gg \omega$, c = speed of light, so that we can neglect the induced electric field and work in the magnetostatic limit of Maxwell's equations. In summary, we have a macroscopic ferromagnetic sample below T_c and we consider behaviour in the long wave limit of self sustained modes in the system which are the macroscopic (semi-classic) solutions of the constitutive torque equation of Chap. 4 (with the adequate effective field acting on each spin) and Maxwell's equations in the magnetostatic limit. We must determine the solutions by imposing the boundary conditions of continuity of the normal component of the magnetic induction field and of the tangential component of the magnetic field at the surface. In this section we shall find convenient to use the letter H for the magnetic field, as there is no confusion with the Hamiltonian. The magnetostatic Maxwell's equations are

$$\nabla \wedge \mathbf{H} = 0$$
$$\nabla \cdot (\mathbf{H} + 4\pi\mathbf{M}) = 0 \tag{7.99}$$

and we also have the constitutive equation

$$\frac{d\mathbf{M}}{dt} = \gamma \mathbf{M} \wedge \mathbf{H} \tag{7.100}$$

neglecting damping. Some authors use the negative sign in (7.100) but the choice of sign will not have any consequence in the present context. The resulting modes, called *magnetostatic modes*, were first experimentally found by White and Solt [3] and Mercenau [4] in ferrite spheres and independently by Dillon [5] in ferrite disks, before they were described theoretically by Walker [6] in his calculations for samples with the shape of ellipsoids of revolution around the z axis. These modes can be excited in a strongly non uniform oscillating magnetic field, which has a wavelength comparable to the sample size. The magnetostatic limit allows us to take $\mathbf{H} = \nabla\psi$ where ψ is a scalar potential. Besides, the system is magnetized to saturation along z, so that the longitudinal component of $\mathbf{M}$ is much larger that the transverse ones, and one writes

$$\mathbf{M} = (m_x, m_y, M_0)$$
$$\mathbf{H} = (h_x, h_y, H) \tag{7.101}$$

The magnetic field must be substituted inside the sample by the effective field:

$$\mathbf{H}_{\text{eff}} = \mathbf{H}_0 + \frac{2A}{M_0^2}\nabla^2\mathbf{M} - 4\pi N_z\mathbf{M} \tag{7.102}$$

The second term above, which is the exchange contribution to the effective field, was obtained in Chap. 4. We are adding the demagnetizing field as calculated in the present chapter for the uniform magnetization case, since the longitudinal component, which is assumed constant, is the dominant one.

We can now substitute (7.102) into (7.100), and linearize the constitutive equation to find:

$$\mathbf{m_k} = \chi(\mathbf{k})\mathbf{h_k} \tag{7.103}$$

where the susceptibility tensor $\chi(\mathbf{k})$ is [7]

$$\chi(\mathbf{k}) = \begin{pmatrix} \chi_k & -iK_k & 0 \\ iK_k & \chi_k & 0 \\ 0 & 0 & 0 \end{pmatrix} \tag{7.104}$$

and

$$\begin{aligned}
\chi_k &= \frac{\omega_M}{4\pi W_k} \\
K_k &= \frac{\omega}{4\pi W_k^2} \\
\omega_M &= 4\pi\gamma M_0 \\
W_k &= \omega_H + \omega_E a^2 k^2 \\
\omega_H &= \gamma(H_0 - 4\pi N_z M_0) \\
\omega_E &= \frac{2A\gamma}{M_0 a^2}
\end{aligned} \tag{7.105}$$

We are neglecting the anisotropy field and the damping terms. The eigenvalues of the susceptibility tensor are $\chi_k \pm K_k$. For $\mathbf{k} \parallel \mathbf{M}_0$ the eigenmodes are the two counter-rotating circularly polarized modes of the transverse magnetization. The difference between the two circular susceptibilities is responsible for the Faraday effect in the microwave region. The magnetic potential outside the sample satisfies Laplace's equation. If one chooses a reference system with the x, y axes along two principal axes of the tensor $\chi(\mathbf{k})$ the potential inside satisfies the differential equation:

$$(1 + 4\pi\chi_0)\left(\frac{\partial^2}{\partial x^2} + \frac{\partial^2}{\partial y^2} \right)\psi + \frac{\partial^2\psi}{\partial z^2} = 0 \tag{7.106}$$

The boundary conditions are the continuity of ψ and of the normal component of $\mathbf{b} = \mathbf{h} + 4\pi\mathbf{m}$ at the surface.

Walker [6] obtained the normal modes for an ellipsoid of revolution around z, which are indexed by three integers n, m, r. n and m are the indices of an associated Legendre polinomial P_n^m in the expansion for ψ, while r numbers the roots of a secular equation.

References

1. Cohen, M. H. and Keffer, F. (1955) *Phys. Rev.* **99**, 1135.

2. Bogoliubov, N.N. (1947) *J. Phys. USSR* **9**, 23; (1958) *Nuovo Cimento* **7**, 794.

3. White, R.L. and Solt, I.H. (1956) *Phys. Rev.* **104**, 56.

4. Mercerau, J.E. (1956) *Bull. Am. Phys. Soc.* **1**, 12.

5. Dillon, J.F. Jr (1956) *Bull. Am. Phys. Soc.* **1**, 125.

6. Walker, L. (1957) *Phys. Rev.* **105**, 390.

7. Keffer, F. (1966) *Hand. der Phys., 2 Aufl.* **XVIII**, 1.

Chapter 8

Coherent States of Magnons

8.1 Introduction

In Chap. 7 we studied states with a given number of magnons excited in a ferromagnet with exchange and dipolar interactions. Therein we remarked that the expectation value of the transverse components of the local spin operator vanishes:

$$\langle n_k \mid S_i^{x,y}(t) \mid n_k \rangle = 0$$

This can be interpreted in the sense that the spins are not rotating coherently: it is as if in a state with a given number of magnons, which we shall herefrom denote as a *"number state"*, a random phase must be assigned to each spin which precesses around the magnetization axis with frequency ω_k . Therefore, number states are not compatible with the semi-classical picture of a magnon as a classical wave as given in Sect. 4.4 . In the search for states that show a classical wave behaviour we must abandon the description in terms of states with a fixed number of magnons, and turn instead to coherent states.

The coherent states we shall presently define in analogy with those of photons [1] behave, in the limit of a large average number of magnons, as a classical spin wave.

We have seen in Chap. 4 that magnons behave like bosons, so that we shall first turn now to the definition and basic properties of a coherent state constructed as a linear combination of bosons.

8.2 Coherent states of bosons

We define a coherent state of bosons [2] as an eigenstate of the annihilation operator:

$$c_k \mid \alpha_k \rangle = \alpha_k \mid \alpha_k \rangle \tag{8.1}$$

where we keep the notation of Chap. 7, since we shall be referring to exchange-dipolar magnons. Notice that we specialize to a particular **k** vector, because our system has by assumption translation invariance, so Bloch's theorem applies. The eigenvalue α_k is a complex number.

We can expand the ket $|\,\alpha_k\rangle$ in the Fock basis of number states

$$|\,\alpha_k\rangle = \sum_{n_k=0}^{\infty} A_{n_k}\,|\,n_k\rangle \tag{8.2}$$

In Eq. (7.52) we expressed the magnetic Hamiltonian with exchange and dipolar interactions as a superposition of independent harmonic oscillators, each of them characterized by a given $\mathbf{k}$. This is true within the lowest order of the Holstein-Primakoff expansion (Sect 4.3). The boson annihilation operator satisfies the relation

$$c_k\,|\,n_k\rangle = \sqrt{n_k}\,\,|\,n_k-1\rangle \tag{8.3}$$

so from Eqs. (8.2) and (8.3) we get the following recursion relation for the expansion coefficients:

$$\alpha_k A_{n_k} = \sqrt{n_k+1}\,\,A_{n_k+1} \tag{8.4}$$

If we assume that $A_0 \neq 0$ we can choose $A_0 = 1$ as the initial condition for solving the recurrence equation and we take care later of the normalization of the coherent state. The expansion then results

$$|\,\alpha_k\rangle = \sum_{n_k=0}^{\infty} \frac{\alpha_k^{n_k}}{\sqrt{n_k}}\,|\,n_k\rangle \tag{8.5}$$

The state with n_k magnons can be obtained upon operating repeatedly on the vacuum state with the creation operator, as:

$$|\,n_k\rangle = \frac{(c_k^{\dagger})^{n_k}}{\sqrt{n_k!}}\,|\,0\rangle \tag{8.6}$$

Substituting (8.6) into (8.5) we get:

$$|\,\alpha_k\rangle = e^{\alpha_k\,c_k^{\dagger}}\,|\,0\rangle \tag{8.7}$$

Exercise 8.1
Prove the following properties of the coherent state:

$$\langle\alpha_k\,|= \langle 0\,|\,e^{\alpha_k^{*}\,c_k} \tag{8.8}$$

$$\langle\alpha_k\,|\,c_k^{\dagger} = \langle\alpha_k\,|\,\alpha_k^{*} \tag{8.9}$$

$$\langle\alpha_k'\,|\,\alpha_k\rangle = e^{\alpha_k'^{*}\,\alpha_k} \tag{8.10}$$

$$c_k^{\dagger}\,|\,\alpha_k\rangle = \frac{\partial}{\partial\alpha_k}\,|\,\alpha_k\rangle \tag{8.11}$$

$$\langle\alpha_k\,|\,c_k = \frac{\partial}{\partial\alpha_k^{*}}\langle\alpha_k\,| \tag{8.12}$$

From (8.10) we get the normalization constant for the coherent states. The normalized ones are defined accordingly as:

$$| \alpha_k \rangle = e^{-\frac{1}{2}|\alpha|^2} e^{\alpha_k c_k^\dagger} | 0 \rangle \tag{8.13}$$

There is no need to use a different notation for the normalized and the original states, since from now on we shall always use the latter. The overlap of two normalized states is:

$$| \langle \alpha | \beta \rangle |^2 = e^{-|\alpha - \beta|^2} \tag{8.14}$$

so although they are never exactly orthogonal, their overlap decreases exponentially as $| \alpha - \beta | \to \infty$.

Exercise 8.2
Prove Eq. (8.14).

8.2.1 Overcompleteness of coherent states basis

In the following we shall drop the subindex k since we shall always deal with Bloch states.

We shall prove the closure relation

$$\mathbf{A} \equiv \int \frac{d\alpha^* \, d\alpha}{2\pi i} \, | \alpha \rangle \langle \alpha | \equiv \mathbf{1} \tag{8.15}$$

First of all we verify that

$$[\hat{c}, \mathbf{A}] = [\mathbf{A}, \hat{c}^\dagger] = 0 \tag{8.16}$$

Exercise 8.3
Prove (8.16).

All operators in Fock's subspace of a given $\mathbf{k}$ vector are polynomials in $\hat{c}_k$, $\hat{c}_k^\dagger$, and therefore also conmute with $\mathbf{A}$, which accordingly satisfies the conditions for the application of Schur's lemma [3], which states that: if a matrix (or an operator on a linear space) $\mathbf{A}$ commutes with every matrix (or operator) of the space, then it is a multiple of the identity:

$$\mathbf{A} = a \cdot \mathbf{1} \tag{8.17}$$

where a is some complex number. In order to determine a we calculate the expectation value of $\mathbf{A}$ in the magnon vacuum Fock state:

$$a = \int \frac{d\alpha^* \, d\alpha}{2\pi i} \langle 0 | \alpha \rangle \langle \alpha | 0 \rangle \tag{8.18}$$

Notice that for the normalized states $\langle 0 \mid \alpha \rangle = e^{-\frac{1}{2}|\alpha|^2}$ so that we have:

$$a = \int \frac{da^* \, da}{2\pi i} e^{-|\alpha|^2} \tag{8.19}$$

This is an integral over the whole complex plane of α. Instead of the variables α and α^* we may choose the real and imaginary parts of $\alpha = x + iy$ as the independent variables. We find that $a = 1$, so that $\mathbf{A} = \mathbf{1}$.

8.3　Magnon number distribution function

Let us now calculate the expectation value and the variance of the number operator $\hat{n} = \hat{c}^\dagger \hat{c}$ in a coherent state. For the expectation value we have:

$$\overline{n} = \langle \alpha \mid c^\dagger c \mid \alpha \rangle = \mid \alpha \mid^2 \tag{8.20}$$

In order to calculate the mean of $\hat{n}^2$ we must first rewrite the product

$$\hat{n}^2 = \hat{c}^\dagger \hat{c} \hat{c}^\dagger \hat{c}$$

in normal order, that is, we must make the necessary permutations so that all creation operators be on the left of the annihilation ones. The result is:

$$\overline{n^2} \equiv \mid \alpha \mid^2 + \mid \alpha \mid^4 = \overline{n} + \overline{n}^2 \tag{8.21}$$

The square of the variance is then

$$(\Delta n)^2 = \langle \alpha \mid (\hat{n} - \overline{n})^2 \mid \alpha \rangle = \mid \alpha \mid^2 = \overline{n} \tag{8.22}$$

so that $\Delta n = \mid \alpha \mid$ and

$$\frac{\Delta n}{\overline{n}} = \frac{1}{\sqrt{\overline{n}}}$$

The internal product $\langle n \mid \alpha \rangle$ is the probability amplitude for finding n magnons in that coherent state:

$$\langle n \mid \alpha \rangle = \frac{\alpha^n}{\sqrt{n!}} e^{-|\alpha|^2/2} \tag{8.23}$$

and the probability distribution of n is the modulus squared of this quantity:

$$P_\alpha(n) = \frac{\mid \alpha \mid^{2n}}{n!} e^{-|\alpha|^2} = \frac{\overline{n}^n}{n!} e^{-\overline{n}} \tag{8.24}$$

which is Poisson's distribution function.

Exercise 8.4
Verify that

$$\sum_{n=0}^{\infty} P_\alpha(n) = 1 \tag{8.25}$$

8.4 Uncertainty relations

We show now that the coherent state is the equivalent, for a harmonic oscillator, of the minimum wave packet for a free particle: the product of the variances of the pair of canonically conjugated operators $\hat{c}, \hat{c}^\dagger$ in a coherent state has the minimum value allowed by Heisenberg uncertainty principle. It is very convenient at this point to introduce the pair of so called "quadrature" operators defined as:

$$\hat{X} = \frac{1}{2}(\hat{c}^\dagger + \hat{c})$$

$$\hat{Y} = \frac{i}{2}(\hat{c}^\dagger - \hat{c}) \tag{8.26}$$

with commutator

$$[\hat{X}, \hat{Y}] = \frac{i}{2} \tag{8.27}$$

so that the quadratic boson Hamiltonian

$$H_q = \hbar\omega_q(\hat{n}_q + \frac{1}{2}) \tag{8.28}$$

can be written as

$$H_q = \hbar\omega_q(\hat{X}^2 + \hat{Y}^2) \tag{8.29}$$

In a number state, we have:

$$\langle n \mid \hat{X}^2 + \hat{Y}^2 \mid n \rangle = (n + \frac{1}{2}) \tag{8.30}$$

while

$$\langle n \mid \hat{X} \mid n \rangle = \langle n \mid \hat{Y} \mid n \rangle = 0 \tag{8.31}$$

The variance of an arbitrary operator A in a given state $\mid s \rangle$ is

$$(\Delta A)^2 \equiv \langle s \mid (A - \langle s \mid A \mid s \rangle)^2 \mid s \rangle \tag{8.32}$$

Then we find that in the number state

$$(\Delta X)^2 = (\Delta Y)^2 = \frac{1}{2}(n + \frac{1}{2}) \tag{8.33}$$

Exercise 8.5
Prove Eq. (8.33).

The form of the uncertainty relation for the quadrature operators in a number state is therefore:

$$\mid \Delta X \mid \mid \Delta Y \mid = \frac{1}{2}(n + \frac{1}{2}) \geq \frac{1}{4} \tag{8.34}$$

On the other hand:

Exercise 8.6
Show that the expectation values of the quadrature operators calculated in a normalized coherent state are:

$$\langle\alpha\mid\hat{X}\mid\alpha\rangle =\mid\alpha\mid\cos\theta$$
$$\langle\alpha\mid\hat{Y}\mid\alpha\rangle =\mid\alpha\mid\sin\theta \qquad (8.35)$$

where θ is the phase of the complex number α.

We can obtain now the variances in the coherent state:

Exercise 8.7
Show that:

$$\langle\alpha\mid(\hat{X}-\langle\alpha\mid\hat{X}\mid\alpha\rangle)^2\mid\alpha\rangle = \langle\alpha\mid(\hat{Y}-\langle\alpha\mid Y\mid\alpha\rangle)^2\mid\alpha\rangle = \frac{1}{4} \qquad (8.36)$$

This result implies that in the coherent state the product of the variances of X and Y has the minimum value allowed by the uncertainty principle. To see this, let us remind that given a pair of non-commuting operators A, B such that $[A, B] = K =$ number, the product of the variances calculated in a given state satisfies:[4]

$$\Delta A\,\Delta B \geq \frac{1}{2}\mid\langle K\rangle\mid \qquad (8.37)$$

The equality is obtained if the states chosen to calculate the expectation values have that special property for the pair of operators considered. Such states, are the analog of the *"minimum uncertainty wave packet"* for a free particle. Eq. (8.36) shows that this is the case for the quadrature operators in any coherent state, independently of the value of α. For a number state this property only obtains, in the vacuum ($n = 0$) case.

8.5   Phase states

We shall follow the theory of the *phase states* of the electromagnetic field [2] and define the *phase operator* through the separation of amplitude and phase of the annihilation operator:

$$\hat{c}_k = (\hat{n}_k + 1)^{1/2}e^{i\hat{\phi}} \qquad (8.38)$$

which is equivalent to

$$e^{i\hat{\phi}} \equiv (\hat{n}_k + 1)^{-1/2}\hat{c}_k \qquad (8.39)$$

Assume $\hat{\phi}^\dagger = \hat{\phi}$. Then, upon taking the hermitian conjugate of Eq. (8.39) we have:

$$e^{-i\hat{\phi}} = c_k^\dagger(\hat{n}_k + 1)^{-1/2} \qquad (8.40)$$

and we also find that

$$e^{i\hat\phi}e^{-i\hat\phi} = 1 \tag{8.41}$$

although we cannot invert the order of the product.

Exercise 8.8
Prove Eq. (8.41)

One verifies the following properties:

$$e^{i\hat\phi} \mid n\rangle = \mid n-1\rangle \ , \ n > 0 \ ;$$
$$e^{i\hat\phi} \mid 0\rangle = 0$$
$$e^{-i\hat\phi} \mid n\rangle = \mid n+1\rangle$$
$$\langle n \mid e^{i\hat\phi} \mid n+1\rangle = \langle n+1 \mid e^{-i\hat\phi} \mid n\rangle = 1 \tag{8.42}$$

We remark that

$$\langle n \mid e^{i\hat\phi} \mid n+1\rangle \neq (\langle n+1 \mid e^{i\hat\phi} \mid n\rangle)^* \tag{8.43}$$

so that the exponential operator is not Hermitian. Instead,

Exercise 8.9
Prove that $\cos\hat\phi$ and $\sin\hat\phi$ are Hermitian, and that

$$\langle n-1 \mid \cos\hat\phi \mid n\rangle \ = \ \frac{1}{2} \ = \ \langle n \mid \cos\hat\phi \mid n-1\rangle$$
$$\langle n-1 \mid \sin\hat\phi \mid n\rangle \ = \ \frac{1}{2i} \ = -\langle n \mid \sin\hat\phi \mid n-1\rangle \tag{8.44}$$

Therefore these are observables, although they cannot be measured simultaneously as shown by their non-vanishing commutator:

Exercise 8.10
Show that

$$[\cos\hat\phi \, , \, \sin\hat\phi] = \frac{1}{2i} \left(\hat{c}^\dagger (\hat{n}+1)^{-1} c - 1 \right) \tag{8.45}$$

and verify that all matrix elements of the commutator vanish, except for the diagonal ground-state matrix element, which is:

$$\langle 0 \mid [\cos\hat\phi \, , \, \sin\hat\phi] \mid 0\rangle = -\frac{1}{2i} \tag{8.46}$$

Other useful commutators are:

$$[\hat{n} \, , \, e^{i\hat\phi}] = -e^{i\hat\phi}$$
$$[\hat{n} \, , \, e^{-i\hat\phi}] = e^{-i\hat\phi}$$
$$[\hat{n} \, , \, \cos\hat\phi] = -i\sin\hat\phi$$

$$[\hat{n}, \sin\hat{\phi}] = \cos\hat{\phi}$$

Since the number and the phase operators don't commute the amplitude and the phase of the transverse components of the local spin operator cannot be simultaneously precisely determined. The corresponding uncertainty relations can be easily obtained from the definition of the variances, following the usual procedure based on Schwartz's inequality[4]:

$$\Delta n \Delta cos\phi \;\geq\; \frac{1}{2}\,|\,\langle \sin\phi \rangle\,|$$

(8.47)

$$\Delta n \Delta sin\phi \;\geq\; \frac{1}{2}\,|\,\langle \cos\phi \rangle\,|$$

(8.48)

8.6   Magnon states of well defined phase

We have seen that $\cos\hat{\phi}$ and $\sin\hat{\phi}$ do not commute. However, there is only one non vanishing matrix element of their commutator in the number basis. This justifies the search for states which are approximately eigenstates of both operators.

We define

$$|\,\phi_s\rangle = \frac{1}{\sqrt{s+1}} \sum_{n=0}^{s} e^{in\phi} \,|\,n\rangle$$

(8.49)

and

$$|\,\phi\rangle = \lim_{s\to\infty} |\,\phi_s\rangle$$

(8.50)

Exercise 8.11
Show that

$$\lim_{s\to\infty} \langle \phi_s \,|\, \phi_s \rangle = 1$$

(8.51)

In the limit defined above, we find that

$$\langle \phi_s \,|\, \cos\hat{\phi} \,|\, \phi_s \rangle \to \cos\phi + \mathcal{O}(s^{-1})$$
$$\langle \phi_s \,|\, \sin\hat{\phi} \,|\, \phi_s \rangle \to \sin\phi + \mathcal{O}(s^{-1})$$

(8.52)

One also finds that to order $s^{-1/2}$,

$$\langle \phi_s \,|\, \cos^2\hat{\phi} \,|\, \phi_s \rangle \to \cos^2\phi$$
$$\langle \phi_s \,|\, \sin^2\hat{\phi} \,|\, \phi_s \rangle \to \sin^2\phi$$

(8.53)

From these results we conclude that the state $|\,\phi_s\rangle$ behaves approximately as a simultaneous eigenstate of $\cos\hat{\phi}$ and $\sin\hat{\phi}$. Besides, as a consequence of Eqs. (8.52) and (8.53), it turns out that the uncertainties of both operators in the phase state vanish in the limit $s \to \infty$.

8.7   Properties of the single-mode number states

Let us consider a single-mode (which implies a specific $\mathbf{k}$) excited in a FM, with a given number n of magnons. Clearly, for such a state the uncertainty in the number, Δn is zero. For the phase operators we find:[2]

$$\langle \hat{n} \mid \cos \hat{\phi} \mid \hat{n} \rangle = \langle \hat{n} \mid \sin \hat{\phi} \mid \hat{n} \rangle = 0 \tag{8.54}$$

while

$$\langle \hat{n} \mid \cos^2 \hat{\phi} \mid \hat{n} \rangle = \langle \hat{n} \mid \sin^2 \hat{\phi} \mid \hat{n} \rangle \quad = \frac{1}{2} \quad \text{for } n \neq 0$$

$$= \frac{1}{4} \quad \text{for } n = 0. \tag{8.55}$$

We find that, leaving aside the $n = 0$ state, the uncertainties are

$$\Delta \cos \hat{\phi} = \Delta \sin \hat{\phi} = \frac{1}{\sqrt{2}} \tag{8.56}$$

which implies that the number state phase is completely undetermined, and can have any value between 0 and 2π.

These results lead to an immediate physical interpretation of the properties of the transverse components of the local spin operator.

Returning to Eq. (4.22),

$$S^+(\mathbf{r}) = \sqrt{\frac{2S}{N}} \sum_k e^{i\mathbf{k}\cdot\mathbf{r}} \hat{b}_k \tag{8.57}$$

where higher order terms of the Hosltein-Primakoff expansion have been neglected, we express $\hat{b}_k$ in terms of the exchange-dipole magnon operators $\hat{c}_k, \hat{c}_k^\dagger$, according to Eq. (7.43). We transform back from polar to cartesian components of $\mathbf{S}$ and finally rearrange the terms in the k summation, to get:

$$S_x(\mathbf{r}) = \sqrt{\frac{S}{2N}} \sum_k (u_k - v_k^*) \left(e^{i\mathbf{k}\cdot\mathbf{r}} \hat{c}_k + e^{-i\mathbf{k}\cdot\mathbf{r}} \hat{c}_k^\dagger \right) \tag{8.58}$$

The expectation value of $S_x^2(\mathbf{r})$ in the single-mode number state $\mid n_k \rangle$ is:

$$\langle n_k \mid S_x^2(\mathbf{r}) \mid n_k \rangle = \frac{S}{N} \left(n_k + \frac{1}{2} \right) (u_k - v_k^*)^2 \tag{8.59}$$

The amplitude of the transverse components is therefore proportional to $n^{1/2}$ for large n.

On the other hand, we have just seen that the phase is completely undefined for this state.

8.8   Properties of a single-mode phase state

We have seen that the uncertainties of the $\cos\hat{\phi}$ and $\sin\hat{\phi}$ operators vanish in a phase state.

Let us now obtain the variance of the number operator. First we calculate:

$$
\lim_{s\to\infty}\langle\phi_s\mid\hat{n}\mid\phi_s\rangle = \lim_{s\to\infty}\sum_{n_1}^{s}\sum_{n_2}^{s}\langle n_1\mid\hat{n}\mid n_2\rangle e^{i(n_1-n_2)\phi} =
$$

$$
= \lim_{s\to\infty}\frac{1}{s+1}\sum_{n}^{s}n = s/2 \tag{8.60}
$$

Let us now calculate the average of $\hat{n}^2$:

$$
\lim_{s\to\infty}\langle\phi_s\mid\hat{n}^2\mid\phi_s = \frac{s(2s+1)}{6} \tag{8.61}
$$

Finally,

$$
(\Delta n)^2 = \langle\hat{n}^2\rangle - (\langle\hat{n}\rangle)^2 = \frac{s^2}{12} \tag{8.62}
$$

For the r. m. s. relative deviation we find

$$
\frac{\Delta n}{\langle\hat{n}\rangle} = \sqrt{\frac{1}{3}} \tag{8.63}
$$

As to the expectation value of the local spin components, we find, before taking the limit $s\to\infty$:

$$
\langle\phi_s\mid S_x(\mathbf{r})\mid\phi_s\rangle = \frac{2}{3}(s+1)^{1/2} \tag{8.64}
$$

Exercise 8.12
Prove Eq. (8.64).

The same result obtains for $S_y(\mathbf{r})$.

In conclusion, we have shown that in a single-mode phase state the transverse components of magnetization have a well defined phase but a divergent amplitude.

8.9   Expectation value of local spin operators in a coherent state

We shall change now to the Heisenberg representation for the transverse local spin operators, in order to display the traveling wave characteristic of these excitations. In this representation, the magnon creation and annihilation operators at time t relate to those at time $t=0$ as:

$$
\hat{c}_k(t) = \hat{c}_k \cdot e^{-i\omega_k t}
$$
$$
\hat{c}_k^{\dagger}(t) = \hat{c}_k^{\dagger} \cdot e^{-i\omega_k t} \tag{8.65}
$$

With the use of Eqs. (4.22) (and its hermitian conjugate) and (7.43) we can express the transverse operators $S_{x,y}(\mathbf{r},t)$ in terms of $\hat{c}_k(t), \hat{c}_k^\dagger(t)$ *(notice that the operators a_k in (4.22) are called b_k in (6.43)* , and then calculate their expectation values in a coherent state $| \alpha \rangle$, to obtain [5]:

$$\langle \alpha_k | S_x(\mathbf{r},t) | \alpha_k \rangle = S_k \cos(\mathbf{r} \cdot \mathbf{k} - \omega_k t + \theta_k)$$
$$\langle \alpha_k | S_y(\mathbf{r},t) | \alpha \rangle = e_k S_k \sin(\mathbf{r} \cdot \mathbf{k} - \omega_k t + \theta_k) \qquad (8.66)$$

where θ_k is the phase angle of α_k, and we have defined:

$$S_k = \sqrt{\frac{2S}{N}}(u_k - v_k) | \alpha_k |$$

$$e_k = \frac{u_k + v_k}{u_k - v_k} = \left(\frac{R_k + | \beta_k |}{R_k - | \beta_k |}\right)^{1/2} \qquad (8.67)$$

We are assuming for simplicity that u_k and v_k are real.

Exercise 8.13
Prove Eq. (8.67).

In conclusion, we have shown that in a coherent state with wave vector $\mathbf{k}$ the transverse components of the spin operator behave as those of a transverse travelling wave with the frequency and the ellipticity already found in Chap.7. The physical possibility of experimentally exciting macroscopic coherent states was discussed in Ref. [5]. These states have since been observed in several experiments.[6]

References

1. Glauber, R. J. (1963) *Phys. Rev.* **131**,2766.

2. Loudon, Rodney (1978) *"The Quantum Theory of Light"*, Clarendon Press, Oxford.

3. Heine, Volker (1977) *"Group Theory in Quantum Mechanics"*, Pergamon Press.

4. Messiah, Albert (1959) *"Mécanique Quantique"* v.1, pg. 379, Dunod, Paris.

5. Zagury, Nicim and Rezende, Sergio M. (1971) *Phys. Rev.* **B4**, 201

6. Tsoi, M. *et.al.* (2000) *Nature* **406**, 46; Melnikov, A. *et.al.* (2003) *Phys. Rev. Lett.* **91** , 227403; Zafar Iqbal M. *et.al.* (1979) *Phys. Rev* **B20**, 4759; Weber, M. C.*et.al.* (2006) *J. App. Phys* **99**, 08J308.

Chapter 9

Itinerant Magnetism

9.1 Introduction

We have considered up to now insulating (or dielectric) systems, where electron spins are well localized. The models we used do not describe therefore systems where all or at least a finite fraction of the electrons are itinerant, as in metals. And yet the most common ferromagnetic materials are metallic, namely Fe, Ni and Co. In this chapter we shall present some of the simplest models for the magnetism of metals. Those metals which display ordered magnetic phases do not in fact lend themselves to a description based exclusively upon extended electron states. As I shall mention later on, realistic models for these systems require a formulation capable of representing their dual extended and localized character.

I shall first describe briefly the simpler problem of the paramagnetism of a metal with a non-degenerate (s state) conduction band.

9.2 Pauli paramagnetic susceptibility

We consider for the time being independent electrons in a single band and derive in what follows the expression for their static, zero field, magnetic susceptibility. [1] In this case the electron states are completely specified by their wave-vector and spin. Under an external applied magnetic field B an electron with wave vector $\mathbf{k}$ has a spin dependent energy:

$$\epsilon_{\mathbf{k}\sigma} = \epsilon_{\mathbf{k}} - \mu_B \sigma B \tag{9.1}$$

and the spin-up and spin-down bands are now relatively shifted by a constant amount.

Let $N(E)$ be the total average number of electrons with energy $\leq E$, that is, the sum of the averages $N_{\pm}$ for both spins. In the presence of B we have:

$$N_{\pm} = \frac{1}{2} N(E_F \pm \mu_B B) \tag{9.2}$$

We verify that this is consistent with the fact that for zero field the numbers of up and down spin electrons are equal. We shall neglect the dependence of E_F on B, since we shall limit ourselves to effects which are of first order in B. Since the Zeemann term of the energy is small compared with E_F we can expand the r. h. s. of Eq.(9.2) in powers of the Zeemann term, and keep only the first:

$$N_\pm \approx \frac{1}{2}N(E_F) \pm \mu_B B \left(\frac{dN}{dE}\right)_{E_F} \tag{9.3}$$

For a degenerate metal all states of spin σ are occuppied up to the chemical potential

$$\mu_\sigma = E_F + \sigma\mu_B B$$

so that the average number of electrons of that spin with energy less or equal to some value E is equal to the number of available states up to that energy, which we call $Z(E)$, which implies

$$N_\pm = \frac{1}{2}Z(E_F) \pm \left(\frac{dZ}{dE}\right)_{E_F} \tag{9.4}$$

If $Z(E)$ is refferred to unit volume, the magnetization is

$$M = \mu_B(N_+ - N_-) = 2\mu_B^2 B \rho(E_F) \tag{9.5}$$

where $\rho(E_F) = \left(\frac{dZ}{dE}\right)_{E_F}$. We find now that Pauli susceptibility is [1]

$$\chi_P = 2\mu_B^2 \rho(E_F) \tag{9.6}$$

9.3   Stoner model of ferromagnetic metals

In order to deal with itinerant electron states it is convenient to consider continuous space distributions of charge and spin. In particular, if we want to calculate the total spin $\Delta\mathbf{S}$ contributed by electrons in a volume ΔV around a point $\mathbf{R}$ in space, we can integrate the spin volume-density operator over ΔV :

$$\Delta\mathbf{S} = \int_{\Delta V} d^3\mathbf{x}\, \vec{\sigma}(\mathbf{R}+\mathbf{x}) \tag{9.7}$$

where the spin density operator for a system with a total number N of electrons is

$$\vec{\sigma}(\mathbf{r}) = \hbar \sum_{i=1}^{N} \mathbf{S}(\mathbf{r}-\mathbf{x_i}) \tag{9.8}$$

and we define the operator

$$\mathbf{S}(\mathbf{r}-\mathbf{x_i}) = \delta^3(\mathbf{r}-\mathbf{x_i})\,\vec{\sigma}_i \tag{9.9}$$

$\{x_i\}$ are electron coordinate operators and $\{\sigma_i^\alpha\}$ are Pauli matrices ($\alpha = +, -, z$). According to the rules for second quantizing the one-particle operator $\mathbf{S}(\mathbf{r}-\mathbf{x_i})$

[3] , we represent $\vec{\sigma}$ as a series of terms, each of which is the product of its matrix element $\langle s \mid \vec{\sigma} \mid t \rangle$ between one particle states s, t in any complete orthonormal set (c. o. s.) which spans the one particle Hilbert space, times the product $a_s^\dagger a_t$ of the second quantized operators which destroy the initial state t and create the final one s:

$$\vec{\sigma}(\mathbf{x}) = \sum_{\nu,\alpha;\mu,\beta} a_{\mu,\alpha}^\dagger a_{\nu,\beta} \, \vec{\sigma}_{\mu,\alpha;\nu,\beta}(\mathbf{x}) \tag{9.10}$$

where the matrix element is

$$\begin{aligned}
\vec{\sigma}_{\mu,\alpha;\nu,\beta}(\mathbf{x}) &= \int d^3 x_{el} \phi_\mu^*(\mathbf{x}_{el}) \langle \alpha \mid \hbar \delta^3(\mathbf{r} - \mathbf{x_{el}}) \vec{\sigma} \phi_\nu(\mathbf{x}_{el}) \mid \beta \rangle \\
&= \hbar \phi_\mu^*(\mathbf{x}) \vec{\sigma}_{\alpha\beta} \phi_\nu(\mathbf{x})
\end{aligned} \tag{9.11}$$

and $\vec{\sigma}_{\alpha\beta}$ are the $\alpha\beta$ matrix elements of the three Pauli matrices. In the sum (9.10) $\alpha, \beta = \pm 1$ corresponding to the two eigenstates of the z component of the spinor. The basis states in Eq. (9.10) are spin-orbitals, products of a wave function ϕ and a spinor $\mid \alpha \rangle$ as explicitly indicated in Eq. (9.11). For a traslationally invariant system it is convenient to use the Bloch basis [4]

$$\phi_{\mathbf{k},\nu,\alpha} = e^{i\mathbf{k}\cdot\mathbf{x}} u_{\mathbf{k},\nu}(\mathbf{x}) \mid \alpha \rangle \tag{9.12}$$

in which case

$$\vec{\sigma}(\mathbf{x}) = \sum_{k,q} \sum_{\mu\nu} \sum_{\alpha\beta} e^{-i\mathbf{q}\cdot\mathbf{x}} u_{\mathbf{k}+\mathbf{q},\mu}^*(\mathbf{x}) \vec{\sigma}_{\alpha,\beta} u_{\mathbf{k},\nu}(\mathbf{x}) \, a_{\mathbf{k}+\mathbf{q},\mu\alpha}^\dagger a_{\mathbf{k},\nu\beta} \tag{9.13}$$

The u factor above is a periodic function of its argument, and the second subindex corresponds to the various (in principle infinite) bands. The summation over k and q above extends over the volume of the first Brillouin zone (BZ). If we use plane waves instead of Bloch waves the factors u are replaced by a constant, determined simply by the normalization condition chosen, while the k space is extended up to infinity. The crystalline structure can be taken into account in this case by imposing that all wave vectors which are not within the first BZ are folded back into it by subtracting the adequate reciprocal vector $\mathbf{G}$ (*empty lattice model*). This procedure yields a unique result for each arbitrary $\mathbf{k}$. In the plane wave representation, we can write

$$\vec{\sigma}(\mathbf{x}) = \sum_{q} e^{i\mathbf{q}\cdot\mathbf{x}} \vec{\sigma}(\mathbf{q}) \tag{9.14}$$

where

$$\sigma^\alpha(\mathbf{q}) = \sum_{pst} a_{\mathbf{p}+\mathbf{q},\, s}^\dagger \sigma_{st}^\alpha a_{\mathbf{p}\, t} \tag{9.15}$$

and s, t are spin indices. The circular components of σ are then

$$\sigma^+(q) = \sum_{p} a_{p+q\uparrow}^\dagger a_{\mathbf{p}\downarrow}$$

$$\sigma^-(q) = \sum_p a^\dagger_{p+q\downarrow} a_{p\uparrow}$$

$$\sigma^z(q) = \frac{1}{2} \sum_p \left(a^\dagger_{p+q\uparrow} a_{p\uparrow} - a^\dagger_{p+q\downarrow} a_{p\downarrow} \right) \tag{9.16}$$

9.4   Hubbard Hamiltonian

In order to proceed, we must now define the Hamiltonian for the metal. For the case of the $3d$ transition metals, Hubbard [5] proposed a simplified form of the electron-electron interaction potential by discarding, as relatively less important, the electrostatic Coulomb repulsion among electrons localized on *different* sites, which is expected to be reasonably screened by the less localized s and p electrons. The orbital degeneracy of the d orbitals demands in principle a representation with several states on each site, but as a first simple approach one works with the single orbital case chosen above. The remaining terms in the Hamiltonian are one electron terms, including the atomic binding energy of one electron and the hopping to the neighbouring ions due to the self-consistent average one-electron potential. In the site representation the Coulomb on-site repulsion is parametrized by a constant U, which is usually fitted by comparison with experiment. The simplest (one band) Hubbard Hamiltonian in the Wannier representation is then [5]

$$H = E_0 \sum_{i\sigma} a^\dagger_{i\sigma} a_{i\sigma} + \sum_{i,j\sigma} t_{ij} \left(a^\dagger_{i\sigma} a^\dagger_{j\sigma} + c.c. \right)$$
$$+ U \sum_i \hat{n}_{i\uparrow} \hat{n}_{i\downarrow} \tag{9.17}$$

where E_0 and the hopping terms t_{ij} are adequate band parameters, which in the tight-binding approximation are taken from atomic parameters. In particular the diagonal term can be approximated by the atomic ground state energy. Changing to the Bloch representation through the unitary transformation

$$a_i = \frac{1}{\sqrt{N}} \sum_k a_k e^{-i\mathbf{k}\cdot\mathbf{R_i}}$$

$$a_k = \frac{1}{\sqrt{N}} \sum_i a_i e^{+i\mathbf{k}\cdot\mathbf{R_i}} \tag{9.18}$$

we have

$$H = \sum_{k\sigma} \epsilon_k a^\dagger_{k\sigma} + \frac{U}{N} \sum_{pkq} a^\dagger_{p+q\uparrow} a_{p\uparrow} a^\dagger_{k-q\downarrow} a_{k\downarrow} \tag{9.19}$$

where

$$\epsilon_p = E_0 + \sum_{R_{ij}} t_{ij} \, e^{i\mathbf{R_{ij}}\cdot\mathbf{P}} \tag{9.20}$$

In the Hartree–Fock approximation, the electrons are independent particles which occupy states with energies

$$E_{p\sigma} = \epsilon_p + \frac{U}{N} \sum_{\mathbf{k}} f_{\mathbf{k},-\sigma} \tag{9.21}$$

The second term on the right hand side of (9.21) is the average self-consistent field acting on each electron due to all the other ones in the metal, within the on-site truncated Hubbard model. The energies $E_{p\sigma}$ are the quasi particle energies in the Hartree–Fock approximation to Hubbard Hamiltonian. In (9.21) $f_{\mathbf{k}\sigma} \equiv \langle a_{\mathbf{k}\sigma}^{\dagger} a_{\mathbf{k}\sigma} \rangle$ is the Fermi distribution

$$f_{\mathbf{k}\sigma} = \frac{1}{e^{\beta(E_{\mathbf{k}\sigma} - \mu)} + 1}$$

for quasiparticles with energy $E_{\mathbf{k}\sigma}$ and spin σ. To impose self-consistency we demand that the Fermi level μ be determined by the condition that the total average number of particles has a given value, assumed known, $n = N_{el}/N$. We shall calculate the average number of electrons with spin σ per atom:

$$\frac{1}{N} \sum_{\mathbf{k}} f_{\mathbf{k}\sigma} = \frac{1}{N} \sum_{\mathbf{k}} n_{\mathbf{k}\sigma} = n_{\sigma} \tag{9.22}$$

N being the total number of atoms in the system. Then we must have

$$n = n_{\uparrow} + n_{\downarrow} \tag{9.23}$$

In the paramagnetic phase, $n_{\uparrow} = n_{\downarrow}$. We may ask ourselves whether this phase is stable with respect to an infinitesimal ferromagnetic breaking of spin-up $\leftrightarrow$ spin-down symmetry [6]. The Hamiltonian (9.19) was chosen to represent the competition between the tendency of the kinetic energy term to de-localize electrons and the opposite effect of the intra-atomic repulsion U. The latter favours states with just one electron, or none, on each atom. Hopping of electrons between different ions clearly requires some double occupancy, so the two terms have opposite effects. In the $3d$ metals the d band width is $W \approx 4 \ eV$, while $U \approx 1$–$3 \ eV$, so $U/W < 1$. In the $4f$ metals the f bandwidth is very small, while $U \approx 5$–$6 \ eV$ so that $U/W > 20$. In the $5f$ metals, W decreases from $2.5 \ eV$ to $0.5 \ eV$ on going from Ac to Bk while U varies between $2 \ eV$ and $10 \ eV$ [8] so that U/W is between 1 and 7. As we see, the application to all these metals of Hubbard model requires its solution not only in the two opposite limits of weak and strong correlations, but also in the intermediate case. The exact solution of Hubbard's Hamiltonian is unknown, except for the $1d$ case [9]. In the rest of this chapter we shall develop the Hartree–Fock solution of (9.19), which is strictly valid for the weak correlation case $U/W < 1$.

9.5 Instability of paramagnetic phase

Let us imagine that we excite the down spin electrons which have an energy contained within a shell of thickness δE below the Fermi level μ and replace them in the unoccuppied states of the up spin band, just above μ. Let us now calculate the energy change per atom due to this operation, which can be thought as an infinitesimal splitting of the bands. The change of kinetic energy per atom is

$$\Delta E_{kin} = \rho(\mu)(\delta E)^2 \qquad (9.24)$$

where $\rho(E)$ is the density of electron states of energy E per atom. The variation of the interaction energy, within the Hartree–Fock approximation, is

$$\begin{aligned} \Delta E_{int}/U &= (n/2 + \rho\delta E)(n/2 - \rho\delta E) - n^2/4 = \\ &\quad -\rho^2(\mu)(\delta E)^2 \end{aligned} \qquad (9.25)$$

The total change in energy per atom is

$$\Delta E = \rho(\mu)(\delta E)^2 \left(1 - U\rho(\mu)\right) \qquad (9.26)$$

The condition for the instability of the non-magnetic state is then [7]

$$U\rho(\mu) \geq 1 \qquad (9.27)$$

It is interesting to remark that Eq. (9.26) is the same for either up or down spin, reflecting the time reversal symmetry of the model. Condition (9.27) is called *Stoner condition*. The main concepts of this model of itinerant ferromagnetism were developed by Slater [10] and Stoner [11] around 1940, and the resulting theory is known as *Stoner's model* of FM metals. In the next section we shall calculate the dynamic transverse susceptibility $\chi^{-+}(\omega, \mathbf{q})$ in the paramagnetic state and we shall find that the static uniform susceptibility in the non-magnetic limit diverges for $U\rho(\mu) = 1$.

Within this model, the energy band, which in the paramagnetic state is the same for both spins, is split into a pair of bands, each containing states of a given spin, which are identical to the original paramagnetic bands, only mutually displaced, so that the lower band contains more electrons, becoming the majority band, thereby creating a state with a finite magnetization. One can distinguish between two cases of split band FM. We define the parameter

$$\Delta_0 = U(n_\downarrow - n_\uparrow) \equiv Um \qquad (9.28)$$

where $m = \langle S^z \rangle$ is the average of the z component of spin per atom and we are assuming that the sample magnetization points down, so that $m > 0$. Δ_0 is called the *exchange splitting* . If $\Delta_0 > \mu$, the minority spin band will be completely empty at low T. When this happens, one speaks of (a) *complete ferromagnetism* , as compared to (b) *partial or incomplete ferromagnetism* in the opposite case. Both cases are schematically shown in Figs. 9.1 and 9.2.

9.6   Magnons in the Stoner model

9.6.1   The RPA susceptibility

Let us suppose that the system has a ferromagnetic ground state. We assume that the temperature is well below Curie's T_c, so that a well defined magne-

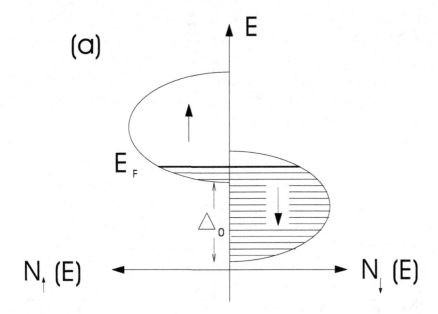

Figure 9.1: *Split-band incomplete ferromagnetic state.*

tization axis exists, and we shall consider a one domain situation, that is, the average magnetization is constant throughout the sample. We shall obtain a description of a magnon in an itinerant ferromagnetic system as a travelling wave of transverse fluctuations of the local magnetization.

Let us assume that in the ground state the majority spins point down. Now consider the retarded spin-flip propagator (see Chap. 6) [13]:

$$
\begin{aligned}
\chi^{-+}(\mathbf{q},t) &= i\theta(t)\langle[\sigma^{-}(\mathbf{q},t),\sigma^{+}(\mathbf{x}=0,t=0)]\rangle \\
&= \sum_{k}\chi^{-+}(\mathbf{k},\mathbf{q},t)
\end{aligned}
\tag{9.29}
$$

where we define

$$
\chi^{-+}(\mathbf{k},\mathbf{q},t) = \langle\langle a^{\dagger}_{\mathbf{k}+\mathbf{q}\downarrow}a_{\mathbf{k}\uparrow}\,;\sigma^{+}(\vec{x}=0,t=0)\,\rangle\rangle
\tag{9.30}
$$

with the notation of Chap. 6. We limit ourselves from now on, for simplicity, to the one-band case and we eliminate accordingly the band index from the operators.

The equation of motion for the propagator defined in (9.29) is

$$
\begin{aligned}
i\hbar\frac{\partial}{\partial t}\chi^{-+}(\mathbf{k},\mathbf{q},t) = {} &-\delta(t)\langle[\,a^{\dagger}_{\mathbf{k}+\mathbf{q}\downarrow}a_{\mathbf{k}\uparrow},\sigma^{+}(\mathbf{x}=0,t=0)\,]\rangle \\
&+ \langle\langle[a^{\dagger}_{\mathbf{k}+\mathbf{q}\downarrow}a_{\mathbf{k}\uparrow},H]\,;\sigma^{+}(\vec{x}=0,t=0)\rangle\rangle
\end{aligned}
\tag{9.31}
$$

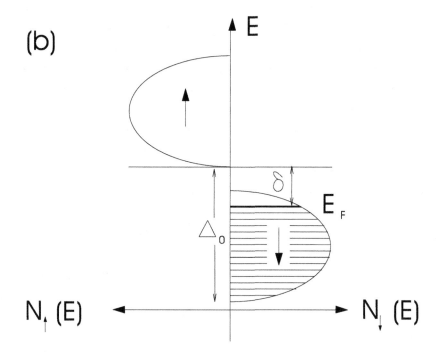

Figure 9.2: *Split-band complete ferromagnetic state.*

where H is Hubbard's Hamiltonian (9.19). Upon substituting H in the commutator of Eq. (9.31) we obtain several terms with four electron operators.

As in the case of the RPA for the Heisenberg Hamiltonian we truncate the corresponding hierarchy of Green's-function integro-differential equations by applying the RPA appropriate for this case. This consists of reducing all four electron operator terms to products of two operators, by substituting every pair of one creation and one annihilation operator by its statistical average. In order to do this factorization consistently one selects all possible different products of one creation and one annihilation operator from each four operator term in all possible ways, and adds all the terms obtained with the sign corresponding to respecting the anticommutation rules of fermionic operators. This is easily done by keeping track of the parity of the number of permutations needed to bring both operators selected to adjacent positions in the product, and assigning to the corresponding term in the sum the sign + for even, − for odd permutations. As a result, Eq. (9.31) becomes:

$$\left[i\hbar \frac{\partial}{\partial t} + (E_{\mathbf{p+q}\uparrow} - E_{\mathbf{p}\downarrow}) \right] \chi^{-+}(\mathbf{p},\mathbf{q},t)$$

$$= -\delta(t)\left[f_{\mathbf{p+q}\uparrow} - f_{\mathbf{p}\downarrow} \right] - (f_{\mathbf{p+q}\uparrow} - f_{\mathbf{p}\downarrow}) \frac{U}{N} \sum_{\mathbf{k}} \chi^{-+}(\mathbf{k},\mathbf{q},t) \quad (9.32)$$

The one-electron energies $E_{\mathbf{k},\sigma}$ appearing in (9.32) are those defined in (9.21). We Fourier transform (9.32) with respect to the time variable, obtaining

$$\chi^{-+}(\mathbf{p},\mathbf{q};\omega) = \frac{(f_{\mathbf{p}\uparrow} - f_{\mathbf{p}+\mathbf{q}\downarrow})\,(1 + U\chi^{-+}(\mathbf{q},\omega))}{\omega + E_{\mathbf{p}+\mathbf{q}\uparrow} - E_{\mathbf{p}\downarrow}} \tag{9.33}$$

where we introduced the quantity

$$\chi^{-+}(\mathbf{q},\omega) = \frac{1}{N}\sum_{\mathbf{p}} \chi^{-+}(\mathbf{p},\mathbf{q},\omega) \tag{9.34}$$

Exercise 9.1
Prove Eq. (9.33).

Equation (9.33) is an inhomogeneous linear integral equation with a separable kernel, which can be easily solved. The quantity we want is $\chi^{-+}(\mathbf{q},\omega)$ defined in (9.34, and we find:

$$\chi^{-+}(\mathbf{q},\omega) = \frac{\chi_0^{-+}(\mathbf{q},\omega)}{1 - U\chi_0^{-+}(\mathbf{q},\omega)} \tag{9.35}$$

where

$$\chi_0^{-+}(\mathbf{q},\omega) = \frac{1}{N}\sum_{\mathbf{p}} \frac{f_{\mathbf{p}+\mathbf{q}\uparrow} - f_{\mathbf{p}\downarrow}}{\omega - \Delta_{\mathbf{p},\mathbf{q}}} \tag{9.36}$$

and

$$\Delta_{\mathbf{p},\mathbf{q}} = E_{\mathbf{p}+\mathbf{q}\uparrow} - E_{\mathbf{p}\downarrow} \tag{9.37}$$

Exercise 9.2
Prove Eq. (9.35).

We have now obtained the Fourier transform of the spin propagator defined in (9.29), within the RPA. Since we want the retarded propagator, we add a small positive imaginary part to ω. If $U \to 0$, (9.35) reduces to the transverse dynamic spin susceptibility of free electrons $\chi_0^{-+}(\mathbf{q},\omega)$ [13]. At very low T, the Fermi distribution reduces to the step function, so that the only contributions to the sum over k space in (9.36) are those terms in which one state is above and the other below the Fermi level. One can describe the energy difference $\Delta_{\mathbf{p},\mathbf{q}}$ in the denominator of (9.36) as individual electron-hole excitation energies, the *Stoner excitations*. Let us assume, without any loss of generality, that the metal in the ferromagnetic phase is polarized with the majority band $\downarrow$. This means that the average z component of spin per atom is $m = n_\downarrow - n_\uparrow > 0$. In the simplest model of independent electrons for a metal one includes the effect of the periodic crystal potential approximately by substituting the free electron mass by an effective mass (effective mass approximation) so that the electron energy is

$$\epsilon_{\mathbf{k}} = \hbar^2 k^2/2m^* \tag{9.38}$$

in the simple case in which the effective mass is isotropic.

In general, $(\mathbf{m}^*)^{-1}$ is a tensor [4]. Adopting the isotropic form (9.38) we write the Stoner excitation energy (9.37) as:

$$\Delta_{\mathbf{k},\mathbf{q}} = \epsilon_{\mathbf{k}+\mathbf{q}\uparrow} - \epsilon_{\mathbf{k}\downarrow} + \Delta_0 \tag{9.39}$$

We keep the spin indices in the free electron energies because the split bands have different Fermi wave vectors.

9.6.2   Singularities of the susceptibility

The poles of $\chi^{-+}(\mathbf{q},\omega)$ determine the excitation spectrum of electron-hole pairs with spin flipping. In the paramagnetic state, the spin change is irrelevant from the point of view of the energy of the excitation, at least as long as there is no external magnetic field. In the ferromagnetic state, there is a change Δ_0 of energy with a spin flip, even though the kinetic energy change be zero in (9.39). According to (9.21), which defines the single particle Hartree–Fock energies, the Fermi wave-vectors for each spin are determined by the conditions

$$\frac{\hbar^2 k_{F\sigma}^2}{2m^*} + U n_{-\sigma} = \mu \tag{9.40}$$

where μ is the Fermi level, and $\sigma = \uparrow$ or $\downarrow$. Then, if the ground state has a majority of down spin electrons, $k_{F\downarrow} > k_{F\uparrow}$.

The propagator $\chi^{-+}(\mathbf{q},\omega)$ has a cut along an interval of the real energy axis where, for fixed q, the electron-hole excitation energy $\Delta_{\mathbf{p},\mathbf{q}}$ varies between its maximum and minimum values, which are respectively

$$\Delta_{max}(q) = \Delta_0 + \frac{\hbar^2}{2m^*}\left(2k_{F\downarrow}q + q^2\right)$$

$$\Delta_{min}(q) = \Delta_0 + \frac{\hbar^2}{2m^*}\left(-2k_{F\downarrow}q + q^2\right) \tag{9.41}$$

The curves for $\Delta_{max,min}(q)$ are represented in Fig. 9.3 for an incomplete FM and in Fig. 9.4 for a complete one.

Exercise 9.3
Verify Eqs. (9.41).

There may be other singularities of χ^{-+} for fixed q. For some $\omega(q)$ outside the cut, the denominator of (9.36) might vanish, which implies that the condition

$$U \chi_0^{-+}(\mathbf{q}, \omega(bfq)) = 1 \tag{9.42}$$

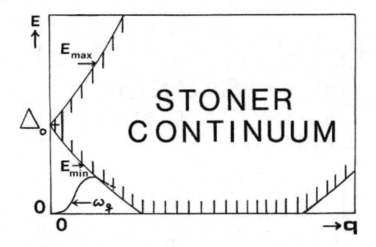

Figure 9.3: *Magnon dispersion relation ω_q and upper (E_{max}) and lower (E_{min}) limiting curves of the spectrum of Stoner excitations in an incomplete ferromagnetic metal within the isotropic one-band effective-mass model, shown schematically as a function of q. The majority band is $\downarrow$. The hatched region is covered as $\mathbf{k}$ spans all its allowed values.*

is satisfied. Solutions of this equation for $w(q)$ are eigenvalues of the Hubbard Hamiltonian in the Hilbert subspace which contains an electron-hole pair excited from the Fermi sea. One may interpret the corresponding eigenstate as a stationary superposition of an electron and a hole of opposite spins, of energy $w(q)$, carrying spin 1 and momentum $\mathbf{q}$. This is a magnon in the present formulation. Eq. (9.42) requires that $\chi_0^{-+}(\mathbf{q}, w(\mathbf{q})) > 0$. Let us consider the small q limit. Then, $\chi_0^{-+}(\mathbf{q}, w) \geq 0$ for $w \to -\infty$ and a look at Fig. 9.5 shows that for some $w > 0$ Eq. (9.42) is satisfied. In that figure we depict schematically the variation of χ_0^{-+} as a function of w for fixed q. There is a series of poles of χ_0^{-+} which in the thermodynamic limit, that is $N \to \infty$, fill the cut contained within the interval

$$\omega_1 = \min\left(\Delta_{min}, 0\right) \leq \omega \leq \omega_2 = \Delta_{max})\qquad(9.43)$$

spanned by all the admissible values of $\mathbf{k}$, for fixed q. Ones sees that there are in general as many intersections of the curve with the horizontal line $\chi_0^{-+} = \frac{1}{U}$, as poles , except for one, which lies at some energy $w(q)$ below the quasi-continuum. Clearly, when $U \to 0$, $w(q) \to \Delta_{min}(q)$ and the discrete state merges into the quasi-continuum. One can prove that

$$\lim_{q \to 0} w(q) = 0\qquad(9.44)$$

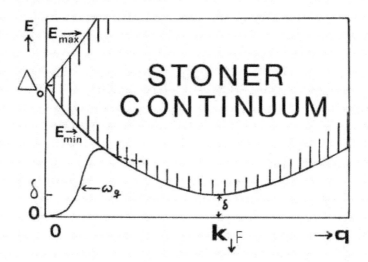

Figure 9.4: *Same as in Fig. 9.3 for a complete FM. There is a minimum exci-tation energy δ at $q = k_{F\downarrow}$ for the creation of a single electron-hole pair.*

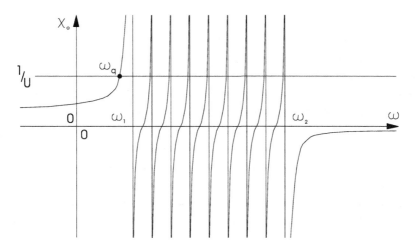

Figure 9.5: *Schematic variation of $\chi_0^{-+}(\mathbf{q}, \omega)$, for fixed q, as a function of ω.*

Eq. (9.44) is a particular case of the application of Goldstone's theorem, and the $q = 0$ magnon is a Goldstone boson. One can prove as well that the magnon dispersion relation is

$$\omega(q) = Dq^2 + O(q^3) \tag{9.45}$$

for small q.

To be more precise:

Exercise 9.4
Prove that

$$\omega(q) = \frac{\hbar^2 q^2}{2M^*} + O(q^3) \tag{9.46}$$

where

$$\frac{M^*}{m^*} = \left[1 - \frac{2}{3\pi^2 m^2 U} \left(n_\downarrow \langle \epsilon_\downarrow \rangle - n_\uparrow \langle \epsilon_\uparrow \rangle \right) \right]^{-1} \tag{9.47}$$

and we assume that the average down magnetization $m = n_\downarrow - n_\uparrow > 0$. *In Eq. (9.46)* $\langle \epsilon_\sigma \rangle$ *is the average kinetic energy in the sub-band of spin* σ.

If the majority spin, as assumed above, is down, the parenthesis in (9.47) is positive, and $M^* > m^*$. So, for a model of one parabolic band, the magnon effective mass, as obtained within the RPA, is larger than the electron one.

It is instructive to look at this problem from another point of view, by appealing to the *equation of motion method*. We are guided in this case by the main idea that there are excitations which consist of the correlated motion of an electron in the spin up (minority) band and a hole in the spin down band, with their center of mass propagating with momentum **q**. We may therefore guess that the operator for the creation of such a presumably well defined excitation, consists, to lowest order, of a sum of products of electron and hole operators which create pairs with the required quantum numbers, taken from all the possible free electron states, with some weight function to be determined later. We construct then the operator [6]

$$B^\dagger(\mathbf{q}) = \sum_k \alpha_\mathbf{q}(\mathbf{k}) a^\dagger_{\mathbf{k}+\mathbf{q}\uparrow} a_{\mathbf{k}\downarrow} \tag{9.48}$$

where $\alpha_q(\mathbf{k})$ is a c-number. The next step, as in the case of the AFM magnons in Chap. 4, is to demand that this operator satisfy the Heisenberg equation

$$\left[H, B^\dagger(\mathbf{q})\right] = \omega(\mathbf{q}) B^\dagger(\mathbf{q}) \tag{9.49}$$

which implies that $B^\dagger(\mathbf{q}) \mid 0\rangle$ would be an approximate excited eigenstate of H with energy $E_0 + \omega(\mathbf{q})$. The vacuum state $\mid 0\rangle$ is the FM ground state of the Fermi sea.

Exercise 9.5
Prove that the function $\alpha_q(\mathbf{k})$ *is:*

$$\alpha_q(\mathbf{k}) = \frac{A(\mathbf{q})}{\omega(\mathbf{q}) - (E_{\mathbf{k}+\mathbf{q}\uparrow} - E_{\mathbf{k}\downarrow})} \tag{9.50}$$

where $A(\mathbf{q})$ *is a normalization constant.*

The operators $B(\mathbf{q}), B^{\dagger}(\mathbf{q})$ have some complicated commutation relations:

Exercise 9.6
Verify that

$$
\left[\, B^{\dagger}(\mathbf{q}), B(\mathbf{q}') \,\right] = \left[\sum_{\mathbf{k}\,\mathbf{k}'} c^{\dagger}_{\mathbf{k}'+\mathbf{q}'\uparrow} c_{\mathbf{k}+\mathbf{q}\uparrow}\delta_{\mathbf{k}+\mathbf{k}'} - \sum_{\mathbf{k}\,\mathbf{k}'} c^{\dagger}_{\mathbf{k}\downarrow} c_{\mathbf{k}'\uparrow}\delta_{\mathbf{k}+\mathbf{q},\mathbf{k}'+\mathbf{q}'}\right]
$$
$$
\times\, \alpha^{*}_{\mathbf{q}}(\mathbf{k})\alpha_{\mathbf{q}'}(\mathbf{k}') \tag{9.51}
$$

However, within the RPA we can substitute the number operators in (9.51) by their corresponding self-consistent averages, and demand that in this approximation the right hand side of (9.51) is $\delta_{\mathbf{q},\mathbf{q}'}$. This condition also determines the unknown factor $A(\mathbf{q})$:

Exercise 9.7
Prove that

$$
A^2(\mathbf{q}) = \frac{1}{(\partial\chi^{-+}/\partial\omega)_{\omega(\mathbf{q})}} \tag{9.52}
$$

Notice that this is precisely U^2 times the residue of χ^{-+} at its pole $\omega(\mathbf{q})$. This is just a verification of the proportionality between the square of the wave function and the residue of the corresponding Green's function [14].

Consideration of the band structure details is necessary to obtain a quantitative description of the dispersion relation of magnons in real metallic ferromagnets [15, 16]. The theoretical results can be compared with data obtained with inelastic neutron scattering[17].

9.7 T_c in Stoner model

Let us consider again a metal with only one band. The average electronic charge and $\langle S^z \rangle$ per atom are :

$$
n = n_{\downarrow} + n_{\uparrow}
$$
$$
2\langle S^z\rangle \equiv m = n_{\downarrow} - n_{\uparrow} > 0 \tag{9.53}
$$

where the average occupation numbers above were defined in (9.22). When T approaches T_c from below $m \to 0$. We can then expand the self consistency conditions (9.22) in powers of m. We first invert the system (9.53), expressing $n_{\downarrow}$ and $n_{\uparrow}$ in terms of n and m:

$$
n_{\sigma} = n/2 - \sigma m/2 \qquad (\sigma = \pm 1) \tag{9.54}
$$

Then

$$
E_{\mathbf{p}\sigma} = \tilde{\epsilon}_p + \sigma U m/2 \tag{9.55}
$$

with $\tilde{\epsilon}_p \equiv \epsilon_p + Un/2 - \mu$. Let us now expand the Fermi function around $\tilde{\epsilon}_p$:

$$
f(E_{p\sigma}) = f(\tilde{\epsilon}_p) + \frac{1}{2}\sigma U m\frac{\partial f}{\partial\tilde{\epsilon}_p} + \mathcal{O}(m^2) \tag{9.56}
$$

We must impose

$$\frac{1}{N}\sum_p f(\tilde{\epsilon}_p) = \frac{n}{2} \tag{9.57}$$

Exercise 9.8
Prove that the critical temperature must satisfy the condition

$$1 = \frac{U}{N}\sum_p \frac{\partial f}{\partial \tilde{\epsilon}_p} \tag{9.58}$$

Since the integral in (9.58) is obviously bounded, and we have assumed that there is an ordered ferromagnetic state at $T = 0$, we cannot have an arbitrarily small U. There is therefore a minimum U compatible with this assumption, as we already know (Stoner condition).

We shall now calculate explicitly the r. h. s. of (9.58). To this end, we take the thermodynamic limit, $N \to \infty$, and substitute consequently the sum over the quasi continuum set of wave vectors by an integration over the continuum:

$$\frac{U}{N}\sum_p \frac{\partial f}{\partial \tilde{\epsilon}_p} = \int_{-\infty}^{\infty} d\epsilon \, \rho(\epsilon) \frac{\partial f}{\partial \epsilon} \tag{9.59}$$

where $\rho(\epsilon)$ is the density of one electron states per atom in the non-magnetic phase and $\epsilon = E - \mu$. Let us recall that

$$\frac{\partial f}{\partial \epsilon} = -\beta \frac{e^x}{(e^x + 1)^2} \tag{9.60}$$

which is an even function of $x \equiv \beta\epsilon$ which decreases rapidly as $|x| > 1$, or $|\epsilon| > k_B T$. For all transition metals the band width W and the Fermi level μ, are much larger than $k_B T$, so that the variation of $\rho(\epsilon)$ is very small within the interval where the function (9.60) is non negligible. This allows us to expand ρ in a Taylor series in ϵ around $\epsilon = 0$ (Fermi level):

$$\rho(\epsilon) = \rho(0) + \epsilon\,\rho'(0) + \frac{1}{2}\epsilon^2\,\rho''(0) + \cdots \tag{9.61}$$

The second moment of the derivative of the Fermi distribution is [18]:

$$\int_{-\infty}^{\infty} \frac{e^x}{(e^x + 1)^2} x^2 dx = \frac{\pi^2}{3} \tag{9.62}$$

so that

$$-\int d\epsilon\,\rho(\epsilon)\frac{\partial f}{\partial \epsilon} = \rho(0) + \rho''(0)\frac{\pi^2}{6\beta_c^2} + \cdots = \frac{1}{U} \tag{9.63}$$

Or

$$\frac{\pi^2}{6}(k_B T_c)^2 \approx \frac{U\rho(0) - 1}{U\rho''(0)} \tag{9.64}$$

We observe that in this approximation Eq. (9.64) can only be satisfied if the Fermi level is near a maximum of the density of states. We can roughly estimate the order of magnitude of T_c as obtained from (9.64) in terms of the bandwidth W, by assuming that $\rho(0) \approx 1/W$ and that $\rho''(0) \approx 1/W^3$. Then we get $k_B T_c \approx W$ if U and W are of the same order, which is supposed to happen in the $3d$ transition metals. This temperature is of the order of several eV, or tens of thousands K, which of course is almost two orders of magnitude larger than the measured Curie temperatures in the ferromagnetic metals. The conclusion is therefore that the simple Stoner model does not predict correctly the value of T_c. The prevalent current view is that the loss of magnetization does not proceed by the overall change into the non magnetic state at each site, as in the Stoner model, but rather that at each site there may persist a non zero net effective magnetic moment, even for T several times T_c. The orientation in space of these local moments would be however random at $T > T_c$, thus destroying the long range magnetic order, and yielding a vanishing total magnetization [19, 20].

9.8   Metals with degenerate bands

Let us consider now a more realistic case in which we extend the basis of Wannier orbitals by including N_d degenerate states on each site. For instance, in a $3d$ transition metal $N_d = 5$, corresponding to the five $d = 2$ orbitals. If the electron annihilation operator in second quantization is expanded in the corresponding spin-orbitals, we get

$$\psi_\sigma(x) = \sum_{\alpha,m} W_m(x - R_\alpha) \mid \sigma\rangle a_{\alpha m \sigma} \tag{9.65}$$

where the wave function $W_m(x - R_\alpha)$ is the Wannier orbital of type m centered at site R_α, and $m = 1, \cdots, N_d$. The two-body interaction in this basis has the form

$$V = \frac{1}{2} \sum_{\sigma\sigma'} \int\int dx_1 dx_2 \psi_\sigma^\dagger(x_1)\psi_{\sigma'}^\dagger(x_2)V(x_1 - x_2)\psi_{\sigma'}(x_2)\psi_\sigma(x_1) \tag{9.66}$$

If we assume as before that the only important terms in the interaction are the intra-atomic ones, we generalize Hubbard's interaction Hamiltonian to

$$V_{Hu} = \frac{1}{2} \sum_{\alpha,\sigma,\sigma'} \sum_{m_1 m_2} \sum_{m_3 m_4}$$
$$(m_1 m_2 \mid V \mid m_3 m_4) a_{\alpha m_1 \sigma}^\dagger a_{\alpha m_2 \sigma'}^\dagger a_{\alpha m_3 \sigma'} a_{\alpha m_4 \sigma} \tag{9.67}$$

where

$$(m_1 m_2 \mid V \mid m_3 m_4) \equiv$$
$$\int\int dx_1 dx_2 \, W_{\alpha m_1}^*(x_1)W_{\alpha m_2}^*(x_2)V(x_1 - x_2)W_{\alpha m_3}(x_2)W_{\alpha m_4}(x_1) \tag{9.68}$$

which is independent of α.

Let us now make some further simplifications:

1. We assume that the diagonal term is the same for all m:

$$(mm \mid V \mid mm) = I \quad ;$$

2. A given pair of different orbitals m, m' gives rise to a direct interaction matrix element

$$(mm' \mid V \mid m'm) \equiv U \qquad (9.69)$$

   which we assume is the same for all possible pairs $m \neq m'$ and

3. an exchange term

$$(mm' \mid V \mid mm') \equiv J \qquad (9.70)$$

   also assumed the same for all pairs $m \neq m'$.

4. Finally we assume that symmetry requires that

$$(mm \mid V \mid m'm') \equiv 0 \qquad (9.71)$$

for $m \neq m'$.

The three parameters so defined cannot be independent. Suppose we perform a symmetry operation such that the two orbitals involved are transformed into two linear combinations thereof, as usual. Since the Hamiltonian is invariant under the transformation the parameters must also be invariant. For example consider $U \to U'$, where the new orbitals are

$$\begin{aligned} \mid M \rangle &= c \mid m \rangle + s \mid m' \rangle \\ \mid M' \rangle &= -s \mid m \rangle + c \mid m' \rangle \end{aligned} \qquad (9.72)$$

with $c^2 + s^2 = 1$, and we take $c, s =$ real. It is easy to show now that

$$(MM' \mid V \mid M'M) = U' = U$$

if

$$I - J = U \qquad (9.73)$$

which we shall assume in the following. Using this relation we can write the interaction term as

$$\begin{aligned} V_{Hu} = \ &\frac{1}{2}(U + J) \sum_{\alpha m \sigma} n_{\alpha m \sigma} n_{\alpha m -\sigma} \\ &+ \ \frac{1}{2}(U + J) \sum_{\alpha, m \neq m', \sigma} \{ n_{\alpha m \sigma} n_{\alpha m' -\sigma} \\ &+ \ (U - J) n_{\alpha m \sigma} n_{\alpha m' \sigma} - J a^{\dagger}_{\alpha m \sigma} a_{\alpha m -\sigma} a^{\dagger}_{\alpha m' -\sigma} a_{\alpha m' \sigma} \} \qquad (9.74) \end{aligned}$$

We remark at this point that the terms $m \neq m'$ can be written as

$$\left(U - \frac{J}{2} \right) \sum_{\alpha, m \neq m', \sigma, \sigma'} n_{\alpha m \sigma} n_{\alpha m', \sigma'} - 2J \sum_{\alpha, m \neq m'} \mathbf{S}_{\alpha m} \cdot \mathbf{S}_{\alpha m'} \tag{9.75}$$

which expresses Hund's rule that favours the maximum possible value of $| \mathbf{S}_{\alpha m} + \mathbf{S}_{\alpha m'} |$. The one-electron hopping term is assumed to connect orbital states on different sites with a matrix element which is independent of the orbital quantum-number and of spin, so that the complete Hubbard Hamiltonian is

$$H_{Hu} = \sum_{\alpha \alpha' \sigma m} \{ t_{\alpha \alpha'} a^{\dagger}_{\alpha m \sigma} a_{\alpha' m \sigma} + h.c. \} + V_{Hu} \tag{9.76}$$

We shall now apply the usual Hartree–Fock approximation. To this end the number operators are expressed identically as

$$n_{\alpha m \sigma} \equiv (n_{\alpha m \sigma} - \langle n_{\alpha m \sigma} \rangle) + \langle n_{\alpha m \sigma} \rangle \tag{9.77}$$

Exercise 9.9
Substitute the identity (9.77) into the interaction potential, and neglect the products of fluctuations to obtain the HF Hamiltonian

$$\begin{aligned}
H_{HF} &= \sum_{\alpha \alpha' \sigma m} \{ t_{\alpha \alpha'} a^{\dagger}_{\alpha m \sigma} a_{\alpha' m \sigma} + h.c. \} + (U + J) \sum_{\alpha m \sigma} n_{\alpha m \sigma} \langle n_{\alpha m - \sigma} \rangle \\
&\quad + U \sum_{\alpha m \neq m' \sigma} n_{\alpha m \sigma} \langle n_{\alpha m' - \sigma} \rangle + (U - J) + \sum_{\alpha m \neq m' \sigma} n_{\alpha m \sigma} \langle n_{\alpha m' \sigma} \rangle
\end{aligned} \tag{9.78}$$

Let us now define the average spin and charge of the ion on site α:

$$S^z_{\alpha} = \frac{1}{2} \sum_{m=1}^{N_d} (\langle n_{\alpha m \uparrow} \rangle - \langle n_{\alpha m \downarrow} \rangle)$$

$$n = \sum_{m=1}^{N_d} (\langle n_{\alpha m \uparrow} \rangle + \langle n_{\alpha m \downarrow} \rangle) \tag{9.79}$$

We are assuming that n is site independent. Then the HF interaction potential can be written as

$$\begin{aligned}
V_{HF} &= \frac{n}{2N_d} [(2N_d - 1)U + (2 - N_d)J] \sum_{\alpha m \sigma} n_{\alpha m \sigma} \\
&\quad - \frac{U + N_d J}{N_d} S^z_{\alpha} \sum_{\alpha m} (n_{\alpha m \uparrow} - n_{\alpha m \downarrow})
\end{aligned} \tag{9.80}$$

As an example let us consider the d states in a transition metal, so that $N_d = 5$.

Exercise 9.10
Show that Stoner's criterium for the FM instability in this case is

$$I_{eff}\rho(\mu) \geq 1 \tag{9.81}$$

where $I_{eff} = U + 5J$ and $\rho(\mu)$ is the density of states per atom at the Fermi level of each of the N_d degenerate (in the present approximation) bands of non-interacting electrons.

We see that the effect of degeneracy, under the present assumptions and simplifications, is to enhance the effective Stoner intra-atomic interaction parameter and accordingly to favour the onset of FM ordering.

9.9 Spin-density wave

Let us turn now to other possibilities than the uniform ferromagnetic instability. Just as in the Heisenberg model, we must look for static ($\omega = 0$) instabilities of the response function for some finite q. To simplify the treatment of the problem we shall return to the case of non-degenerate bands.

We first calculate the free electron susceptibility $\chi_0^{-+}(q, 0)$. From Eq. (9.36) we get

$$\chi_0^{-+}(q, 0) = -\frac{1}{N} \sum_k \frac{f_k - f_{k+q}}{E_k - E_{k+q}} \tag{9.82}$$

We consider the paramagnetic phase and take the limit $N \to \infty$, so that the sum tends to the integral

$$\chi_0^{-+}(q, 0) = \frac{2V}{8\pi^3 N} \int d^3k \frac{f_k}{E_{k+q} - E_k} \tag{9.83}$$

which in the paramagnetic phase coincides with the expression for the static dielectric polarizability of an electron gas [21] and can be written as

$$\chi_0^{-+}(q, 0) = \frac{3z}{4\mu} F(q) \tag{9.84}$$

where

$$F(q) = 1 + \left(\frac{4k_f^2 - q^2}{4qk_F} \right) \log \left| \frac{2k_F + q}{2k_F - q} \right| \tag{9.85}$$

and we used the relation

$$\frac{V}{N} k_F^3 = 3\pi^2 z \tag{9.86}$$

valid for a metal of total volume V with N atomic cells, one atom per atomic cell and z electrons per atom.

The RPA dielectric function for an electron gas was first obtained by J. Bardeen [22] and independently by J. Lindhard [23] and it is known as Lindhard dielectric function. Its calculation leads to Eq. (9.84) for the polarizability. One can easily show that $F(q)$ defined in Eq. (9.85) is monotonically decreasing in the interval $0 \leq q < \infty$, so its maximum is at q=0, where $F(0) = 2$. Then the condition

$$\chi_0^{-+}(q,0)U = 1 \qquad (9.87)$$

is first satisfied, as U increases from 0, at $q = 0$, which leads again to the uniform ferromagnetic phase. For $q = 0$ we Eq. (9.84) reduces to the Pauli static susceptibility of the free electron gas

$$\chi_P = \frac{3z}{2\mu} \qquad (9.88)$$

as obtained in Sect. 9.2. The paramagnetic contribution in a ferromagnetic metal may have some quantitative effect upon comparison of theory and experiment, and it must be taken into account when one needs to fit some parameters of a model to measurements.

The description of the possible phase transitions changes considerably if a realistic band structure is considered. Take for instance a simple cubic lattice, where the kinetic part of the energy has the form

$$\epsilon_k = -\epsilon_0(\cos k_x a + \cos k_y a + \cos k_z a) \qquad (9.89)$$

which is a band of width $W = 6\epsilon_0$, and with its minimum at the Γ point $k = 0$. The density of states of this band is symmetric around $\epsilon_k = 0$ [24], so that if the band is half full at $T = 0$ the Fermi energy is $\mu = 0$. Consider now the special wave vector $Q = \frac{\pi}{a}(1,1,1)$. We verify from Eq. (9.89) that

$$\epsilon_{k+Q} = -\epsilon_k \quad \forall \ k$$

and for the special case $\epsilon_k = 0$, that also implies

$$\epsilon_{k+Q} = \epsilon_k \ .$$

This means that different points of the energy surface $\epsilon_k = 0$, are connected by Q, which is half a reciprocal lattice vector. For this special Q we can calculate $\chi_0^{-+}(Q,0)$:

$$\chi_0^{-+}(Q,0) = -\frac{V}{8\pi^3 N} \int_{-3\epsilon_0}^{\mu} d\epsilon \ \frac{\rho(\epsilon)}{\epsilon} \qquad (9.90)$$

which has a logarithmic singularity as $\mu \to 0$. Then the instability for a perturbation of the magnetization with wave vector Q will occur even for $U = 0$ [25]. This is a spin-density-wave (SDW) instability. This situation, in which different finite portions of the Fermi surface are connected by a particular wave vector, is called *nesting*.

Ground states of metals with a static spin-density-wave were predicted by A. W. Overhauser [26], who concluded that for a metal described within the

Hartree–Fock approximation with an unscreened Coulomb interaction (although not for a real Fermi liquid at normal metallic densities), the susceptibility $\chi(q)$ diverges at $q = 2k_F$.

The magnetization in the SDW phase that results from the instability described in the example above oscillates in space as

$$\mathbf{m}(\mathbf{r}) = \mathbf{m}_0 \, e^{i\mathbf{Q}\cdot\mathbf{r}} \qquad (9.91)$$

and for the special s. c. case it takes opposite signs on neighbouring sites in the lattice, in an AF1 spin structure, with spins aligned parallel on each plane of the (111) set, perpendicular to $\mathbf{Q}$, while spins on adjacent planes are anti-parallel. As we mentioned above, in realistic cases a finite U will be necessary to produce such an instability.

9.10 Hartree–Fock description of SDW

Let us return to the Hubbard Hamiltonian in the site representation

$$H = \sum_{i,j,\sigma} t_{ij} a_{i\sigma}^\dagger a_{j\sigma} + U \sum_i n_{i\uparrow} n_{i\downarrow} \qquad (9.92)$$

We can express the number operator for given spin in a form which displays the local z component of spin:

$$n_{i\sigma} \equiv \frac{1}{2} n_i + \sigma S_i^z$$

$$n_i \equiv \sum_\sigma n_{i\sigma}$$

$$S_i^z \equiv \sum_\sigma \frac{1}{2} \sigma n_{i\sigma} \qquad (9.93)$$

In the HFA we neglect as usual terms of second order in the fluctuations, leading to

$$H = \sum_{i,j,\sigma} t_{ij} a_{i\sigma}^\dagger a_{j\sigma} + U \sum_{i\sigma} n_{i\sigma} \langle n_{i-\sigma} \rangle - U \langle n_{i\uparrow} \rangle \langle n_{i\downarrow} \rangle \qquad (9.94)$$

We now use Eq. (9.93) to rewrite the efective potential:

$$U \sum_{i\sigma} n_{i\sigma} \langle n_{i-\sigma} \rangle = U \sum_{i\sigma} n_{i\sigma} \langle \frac{1}{2} n_i + \sigma S_i^z \rangle \qquad (9.95)$$

At this point we make the following two assumptions:
1) The charge density is uniform, so:

$$\langle \frac{1}{2} n_i \rangle = \frac{n}{2} \ , \quad \forall i \ .$$

2) The z component of the spin-density oscillates with a fixed wave vector:

$$\langle S_i^z \rangle = S_Q (e^{i\mathbf{Q}\cdot\mathbf{R_i}} + e^{-i\mathbf{Q}\cdot\mathbf{R_i}})$$

We return to the running-wave representation and find the following form for Hubbard's Hamiltonian in the HFA:

$$
\begin{aligned}
H \;=\; & \sum_{k\sigma} E_{k\sigma} a^{\dagger}_{k\sigma} a_{k\sigma} \\
& - U S_Q \sum_{k\sigma} \sigma \left(a^{\dagger}_{k\sigma} a_{\mathbf{k}+\mathbf{Q}\sigma} + a^{\dagger}_{\mathbf{k}+\mathbf{Q}\sigma} a_{k\sigma} \right) \\
& - \sum_i \frac{U n^2}{2} + U \sum_i \langle S^z_i \rangle^2
\end{aligned}
\tag{9.96}
$$

where the quasi-particle energy is

$$
E_{k\sigma} = \epsilon_{k\sigma} - \sigma U S_Q
\tag{9.97}
$$

and we have the relations

$$
\langle S^z_i \rangle = \langle S^z_Q \rangle e^{i\mathbf{Q}\cdot\mathbf{R_i}} + \text{c. c.}
$$

$$
S^z_Q \equiv \frac{1}{2} \sum_{k\sigma} \sigma a^{\dagger}_{\mathbf{k}\sigma} a_{\mathbf{k}+\mathbf{Q}\sigma}
$$

$$
\langle S^z_Q \rangle \equiv S_Q
\tag{9.98}
$$

In summary, we have added a molecular field to the kinetic energy each electron had in the paramagnetic phase and we obtained a coupling of pairs of electrons with wave vectors $\{\mathbf{k}, \mathbf{k}+\mathbf{Q}\}$. Before we continue it will be useful to simplify the model somehow by assuming that

$$
2\mathbf{Q} = \mathbf{G}
\tag{9.99}
$$

where $\mathbf{G}$ is a reciprocal lattice basis vector. Then, we consider a domain within the first BZ containing, and symmetrical with respect to, the origin, that we call RBZ (Reduced Brillouin Zone) with the property that

$$
\forall \mathbf{p} \in RBZ , \quad \mathbf{p}+\mathbf{Q} \in BZ
\tag{9.100}
$$

in such a way that the whole set of points $\{\mathbf{k}\}$ inside the BZ is the direct sum of the set $\{\mathbf{p}\}$ and the set $\{\mathbf{p}+\mathbf{Q}\}$. In the particular case $\mathbf{Q} = (Q,0,0)$, we can take $\{\mathbf{p} :\mid p_x \mid \le Q/2\}$ for the RBZ. One easily verifies that for any such $\mathbf{p}$ either $\mathbf{p}+\mathbf{Q}$ or $\mathbf{p}+\mathbf{Q}-\mathbf{G}$ lies inside the BZ. Besides, due to the condition (9.99) we have $a_{\mathbf{p}+2\mathbf{Q},\,\sigma} = a_{\mathbf{p}\sigma}$, *etc.* Let us now write the sums over the BZ using this decomposition, where p is always inside the RBZ:

$$
\begin{aligned}
H \;=\; & \sum_{p\sigma} E_{p\sigma} a^{\dagger}_{p\sigma} a_{p\sigma} \\
& + \sum_{p\sigma} E_{p+Q\sigma} a^{\dagger}_{p+Q\sigma} a_{p+Q\sigma} \\
& - 2 U S_Q \sum_{p\sigma} \sigma \left(a^{\dagger}_{p\sigma} a_{p+Q\sigma} + a^{\dagger}_{p+Q\sigma} a_{p\sigma} \right) + E_0
\end{aligned}
\tag{9.101}
$$

where E_0 contains the c-number part of H. Diagonalizing H so written is now a familiar problem, and we naturally proceed to perform the necessary Bogoliubov transformation from the pair of operators $a_{p\sigma}, a_{(p+Q)\sigma}$ to a new pair $c_{1p\sigma}, c_{2p\sigma}$. The transformation is parametrized conveniently as a rotation:

$$
\begin{aligned}
a_{p\sigma}^\dagger &= c_{1p\sigma}^\dagger \cos\theta_{p\sigma} - c_{2p\sigma}^\dagger \sin\theta_{p\sigma} \\
a_{(p+Q)\sigma}^\dagger &= c_{1p\sigma}^\dagger \sin\theta_{p\sigma} + c_{2p\sigma}^\dagger \cos\theta_{p\sigma}
\end{aligned}
\tag{9.102}
$$

Exercise 9.11
Prove that the anticommutation relations

$$
\{c_{\alpha p\sigma}, c_{\beta p'\sigma'}^\dagger\} = \delta_{\alpha\beta}\delta_{pp'}\delta_{\sigma\sigma'}
\tag{9.103}
$$

are preserved, that the non-diagonal terms of the form $c_1 c_2^\dagger$, etc. have vanishing coefficients if:

$$
\tan 2\theta_{p\sigma} = \frac{2US_Q\sigma}{E_{p+Q} - E_p}
\tag{9.104}
$$

and that if $\theta_{p\sigma}$ is so chosen the resulting transformed Hamiltonian is:

$$
H = \sum_{\alpha p\sigma} E_{\alpha p\sigma} c_{\alpha p\sigma}^\dagger c_{\alpha p\sigma}
\tag{9.105}
$$

where $\alpha = (1,2)$ and the new quasi-particle energies are:

$$
E_{1(2)p\sigma} = \frac{1}{2}\Big[\, \epsilon_p + \epsilon_{p+Q} + Un
$$
$$
\pm \sqrt{(\epsilon_p - \epsilon_{p+Q})^2 + (2US_Q)^2}\,\Big]
\tag{9.106}
$$

where $1(2) \to -(+)$.

We get therefore two bands, but they are not distinguished, as in Stoner ferromagnet, by spin, since as we see the quasiparticle energies are spin independent, and two electrons, with both spins, will occupy each of these states in the Fermi vacuum, up to the Fermi level. Since the Bogoliubov transformation is unitary the operator for the total number of particles takes the same form in both representations:

$$
\sum_{\alpha p\sigma} c_{\alpha p\sigma}^\dagger c_{\alpha p\sigma} = \sum_{k\sigma} a_{k\sigma}^\dagger a_{k\sigma}
\tag{9.107}
$$

and as a consequence

$$
\int_{E_{1l}}^{E_{2u}} dE\, N(E) = \int_{-W/2}^{W/2} d\epsilon\, \rho(\epsilon)
\tag{9.108}
$$

where E_{1l}, E_{2u} are respectively the lower limit of the lower sub-band and the upper limit of the upper sub-band, $N(E)$ is the density of states per atom per spin in the AFM phase and $\rho(\epsilon)$ the corresponding function for the free electron (U=0) problem, assumed to be symmetric and of width W. Remark that only RBZ wave-vectors are involved on the l. h. s. of Eq. (9.107), while k vectors on the r. h. s. span the whole (atomic) BZ. One can use the periodicity of E_p,

$$E_{p+G} = E_p$$

to plot both AFM sub-bands as one band in the extended RBZ, which coincides with the original BZ, or plot both sub-bands within the RBZ.

In special cases we find a gap separating both sub-bands. Let us consider the *perfect nesting* case where

$$\epsilon_{p+Q} = -\epsilon_p \ , \ \ \forall \, p \tag{9.109}$$

This is also called the *folding* condition. In the case of only first n. n. hopping, one can find some vectors Q which fulfill this condition in the s. c. and b. c. c. lattices, but not in the f. c. c. one. If (9.109) holds, the sub-band energies of the AFM phase are

$$E_{1,2} = \frac{Un}{2} \mp \sqrt{\epsilon_p^2 + (US_Q)^2} \tag{9.110}$$

where 1(2) refers to the lower (upper) sub-band.

Let us now calculate the density of states per atom, assuming $\rho(\epsilon)$ is known. By definition, and using Eq. (9.110) explicitly, we have

$$N_{1,2}(E) = \frac{2}{N} \sum_p \delta \left(E - \frac{Un}{2} \pm \sqrt{\epsilon_p^2 + (US_Q)^2} \right) \tag{9.111}$$

Let us simplify the notation and call $US_Q \equiv \Gamma$. Now we take the thermodynamic limit and substitute the sum by an integral:

$$N_{1,2}(E) = 2 \int d\epsilon \, \rho(\epsilon) \delta \left(\left(E - \frac{Un}{2} \pm \sqrt{\epsilon^2 + \Gamma^2} \right) \right) \tag{9.112}$$

The argument of the δ distribution above has the form

$$X(E, \epsilon) = E - \frac{Un}{2} + f(\epsilon)$$

and for fixed E the only contributions to the integral arise from the set $\{\epsilon_i\}$ of the roots of X. Then,

$$\delta(X) = \sum_i \frac{\delta(\epsilon - \epsilon_i)}{|\partial f/\partial \epsilon|_{\epsilon_i}} \tag{9.113}$$

The roots of X are

$$\epsilon_\pm = \pm \sqrt{(E - \frac{Un}{2})^2 - \Gamma^2} \tag{9.114}$$

We notice that there are no real roots in the interval of E

$$\frac{Un}{2} - \Gamma \le E \le \frac{Un}{2} + \Gamma \tag{9.115}$$

which therefore constitutes a gap of width 2Γ.

Exercise 9.12
Show that the density of states per atom for both sub-bands is

$$N_{1,2}(E) = 2\,\frac{\rho(\epsilon_\pm)\,|\,E - \frac{Un}{2}\,|}{\sqrt{(E - \frac{Un}{2})^2 - \Gamma^2}} \tag{9.116}$$

We can write (9.116) as

$$N_{1,2}(E) = \mathcal{N}(\omega) = 2\,\frac{\rho(\pm\epsilon(\omega))\,|\,\omega\,|}{\epsilon(\omega)} \tag{9.117}$$

in terms of

$$\omega \;\equiv\; E - \frac{Un}{2}$$
$$\epsilon(\omega) \;\equiv\; \sqrt{\omega^2 - \Gamma^2} \tag{9.118}$$

We obtained a mapping of $\rho(\epsilon)$ onto $\mathcal{N}(\omega)$ which yields a symmetric distribution in ω if ρ is symmetric.

At both borders of the gap, $\omega = \pm\Gamma$, the denominator $\epsilon(\omega) = 0$, so that the AFM density of states has a square root divergence (see Fig. 9.6).

Exercise 9.13
Verify that the perfect nesting condition

$$\epsilon_{\mathbf{p+Q}} = -\epsilon_{\mathbf{p}}$$

holds for:

1. *The s.c. lattice with*
$$\mathbf{Q} = \frac{\pi}{a}(\pm 1, \pm 1, \pm 1)$$
   resulting in the AF1 ordering;

2. *The b.c.c. lattice with*
$$\mathbf{Q} = \frac{\pi}{2a}(\pm 1, 0, 0)$$
   leading to AF1 ordering;

3. *The b.c.c. lattice with*
$$\mathbf{Q} = \frac{2\pi}{a}(\pm 1, \pm 1, \pm 1)$$
   resulting in AF ordering with two sublattices.

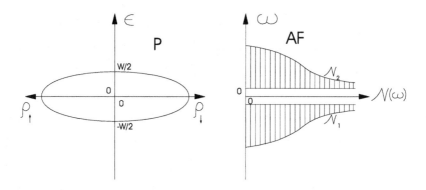

Figure 9.6: *On the l.h.s. of the figure we plot the density of states functions $\rho_\uparrow$ and $\rho_\downarrow$ for the non-interacting-electron ($U = 0$) band vs. the kinetic energy ϵ; on the r.h.s. we show the schematic density of states of the split band as a function of the quasi-particle energy ω as defined in Eq. (9.118) for the AFM phase.*

In order to complete the solution for the AFM split band case we need to find the conditions for the self-consistent evaluation of the order parameter S_Q. From Eq. (9.98), and upon substituting the a_k, a_{k+Q} operators by c_1, c_2 from Eq. (9.102) we find

$$S_Q = \frac{1}{2} \sum_{p,\sigma} \sigma \sin\theta_{p\sigma} \cos\theta_{p\sigma} \left(c_{1p\sigma}^\dagger c_{1p\sigma} - c_{2p\sigma}^\dagger c_{2p\sigma} \right) \qquad (9.119)$$

We now calculate the thermodynamic average of the operator in Eq. (9.119) and we substitute $\sin\theta_{p\sigma}, \cos\theta_{p\sigma}$ by their values obtained from Eq. (9.104), to obtain the result:

$$\langle S_Q^z \rangle \equiv S_Q = \frac{1}{2} \sum_{p\sigma} \frac{U S_Q \langle\langle\, n_{1p\sigma} - n_{2p\sigma}\, \rangle\rangle}{\sqrt{\epsilon_p^2 + (U S_Q)^2}} \qquad (9.120)$$

where 1(2) denotes the lower(upper) sub-band and

$$\langle n_{\alpha p\sigma} \rangle = (e^{\beta(E_{\alpha p\sigma} - \mu)} + 1)^{-1} \ . \qquad (9.121)$$

For a metal with one electron per atom

$$\int_{-\infty}^{\infty} d\epsilon \, \rho(\epsilon) = 1$$

and $n = 1$ is the half-filled-band case, once spin is taken into account, so for this case the Fermi level would be exactly in the middle of the band, that is, at $\epsilon = 0$ if the band is symmetric. In this case, as we have seen, the split-band density of states is also symmetric in ω, so for $n < 1$ the lower band is not completely full of electrons of both spins. Notice that the integrand in (9.120) is spin independent, since so is the spectrum, so that the sum over spin gives just a factor 2. In order to solve completely the problem we must determine the only parameter still unknown, namely the Fermi level μ, from the condition that the total average number of electrons per atom must coincide with the input value n. Using Eq. (9.116) we write this condition for the case in hand as

$$2 \int d\omega \frac{\rho(\sqrt{\omega^2 - \Gamma^2}) \, |\omega|}{\sqrt{\omega^2 - \Gamma^2}} = n \tag{9.122}$$

The system of Eqs. (9.120) and (9.122) determines the two parameters S_Q and μ for given $n < 1$. We remark that

$$\frac{|\omega|}{\sqrt{\omega^2 - \Gamma^2}} = \left| \frac{d\epsilon}{d\omega} \right| \tag{9.123}$$

which is just the jacobian for the change of variables from ω to ϵ, so one can simply rewrite Eq. (9.122) as

$$2 \int_{-W/2}^{\epsilon_F} d\epsilon \, \rho(\epsilon) = n \tag{9.124}$$

where

$$\epsilon_F = -\sqrt{(\mu - \frac{Un}{2})^2 - \Gamma^2} \tag{9.125}$$

and simplify also Eq. (9.120) as

$$\frac{1}{U} = \int_{-W/2}^{\epsilon_F} d\epsilon \frac{\rho(\epsilon)}{\sqrt{\epsilon^2 + \Gamma^2}} \quad . \tag{9.126}$$

If $n > 1$ the lower sub-band is full, and the last two equations are substituted by

$$2 \int_0^{\epsilon_F} d\epsilon \, \rho(\epsilon) = n - 1 \tag{9.127}$$

and

$$\frac{1}{U} = \int_{-W/2}^{0} d\epsilon \frac{\rho(\epsilon)}{\sqrt{\epsilon^2 + \Gamma^2}} - \int_0^{\epsilon_F} d\epsilon \frac{\rho(\epsilon)}{\sqrt{\epsilon^2 + \Gamma^2}} \quad . \tag{9.128}$$

with

$$\epsilon_F = \sqrt{(\mu - \frac{Un}{2})^2 - \Gamma^2} \tag{9.129}$$

We remark that the first integral in (9.128) is independent on n.

More refined treatments of the effect of nesting which include considering several bands and a more realistic description of the Fermi surface can be found in the literature [29].

Among the pure $3d$ metals only Cr and $\alpha - Mn$ are AF. Body centered Cr has a sinusoidal spin-wave with $Q \approx 0.96 \times 2\pi/a$ below $T_c = 310 \ K$. The polarization is longitudinal, *i.e.* $\parallel \mathbf{Q}$ for $T < 115\ K$ and transverse ($\perp \mathbf{Q}$) for $T > 115\ K$. The maximum value of the magnetization along the wave at low T is $0.59\ \mu_B$. Theoretical analysis based on the itinerant model, including detailed information on the Fermi surface has been reasonably succesful in describing the SDW in Cr [28, 27]. In real cases, the overlap of different bands will not allow in general the presence of an AF gap over the whole BZ. In a realistic calculation for Cr in the SDW phase the DOS does not vanish anywhere near the Fermi surface, although it shows a pronounced dip therein.

For further detailed treatments of itinerant magnetism the reader is referred to several reviews in the literature [28, 30, 31].

9.11   Effects of correlations

Let us return to the Hubbard Hamiltonian in the Wannier representation. We want to consider the limit $0 <\mid t \mid /U \ll 1$, where we shall see that in the half-filled band case $n = 1$ the ground state is an AF insulator.[33, 34, 35] In this limit the hopping term in H can be treated as a perturbation. The HF solution yields in the paramagnetic case just one band centered at $E_0 + Un/2$. Hubbard [5] used a Green's function decoupling approximation. In this approach one defines the usual retarded Green's function

$$G_{ij}^{\sigma}(\omega) = \langle\langle\ c_{i\sigma}; c_{j\sigma}^{\dagger}\ \rangle\rangle_{\omega} \tag{9.130}$$

which obeys the equation

$$\omega G_{ij}^{\sigma}(\omega) = \delta_{ij} + \sum_{l} t_{il} G_{lj}^{\sigma}(\omega) + U\ \Gamma_{ij}^{\sigma}(\omega) \tag{9.131}$$

introducing the new Green's function

$$\Gamma_{ij}^{\sigma}(\omega) = \langle\langle\ n_{i\sigma} c_{i\sigma}; c_{j\sigma}^{\dagger}\ \rangle\rangle_{\omega} \tag{9.132}$$

The equation for Γ is now approximated by factoring out $\langle n_{i-\sigma}\rangle$. Fourier transforming to k space one gets the pair of equations

$$\begin{aligned}
(\omega - \epsilon_k)G_k^{\sigma}(\omega) - U\ \Gamma_k^{\sigma}(\omega) &= 1 \\
-\langle n_{-\sigma}\rangle \epsilon_k G_k^{\sigma}(\omega) + (\omega - U)\Gamma_k^{\sigma}(\omega) &= \langle n_{-\sigma}\rangle
\end{aligned} \tag{9.133}$$

Now we find two real poles for each σ:

$$\omega_{\pm}^{\sigma} = \frac{\epsilon_k + U}{2} \pm \sqrt{(\frac{\epsilon_k - U}{2})^2 + n_{-\sigma} U \epsilon_k} \tag{9.134}$$

In the PM phase one gets two sub-bands. In the narrow band limit $U \gg \epsilon_k$ the expression for G simplifies:

$$G_k^{\sigma}(\omega) \approx \frac{1 - n_{-\sigma}}{\omega - \epsilon_k(1 - n_{-\sigma})} + \frac{n_{-\sigma}}{\omega - U - n_{-\sigma}\epsilon_k} \tag{9.135}$$

For a spin σ electron , the original non-interacting band is split into two sub-bands. The lower one can be interpreted as originating in the electrons with spin σ which itinerate occupying succcessively ions which are empty. The width of this sub-band is renormalized by the factor $(1 - n_{-\sigma})$, reflecting the average reduction of the hopping probability due to the exclusion of a fraction $n_{-\sigma}$ of available sites.

The upper sub-band, on the other hand, contains itinerant states in which the electron occupies precisely those ions already singly occupied (by a $-\sigma$ spin electron), so the factor $n_{-\sigma}$ renormalizes the effective width of this sub-band. Since all sites contributing to this band are doubly occupied, their energy is shifted up by an amount equal to the intra-atomic Coulomb repulsion U.

If $U \to \infty$ one expects that the ground state of Hubbard's Hamiltonian be FM.

In this respect we have the Nagaoka–Thouless theorem [32].
Nagaoka makes the following hypotheses:

1. The hopping term has only first n. n. range;

2. The band is almost half filled, so that if N_e is the total number of electrons and N that of atoms in the sample, he assumes that $| N_e - N | \ll N$.

Let us consider separately two cases depending on the crystal structure:

A) s. c. and b. c. c.; f. c. c. and h. c. p. with $N_e > N$;

B) f. c. c. and h. c. p. with $N_e < N$.

Then, for the special case $N_e = N - 1$, that is, exactly one hole in the band, in the limit $U \to \infty$ he finds that:

1. for case A the saturated (also called complete) FM state is the ground state;

2. for case B the one above is not the ground state, which is instead an un-saturated (incomplete) FM.

We see that the $U \to \infty$ limit is quite different from the large but finite U case. In the latter case, the non-interacting ($U = 0$) band splits into two well separated bands. If $n = 1$, at very low T the lower band is completely full and the upper one completely empty, with a gap in between, so that we have an insulator. It can be proven that there is an effective AF coupling which results in an insulating AF , which is known as a Mott insulator, being the ground state in this limit. The AF effective exchange interaction has the structure of the Heisenberg spin Hamiltonian with an effective exchange constant of the order of t^2/U. This is known as the "kinetic exchange" interaction, which we shall now discuss.

9.11.1   Kinetic exchange interaction

We consider again the narrow band limit $t/U \ll 1$, in which the hopping term in H can be considered as a perturbation.

We shall perform a canonical transformation on the Hilbert space that will eliminate the transformed hopping terms to first order and such that the ground state and the low excited ones belong to the subspace with single or zero site occupation. This is to be expected because every doubly occupied site adds a large positive contribution U to the energy [36]. First of all we shall describe the Hamiltonian as a sum of bond (instead of site) terms, so we write

$$H = \sum_{\langle ij \rangle} H^{ij} \tag{9.136}$$

where

$$H^{ij} = t_{ij} \sum_\sigma a_{i\sigma}^\dagger a_{j\sigma} + \text{h.c.} + \frac{U}{z}(\hat{n}_{i\uparrow}\hat{n}_{i\downarrow} + \hat{n}_{j\uparrow}\hat{n}_{j\downarrow}) \tag{9.137}$$

We have divided the interaction term by the coordination number z to compensate the overcounting, since each site contributes to z bonds. The advantage of this site $\leftrightarrow$ bond transformation is that we can start by looking at the simpler problem of a two-site, or one-bond, system first and then extend the result to the whole system.

Let us take two ions i, j and assume $t_{ij} = $ real, so that the bond Hamiltonian is

$$H^{(12)} = t \sum_\sigma (a_{1\sigma}^\dagger a_{2\sigma} + a_{2\sigma}^\dagger a_{1\sigma}) + \frac{U}{z} \sum_{i=1}^2 \hat{n}_{i\uparrow}\hat{n}_{i\downarrow} \tag{9.138}$$

Let us now perform the canonical transformation

$$\tilde{H}^{(12)} = e^{-i\epsilon S} H^{(12)}(\epsilon) e^{i\epsilon S} \tag{9.139}$$

where we define

$$H^{(12)}(\epsilon) = H_0 + \epsilon H_1 \tag{9.140}$$

with a decomposition of $H^{(12)}$ to be defined below. As usual in the perturbation formalism, the parameter ϵ will be substituted by 1 at the end of the calculation, and it only serves the purpose of keeping track of the order of each contribution to the energy. As a preliminary of what follows we shall define several projection operators, that will enable us to separate single from double occupation states.

Let us consider the polynomial in the variable x

$$\prod_\sigma ((1 - \hat{n}_{i\sigma}) + x\hat{n}_{i\sigma}) = \sum_{m_i=1}^2 p(i, m_i) x^{m_i} \tag{9.141}$$

It is easy to see that the operators $p(i, m_i)$ project a state onto the subspace in which $\sum_\sigma n_{i\sigma} = m_i$.

These operators have the following algebraic properties:

$$\sum_{m_i=1}^{2} p(i, m_i) = 1 \tag{9.142}$$

$$p(i, m_i)p(i, n_i) = p(i, n_i)\delta_{m_i, n_i} \tag{9.143}$$

Exercise 9.14
Obtain the explicit form of $p(i, m_i)$ and prove properties (9.142) and (9.143).

For the two-atom problem we define now the projection operators over the zero plus single occupation subspace, P_1, and on the double occupation subspace, P_2, as:

$$P_1 = p(1,0)p(2,0) + p(1,1)p(2,1) + \sum_{i \neq j} p(i,1)p(j,0) \tag{9.144}$$

and

$$P_2 = p(1,2)p(2,2) + \sum_{i \neq j} p(i,2) \left(p(j,0) + p(j,1) \right) \tag{9.145}$$

with the algebra

$$\begin{aligned} P_1 + P_2 &= 1 \\ P_i P_j &= P_i \, \delta_{i,j} \end{aligned} \tag{9.146}$$

We call $H^{(12)} = H$ in the following, since there is no possibility of confusion with the complete H. We study now the effect of these projection operators on H :

Exercise 9.15
Show that when these projection operators are applied on H from both sides we get:

$$P_1 H P_1 = t \sum_{\sigma; i \neq j} (1 - \hat{n}_{i-\sigma}) a_{i\sigma}^\dagger a_{j\sigma} (1 - \hat{n}_{j-\sigma})$$

$$P_1 H P_2 = t \sum_{\sigma; i \neq j} (1 - \hat{n}_{i-\sigma}) a_{i\sigma}^\dagger a_{j\sigma} \hat{n}_{j-\sigma}$$

$$P_2 H P_1 = t \sum_{\sigma; i \neq j} \hat{n}_{i-\sigma} a_{i\sigma}^\dagger a_{j\sigma} (1 - \hat{n}_{j-\sigma})$$

$$P_2 H P_2 = t \sum_{\sigma; i \neq j} \hat{n}_{i-\sigma} a_{i\sigma}^\dagger a_{j\sigma} \hat{n}_{j-\sigma} + \frac{U}{z} \sum_{i} \hat{n}_{i\uparrow} \hat{n}_{i\downarrow} \tag{9.147}$$

We interpret $P_1 H P_1$ as generating the lower sub–band, since it does not contain any intra-atomic contribution to the energy. An electron with spin σ travels only on ions which contain no $-\sigma$ electron. Conversely, $P_2 H P_2$ represents the upper band, since in this case that electron will visit ions which already contain another one with opposite spin, which contributes U to the energy. Clearly, $P_i H P_{j \neq i}$ are cross-sub-band terms.

In the atomic limit (small ϵ) we expect that the upper and lower sub-bands do not mix, so we define the terms in the decomposition of Eq. (9.140) as

$$
\begin{aligned}
H_0 &= P_2 H P_2 + P_1 H P_1 \\
H_1 &= \sum_i P_i H P_{j \neq i}
\end{aligned}
\tag{9.148}
$$

Upon expanding the exponential operator $e^{\pm i \epsilon S}$ in Eq. (9.139) in series of powers of ϵ we find:

$$
\tilde{H}(\epsilon) = H_0 + \epsilon \left(H_1 + i \left[H_0, S \right] \right) +
$$
$$
\frac{1}{2} \epsilon^2 \left(2i \left[H_1, S \right] - \left[\left[H_0, S \right], S \right] \right) + \cdots
\tag{9.149}
$$

Exercise 9.16
Verify Eq. (9.149).

The linear term in (9.149) vanishes if

$$
H_1 + i \left[H_0, S \right] = 0
\tag{9.150}
$$

If we substitute this condition in Eq. 9.149 and let $\epsilon = 1$ we get:

$$
\tilde{H} \equiv \tilde{H}(1) = H_0 + \frac{i}{2} \left[H_1, S \right]
\tag{9.151}
$$

We now substitute the definitions of H_0 and H_1 from (9.148) into Eqs. (9.150) and (9.151), and obtain the following equations:

$$
P_1 H P_2 + P_2 H P_1 + i \left[\left(P_1 H P_1 + P_2 H P_2 \right), S \right] = 0
\tag{9.152}
$$

and

$$
\tilde{H} = P_1 H P_1 + P_2 H P_2 + \frac{i}{2} \left[\left(P_1 H P_2 + P_2 H P_1 \right), S \right]
\tag{9.153}
$$

Applying the projectors on both sides of (9.152) we find

$$
P_\mu H P_\nu (1 - \delta_{\mu\nu}) + i P_\mu H P_\mu P_\mu S P_\nu - i P_\mu S P_\nu P_\nu H P_\nu = 0
\tag{9.154}
$$

Assume

$$
P_\nu S P_\nu = \gamma P_\nu
\tag{9.155}
$$

where γ is a c-number. This satisfies Eq. (9.154) for the diagonal terms. For $\mu \neq \nu$ we have

$$P_\mu H P_\nu + i P_\mu H P_\mu P_\mu S P_\nu - i P_\mu S P_\nu P_\nu H P_\nu = 0 \qquad (9.156)$$

Let us define the quantities E_u, E_l as the averages of respectively the upper and lower sub-band energies. In the narrow band (large U/t) limit we have

$$E_u - E_l \approx U/z \ .$$

Upon substituting the diagonal terms $P_\nu H P_\nu$, $(\nu = 1, 2)$ in Eq. (9.156) by their averages one finds approximately

$$P_1 S P_2 \ = \ -i\frac{z}{U} P_1 H P_2 \qquad (9.157)$$

$$P_2 S P_1 \ = \ i\frac{z}{U} P_2 H P_1 \qquad (9.158)$$

If we now substitute the diagonal and non-diagonal terms involving S from (9.155) and (9.158) into (9.151) we get for the transformed Hamiltonian of the two-site system:

$$\tilde{H} = P_1 H P_2 + P_2 H P_1 - \frac{z}{U} \left(P_1 H P_2 H P_1 - P_2 H P_1 H P_2 \right) \qquad (9.159)$$

which, upon writing explicitly the expressions for the projectors P_i yields

$$\tilde{H} = t \sum_{i \neq j, \sigma} (1 - \hat{n}_{i-\sigma}) a_{i\sigma}^\dagger a_{j\sigma} \hat{n}_{j-\sigma})$$

$$+ t \sum_{i \neq j, \sigma} \hat{n}_{i-\sigma} a_{i\sigma}^\dagger a_{j\sigma} (1 - \hat{n}_{j-\sigma} + i\frac{U}{z} \sum_i \hat{n}_{i\uparrow} \hat{n}_{i\downarrow}$$

$$- \frac{z\, t^2}{U} \sum_{i \neq j, \sigma} [\, \hat{n}_{i\sigma} (1 - \hat{n}_{i-\sigma}) \hat{n}_{j-\sigma} (1 - \hat{n}_{j\sigma})$$

$$- \hat{n}_{i\sigma} \hat{n}_{i-\sigma} (1 - \hat{n}_{j\sigma}) (1 - \hat{n}_{j-\sigma}) \,]$$

$$+ \frac{z\, t^2}{U} \sum_{i \neq j} \left(S_i^+ S_j^- + S_j^+ S_i^- - 2 a_{i\uparrow}^\dagger a_{i\downarrow}^\dagger a_{j\uparrow} a_{j\downarrow} \right) \qquad (9.160)$$

If we sum the above two-site (or bond) Hamiltonian over all bonds we get a good approximation to the total H. If one returns to the sum over sites instead of bonds the form (9.160) can be extended to all sites. As we sum Eq. (9.160) over all bonds, there appear some terms which involve three-center hoppings, which are longer range anyway, and should be neglected consistently with the approximation of first n.n. hopping [36]. In the large U limit at low T the upper sub-band is empty, so that we can neglect the terms $n_{i\sigma} n_{i-\sigma}$. The final result for the canonically transformed Hamiltonian in this case can be conveniently expressed introducing new fermion operators

$$b_{i\sigma} = (1 - \hat{n}_{i-\sigma}) a_{i\sigma} \qquad (9.161)$$

in terms of which the transformed Hamiltonian reads:

$$\tilde{H} = t \sum_{i \neq j, \sigma} b_{i\sigma}^{\dagger} b_{i\sigma} + \frac{z\,t^2}{U} \sum_{i \neq j} (S_i^+ S_j^- + S_i^- S_j^+) - \frac{z\,t^2}{2U} \sum_{i \neq j, \sigma} \hat{\nu}_{i\sigma} \hat{\nu}_{j-\sigma} \qquad (9.162)$$

where $\hat{\nu}_{i\sigma} = b_{i\sigma}^{\dagger} b_{i\sigma}$. The operators $b_{i\sigma}$ and $b_{i\sigma}^{\dagger}$ satisfy the anti-commutation relations

$$\{ b_{i\sigma}, b_{j\tau}^{\dagger} \} = (1 - \hat{n}_{i-\sigma}) \delta_{ij} \delta_{\sigma\tau}$$

$$\{ b_{i\sigma}, b_{j\tau} \} = \{ b_{i\sigma}^{\dagger}, b_{j\tau}^{\dagger} \} = \delta_{ij} \delta_{\sigma\tau} \qquad (9.163)$$

It is easy to verify that

$$\sum_{i \neq j, \sigma} \hat{\nu}_{i\sigma} \hat{\nu}_{j-\sigma} = \sum_{i \neq j, \sigma} \hat{\nu}_{i\sigma} \hat{\nu}_{j\sigma} - \sum_{i \neq j} S_i^z S_j^z \qquad (9.164)$$

In the case of an exactly half-filled band, where

$$\hat{n}_{i\uparrow} + \hat{n}_{i\downarrow} = 1 \qquad (9.165)$$

we have the Hamiltonian for Mott's insulator,

$$H_{eff} = t \sum_{i \neq j} b_{i\sigma}^{\dagger} b_{j\sigma} + \frac{z\,t^2}{U} \sum_{i \neq j} \mathbf{S}_i \cdot \mathbf{S}_j \qquad (9.166)$$

also known as the "t-J" Hamiltonian. Comparing Eq. (9.166) with the Heisenberg hamiltonian, we see that the effective exchange interaction constant is negative, leading to AF behaviour at low T.

The problem of strongly correlated Fermion systems has particular relevance since the discovery of the high-T_c superconducting compounds. See Refs. [37, 38] for reviews and further references.

9.12   Paramagnetic instability and paramagnons

There are many metallic systems which are close to satisfying Stoner's condition (9.27), that is, for which

$$\alpha \equiv U\rho(E_F) \approx 1 \qquad (9.167)$$

with $\alpha < 1$, so that the local Coulomb repulsion is not large enough to induce the transition to the FM ground state. It can be shown that as α increases and approaches 1 some properties exhibit singular behaviour. Doniach [13, 39] calculated the enhancement of the inelastic neutron scattering cross-section as a result of this paramagnetic instability. It is also found that the effective mass of the conduction electrons, as obtained from the low-temperature specific heat, diverges at the critical point $\alpha = 1$, while near the critical point the effective mass renormalization is large and the temperature dependence of the specific heat has a logarithmic singular term.[40] This effects are important in near ferromagnetic metals such as Pd and also in liquid He^3[13, 40, 46].

We shall now calculate the low temperature specific heat correction due to Hubbard U interaction. This requires the evaluation of the correction to the free energy, which we shall perform now.

First we remind ourselves of the so called *Feynmann–Helmann theorem*[1]. Consider a Hamiltonian containing a non-interacting part H_0 and an interaction potential which is multiplied by a parameter λ as customary in perturbation theory:

$$H(\lambda) = H_0 + \lambda V \tag{9.168}$$

with eigenkets and eigenvalues

$$H(\lambda) |n \ \lambda\rangle = E(n \ \lambda) |n \ \lambda\rangle \tag{9.169}$$

Call $E(0 \ \lambda) \equiv E(\lambda) =$ ground state energy. We calculate now the derivative

$$\frac{\partial E(\lambda)}{\partial \lambda} \equiv \frac{\partial \langle 0 \ \lambda | H(\lambda) | 0 \ \lambda \rangle}{\partial \lambda} =$$
$$E(\lambda) \frac{\partial \langle 0 \ \lambda | 0 \ \lambda \rangle}{\partial \lambda} + \langle 0 \ \lambda \ | \ V \ | \ 0 \ \lambda \rangle \tag{9.170}$$

and the first term in the second line of (9.170) is $\equiv 0$ since by assumption the eigenkets are normalized to unity for all λ. Therefore,

$$\Delta E = \int_0^1 \langle V \rangle_\lambda \, d\lambda \equiv \int_0^1 \frac{\langle \lambda V \rangle_\lambda}{\lambda} \, d\lambda \tag{9.171}$$

is the change of the ground state energy after the full interaction is switched on.

A similar formula can be obtained for the change of the free energy due to the interaction. To see this consider the definition of Helmholtz free energy (at constant volume):

$$F(\lambda) = -k_B T \ln Tr \hat{\rho}(\lambda) \tag{9.172}$$

where

$$\hat{\rho}(\lambda) = e^{-\beta(H_0 + \lambda V)}$$

Now we differentiate (9.172) with respect to λ and we find, by the same token as for the ground state energy, that:

$$\frac{\partial F(\lambda)}{\partial \lambda} = \langle V \rangle_\lambda \tag{9.173}$$

where $\langle \cdot \rangle_\lambda$ is the statistical average with the density matrix $\hat{\rho}(\lambda)$. Remark that there is no ambiguity with respect to the order of the operators in (9.173) because we are taking the trace.

Upon integrating (9.173) we find :

$$F(1) - F(0) \equiv \Delta F = \int_0^1 \frac{d\lambda}{\lambda} \langle \lambda V \rangle_\lambda \tag{9.174}$$

---

[1] According to V. Weisskopf, as cited by D. Pines in *"The many-body problem"*, this formula is due to W. Pauli.

We are interested in the case where V is Hubbard local repulsion as in Sect. 9.2. We can now relate the average in (9.174) to the particle-hole propagator we defined in (9.35) :

Exercise 9.17
Prove, with the help of the fluctuation-dissipation theorem, that

$$\langle \lambda V \rangle_\lambda = \frac{-2}{N} \sum_q \int \frac{d\omega}{2\pi} \frac{\lambda U \mathcal{I}m(\chi^{-+}(\mathbf{q}, \omega))}{e^{\beta\omega} - 1} \tag{9.175}$$

Upon substitution in Eq. (9.175) of the expression for $\chi^{-+}(\mathbf{q}, \omega)$ which was given within the RPA in (9.36), we are led to calculate the function

$$\Phi(q, \omega) = \mathcal{I}m \int_0^1 d\lambda \left(\frac{\chi_0^{-+}(q, \omega) U}{1 - U \lambda \chi^{-+}(q, \omega)} \right) =$$
$$\mathcal{I}m \ln \left(1 - U \chi_0^{-+}(q, \omega) \right) \tag{9.176}$$

where

$$\chi_0^{-+}(q, \omega) = \frac{1}{N} \sum_\mathbf{p} \frac{f_{\mathbf{k+q}\uparrow} - f_{\mathbf{p}\downarrow}}{\omega + i\eta - E_{\mathbf{k+q}\uparrow} + E_{\mathbf{k}\downarrow}} \tag{9.177}$$

Notice that we added an infinitesimal positive imaginary part to ω in (9.177) to reproduce the correct analytical properties of a retarded propagator.
We need now to calculate $\chi_0^{-+}(q, \omega)$, which is proportional to the transverse susceptibility of the non-interacting Fermi gas (Sect. 9.5.1). We are interested in the low T limit, and this implies that it suffices to consider the small q and small ω limit. Therefore we can expand $f_{\mathbf{k+q}}$ in a series in powers of

$$\Delta_{k,q} \equiv E_{\mathbf{k+q}} - E_{\mathbf{k}} \ .$$

Let us call

$$\zeta = \mathbf{k} \cdot \mathbf{q} / |\mathbf{k}| |\mathbf{q}|$$

which is the cosine of the angle $\mathbf{k}$ makes with the direction of $\mathbf{q}$. Then,

$$\Delta_{k,q} = E_q + 2\zeta \sqrt{E_k E_q}$$

We shall ignore the spin dependence of the electron energies since we are assuming that the system is in the paramagnetic phase, so that the exchange gap is zero. We have:

$$\chi_0^{-+}(q, \omega) = -\frac{1}{N} \sum_k \frac{1}{\Delta - \omega} \left(f' \Delta + \frac{1}{2} f'' \Delta^2 + \frac{1}{6} f''' \Delta^3 + \cdots \right) \tag{9.178}$$

We must consider up to the third order term above in order to get correctly the contribution to the free energy to second order in T at low T. We can rewrite the first three terms of Eq. (9.178) as:

$$
\begin{aligned}
\chi_0^{-+}(q,\omega) &= -\int_0^\infty dE\, \rho(E)\, \frac{1}{2}\int_{-1}^1 d\zeta \left[f'(E)\left(1+\frac{\omega}{\Delta-\omega}\right)\right.\\
&\quad + \frac{1}{2}f''\left(\Delta-\omega+\frac{\omega^2}{\Delta-\omega}\right)\\
&\quad \left.+ \frac{1}{6}f'''\left(\Delta^2+\Delta\omega+\omega^2+\frac{\omega^3}{\Delta-\omega}\right)+\cdots\right]
\end{aligned}
\tag{9.179}
$$

The factor $\frac{1}{2}$ in front of the ζ integral above is due to the fact that in the definition of the band density of states ρ that integral was already performed. In order to proceed we perform first the ζ integral, to find:

$$
\begin{aligned}
\chi_0^{-+}(q,\omega) &= -\int_0^\infty dE\, \rho(E)\Bigg[f'(E)\\
&\quad \times \left(1+\frac{\omega}{4\sqrt{E\,E_q}}\ln\left(\frac{2\sqrt{E\,E_q}+E_q-\omega-i\eta}{-2\sqrt{E\,E_q}+E_q-\omega-i\eta}\right)\right)\\
&\quad + \frac{1}{2}f''(E)\left(E_q+\omega+\mathcal{O}(\omega^2)\right)\\
&\quad + \frac{2}{9}f'''(E)\left(E\,E_q+\mathcal{O}(E_q^2,\omega E_q,\omega^2)\right)+\cdots\Bigg]
\end{aligned}
\tag{9.180}
$$

At low T we approximate

$$
f'(E) \approx -\delta(E-E_F)
\tag{9.181}
$$

which neglects corrections of the order of $(k_B T/E_F)^2$, and we use the definition of the derivatives of Dirac's δ distribution:

$$
\int dx\, \delta^{(n)}(x)g(x) = (-1)^n\, g^{(n)}(0)
\tag{9.182}
$$

With some algebraic operations [41] we arrive at:

$$
\begin{aligned}
\chi_0^{-+}(q,\omega) &= \chi_P\Bigg[1+\frac{\omega}{4\sqrt{E\,E_q}}\left(i\pi+\mathcal{O}(\frac{E_q-\omega}{\sqrt{E\,E_q}})\right)\\
&\quad -\frac{1}{2}\frac{\rho'(E_F)}{\rho(E_F)}\left(E_q+\omega+\mathcal{O}(\omega^2)\right)\\
&\quad +\frac{4}{9}\frac{\rho'(E_F)}{\rho(E_F)}\left(E_q+\mathcal{O}(E_q^2,\omega E_q,\omega^2)+\cdots\right)\Bigg]
\end{aligned}
\tag{9.183}
$$

The absolute value of $\chi_0^{-+}(q,\omega)$ is maximum for

$$
q^2 \approx \frac{\omega}{\sqrt{2\alpha}}
$$

when α and q/k_F are both very small. Therefore we can limit the calculations to the range $\omega \ll q k_F \ll 1$. Then Eq.(9.183) can be written as:

$$\chi_0^{-+}(q,\omega) \approx \chi_P \left(1 - \frac{E_q}{36 E_F} + i \frac{\pi m \omega}{2 q k_F} \right) \tag{9.184}$$

We can now substitute this expression in Eq. (9.176)for $\Phi(q,\omega)$ and finally insert the result in the equation for the change in free energy , Eq.(9.174):

$$\Delta F = \frac{2}{N} \sum_q \int \frac{d\omega}{2\pi} n(\omega) \arctan \phi(q,\omega) \tag{9.185}$$

where

$$\phi(q,\omega) = \frac{U \, \mathcal{I}m \chi_0^{-+}(q,\omega)}{1 - U \mathcal{R}e \chi_0^{-+}(q,\omega)} = \frac{K \frac{\omega}{q \, v_F}}{\alpha + U \rho(E_F) \left(\frac{q}{6 k_F} \right)^2} \tag{9.186}$$

and the coefficient K is

$$K = \frac{U \pi \rho(E_F)}{2}$$

9.12.1   Paramagnon contribution to the specific heat

We differentiate Eq. (9.185) with respect to T, to obtain: [13, 41]

$$\Delta C(T) = \frac{\partial \Delta F(T)}{\partial T} = \frac{2}{N} k_B \sum_q \int \frac{d\omega}{2\pi} \arctan \phi(q,\omega) \, \beta^2 \, \omega \, n(\omega) \, (1 + n(\omega)) \tag{9.187}$$

For low T, which implies small q and ω, we can expand $\arctan \phi$ in a Taylor series in powers of ϕ , and keep up to the second non vanishing term:

$$\Delta C(T) \approx \frac{2}{N} k_B \sum_q \int \frac{d\omega}{2\pi} \left(\phi(q,\omega) - \frac{1}{3} \phi(q,\omega)^3 \right) \beta^2 \, \omega \, n(\omega) \, (1 + n(\omega)) \tag{9.188}$$

Let us perform first the ω integral arising from the linear term in ϕ: , which we call $I_\omega^{(1)}$:

$$I_\omega^{(1)} = \left(\frac{k_B^2 K}{2 \pi q v_F} \right) \frac{1}{\alpha + U \rho(E_F)(q/6 k_F)^2} \int_{-\infty}^{\infty} d x \, x^2 n(x) \, (1 + n(x)) \tag{9.189}$$

The integral above can be obtained from the formula [42]:

$$\int_0^{\infty} d x \, x^2 \frac{e^x}{(e^x - 1)^2} = 2 \Gamma(2) \zeta(2) \tag{9.190}$$

Therefore, the first order correction to $C(T)$ is

$$\Delta C^{(1)}(T) = \frac{2}{N} \sum_q \frac{A T}{q} \frac{1}{\alpha + U \rho(E_F)(\frac{q}{6 k_F})^2} \tag{9.191}$$

We impose an upper cutoff on the q integral, since q is restricted to the range $q \leq 6k_F$. Since $\alpha \to 0$ the dominant term in (9.191) is

$$\Delta C^{(1)}(T) \propto T|\ln \alpha| \tag{9.192}$$

to be compared with the specific heat of non-interacting electrons at low T,[18]

$$C^0(T) = \gamma_0 \frac{m^*}{m} T \tag{9.193}$$

where m^*, m being respectively the effective band mass and the free mass of the electron. The coefficient in (9.193) is:

$$\gamma_0 = \frac{k_B^2}{3}(6\pi^2)^{1/3}n^{-2/3} m \tag{9.194}$$

so that the effect of the interaction near the quantum critical point $\alpha = 0$ is a logarithmic singularity of the renormalization coefficient of the electron effective mass.

We consider now the contribution of the cubic term in (9.187):

$$\Delta C^{(3)}(T) = B \sum_q \int \frac{d\omega}{2\pi} \hbar \omega^4 \frac{1}{q^3} \left[\alpha + U\rho(E_F)/2(q/v_F)^2\right] \tag{9.195}$$

We remind ourselves of the condition $\omega/qv_F \ll 1$, which sets a lower cutoff for the q integral:

$$\frac{1}{N} \sum_q \frac{1}{q^3} \frac{1}{\alpha + U\rho(E_F)/2(q/v_F)^2} \approx -\frac{1}{2\pi^2 n} \ln \frac{\hbar\omega}{k_F v_F} \tag{9.196}$$

Inserting this into (9.195) the dominant term at low T has the form [41]:

$$\Delta C^{(3)}(T) \approx Const \cdot T^3 \ln k_B T/\hbar k_F v_F \tag{9.197}$$

Collecting both terms we have

$$\Delta C(T) = C_0(T)(F + g(T)) \tag{9.198}$$

here $F \propto |\ln \alpha|$ is the renormalizatrion factor we found above (see Eq. (9.192)), while at low T:

$$g(T) = A(T/T_F)^2 \ln T/T_F + B(T/T_F)^2 + C(T/T_F)^4 + \cdots \tag{9.199}$$

The quadratic and further terms arise from the 5th and successive powers in the series in (9.187).

Equation (9.199) can be compared with the corresponding correction to the electronic specific heat arising from their interaction with phonons. It was found [43, 44, 45] that, up to second order in the interaction, the correction has also the form of Eq. (9.198), where F is a mass renormalization coefficient and the function $g(t)$ has exactly the same form, with different coefficients and with T_F substituted by $\Theta =$ Debye temperature.

As in the phonon case we can interpret this effect as the result of the virtual emission and reabsorption by an itinerant electron of a paramagnon (in the place of a phonon), which results in a correction to the quasi-particle energy, and consequently in a renormalization of the effective electron mass. Alhough the paramagnon is not a well defined excitation of the system, χ_0 can be interpreted as the paramagnon propagator, as we have seen, in the small ω, small q region, with $\omega/(qv_F) \ll 1$ where it has the maximum amplitude. Near this peak the decay time of the paramagnon is long [13].

In order to study how real systems approach the QCP it is necessary to study the variation of some key properties, like the low-T specific heat, as α changes. This can be done either by changing the composition of a metallic alloy or by applying pressure to a system with a fixed composition [47].

9.13   Beyond Stoner theory and RPA

Spin-fluctuation theories have appreciably improved the understanding of itinerant magnetism in transition metals. However, the RPA approximation described above does not lead to a Curie-Weiss law. *Murata* and *Doniach* [48] and also Moriya and *Kawabata* [49] obtained this law by extending the RPA to include mode-mode coupling among spin-fluctuations (See the review by *Moriya*) [31].

Experiments point to the permanence of local moments above T_C in transition metal ferro- and antiferromagnets. This has led to formulations with a basis of both $3d$ and $4s$ states, so that both the local and the itinerant character of the electron states are present from the start. Any theory of itinerant magnetism must tackle the simultaneous consideration of correlated charge and spin fluctuations. Since the 70's the approach to this problem was based on functional integral algorithms for calculating the partition function [50, 51].

Ongoing research is oriented to the application to itinerant magnetism of extensions of the density functional theory. We refer the reader to some recent papers on the magnetism of transition metals and to the references therein [52, 59, 61, 62].

9.14   Magnetism and superconductivity

Some early theoretical work led to the conclusion that strong FM spin-fluctuations would suppress the transition to a superconducting state.[46] However it has been found recently that both phases can coexist, as for instance in UGe_2[54, 55], $ZrZn_2$[56] or $URhGe$[57].

The paramagnon mediated model of superconductivity [58, 59] is the main current theory of the coexistence of both phenomena and it has predicted its ocurrence in $ZrZn_2$ [60].

References

1. Pauli, W. (1926) *Z. Phys.* **41**, 81.

2. Abrikosov, A. A., Gorkov, L. P. and Dzyaloshinski, I. E. (1975) *"Methods of Quantum Field Theory in Satistical Physics"*, Dover Publications Inc., New York.

3. Landau, L. D. and Lifchitz, E. (1966) *"Mécanique Quantique"*, MIR Editions, Moscow, Chap. IX.

4. Callaway, Joseph (1976) *"Quantum Theory of the Solid State"*, Academic Press, New York.

5. Hubbard, J. (1963) *Proc. Roy. Soc.(London)* **A276**, 238.

6. Blandin, A. (1968) *"Theory of Condensed Matter"*, International Atomic Energy Agency,Vienna, p. 691.

7. Stoner, E. C. (1948) *Repts. Prog. in Phys.* **11**, 43; (1950)**13**, 83.

8. Herbst, F. *et al.* (1976) *Phys. Rev.* **B14**, 3265.

9. Lieb, E. and Mattis, D. (1962) *J. Math. Phys.* **3**, 749.

10. Slater, J. C. (1936) *Phys. Rev.* **49**, 537

11. Stoner, E. C. (1938) *Proc. Roy. Soc. London* **A165**, 372.

12. Izuyama, T. *et al.* (1963) *J. Phys. Soc.(Japan)* **18**, 1025

13. Doniach, S. and Sondheimer, E.H. *"Green's Functions for Solid State Physicists"*, W. A. Benjamin, Inc., Reading, Massachusetts.

14. Economou, E. N., *"Green's Functions in Quantum Physics"*, (1983), Springer-Verlag, Berlin.

15. Edwards, D. M. and Bechara Muniz, R. (1985) *J. Phys. F: Met. Phys.* **15**, 2339.

16. Bechara Muniz, R., Cooke, J. F. and Edwards, D. M. (1985) *J. Phys. F: Met. Phys.* **15**, 2339.

17. Lovesey, S. W. (1984) *"The Theory of Neutron Scattering from Condensed matter"*, Oxford University Press.

18. Ziman, J. M. (1964) *"Principles of the Theory of Solids"*, Cambridge at the University Press, pp.116–119.

19. Hubbard, J. (1979) *Phys. Rev.* **B19**, 2626; *idem* **B20**, 4584.

20. Hasegawa, H. (1980) *J. Phys. Soc. Japan* **49**, 963.

21. Callaway, J., *loc. cit.*, p. 586.

22. Bardeen, J. (1937) *Phys. Rev.* **52**, 688.

23. Lindhard, J (1954) *Kgl. Danske Videnskab. Selskab, Mat-Fys. Medd* **28**, No. 8.

24. Jelitto, Rainer J. (1969) *J. Phys. Chem. Solids* **30**, 609.

25. Penn, W. (1966) *Phys. Rev.* **142**, 350.

26. Overhauser, A. W. (1962) *Phys. Rev.* **128**, 1437.

27. Shirane, G. and Takei, W. J. (1962) *J. Phys. Soc. Japan Suppl.* **BIII 35**, 17.

28. Herring, Conyers (1966) in *"Magnetism"*, Edited by Rado, G. T. and Suhl, H., **Vol. IV**, Academic Press, New York and London.

29. Rice, T. M. (1970)*Phys. Rev.* **B2**, 3619.

30. Acquarone, Marcello (1986) in *Proc. National Summer School, Villa Gualino, Turin, Italy* (Borsa, F. and Tognetti, V., Editors), World Scientific, Singapore, pp. 109–153.

31. Moriya, T. (1985) *"Spin Fluctuations in Itinerant Electron Magnetism"*, Springer Ser. Solid-State Sci.,Vol. 56, Springer, Berlin, Heidelberg.

32. Nagaoka, Yosuke (1966) *Phys. Rev.* **147**, 392.

33. Gutzwiller, M. (1963) *Phys. Rev. Lett.* **10**, 159.

34. Lieb, E. (1989) *Phys. Rev. Lett.* **62**, 1927.

35. Rasetti, M., ed. (1991) *"The Hubbard model"*, *Int. J. Mod. Phys.* **B 5** nos. 6 & 7.

36. Chao, K. A., Spałek, J. and Oleś, A. M. (1977) *J. Phys. C: Sol. State Phys.* **10**, L271.

37. Baskaran, G. *et al.* (eds.), (1989) *"Proceedings of the Anniversary Adriatico Conference and Workshop on Strongly correlated electron systems"*, World Scientific, Singapore.

38. Anderson, P. W. (1997) *"The theory of superconductivity in the high-T_c cuprates"*, Princeton University Press, Princeton.

39. Doniach, S. (1967) *Proc. Phys. Soc. (London)* **91**, 86.

40. Doniach, S. and Engelsberg, S. (1966) *Phys. Rev. Lett.* **17**, 750.

41. Enz, Charles P. (1992) *"A Course on Many-Body Theory Applied to Solid State Physics"*, World Scientific, Singapore.

42. Gradshteyn, I. S. and Ryzhik, I. M., *"Table of Integrals, Series and Products"*, Jeffrey, A., Editor; Academic Press, fifth edtn., 1994.

43. Buckingham, M. J. and Schafroth, M. R. (1954) *Proc. Phys. Soc.(London)* **A67**, 828.

44. Grimval, G. (1981) *"The Electron-Phonon Interaction in Metals"*, North-Holland, Amsterdam.

45. Majlis, Norberto, (1965) , Ph. D. Thesis, Birmingham University, England (Unpublished).

46. Berk, N. F. and Schrieffer, J. R., (1966) sl Phys. Rev. Lett. **17**, 433.

47. Smith, T. F. *et al.* (1971) *Phys. Rev. Lett.* **27**, 1732; Huber, J. G. *et al.*, (1975) *Solid State Commun.* **16**, 211; Lo, I. *et al.* (1989) *Phys. Rev. Lett.* **62**, 2555.

48. Murata, K. K. and Doniach, S. (1972) *Phys. Rev. Lett.* **29** 285.

49. Moriya, T. and Kawabata, A. (1973) *J. Phys. Soc. Jpn.* **34** 639.

50. Hubbard, J. (1959) *Phys. Rev. Lett.* **3** 77.

51. (1971)Schrieffer, J. R. *et al.* *J. Phys.* *(Paris)* **32** C1; Schrieffer J. R., (1969) *"Lectures on Magnetism in Metals"*, C. A. P. Summer School, Banff (unpublished)

52. Katsnelson, M. I. and Liechtenstein, A. I. (2004) *J. Phys. : Cond. Matt.* **16** 7439.

53. Karchev, Naoum (2003) *J. Phys. : Cond. Matt.* **15** 2797.

54. Saxena, S. *et al* (2000) *Nature* **406**, 587.

55. Huxley, A. *et al* (2001) *Phys. Rev.* **B 63**, 144519.

56. Pfleiderer, C. *et al* (2001) *Nature* **412**, 58.

57. Aoki, D. *et al* (2001) *Nature* **413**, 613.

58. Scalapino, D. J. *arXiv:cond-mat/9908287 v2* 30 Aug 1999.

59. Karchev, Naoum *arXiv:cond-mat/0405371 v2* 15 Jun 2004.

60. Enz, C. P. and Matthias, B. T. (1978) *Science* **201**, 828; (1979) *Z. Phys.* **B33** 129.

61. Georges, Antoine *et al.* (1996) Rev. Mod. Phys. **68**, 13.

62. Liechtenstein , A. I. *et al.* (1986) *J. Mag. Mag. Mat.* **67**, 65.

Chapter 10

Indirect Exchange

10.1   Introduction

Let us now turn our attention to other systems, metals or semiconductors, in which spins can be considered well localized on some or all the sites of the lattice, while a part of the electrons are in delocalized states belonging to one or more conduction bands.

As the first example, we have the rare earths, with incomplete $4f$ shells, and the actinides, with incomplete $5f$ shells.

The *diluted magnetic alloys* constitute a second family of such systems, in which a non-magnetic host metal contains a small atomic percentage of substitutional impurities, consisting of atoms with some incomplete shell, with unpaired spins, which maintain, even when dissolved in the host metal, a spontaneous magnetic moment.

The third family to consider are the magnetic semiconductors.

In the first two cases, the f shell wave functions are tightly bound, and consequently well localized in space. Neutron studies reveal that the mean radius, for trivalent Nd^{3+} and Er^{3+} is $\sim$ 0.35 Å[1].

In the whole series of the trivalent lanthanide (rare earth) group, only Gd has $L = 0$, an S ground state term. For all other ions in the series $L \neq 0$ and we can adopt, as discussed in Chap. 1, the Russel-Saunders coupling of L to S. As already mentioned in Chap. 1, crystal field effects are much smaller than spin-orbit effects in these elements, so that as a first approximation we consider a ground level degenerate multiplet determined by Hund's rules, with degeneracy $2J + 1$ for given J. The higher multiplets in the rare earths are sufficiently separated in energy to be neglected. A typical value of the spin-orbit splitting parameter $A\hbar^2$ is 0.01 eV [2]. Neglecting crystal field effects then, we have

$$\mu = g_J \mu_B J \tag{10.1}$$

For the second half of the lanthanide series, that is for Tb, Dy, Ho, Er, Tm and Yb, parallel coupling of L and S is required by Hund's rules, that is $J = L + S$, and then

$$g_J = \frac{L + 2S}{L + S} \tag{10.2}$$

which implies

$$S = (g_J - 1)J \tag{10.3}$$

This has been experimentally verified for all ions listed above [3]. Static susceptibility measurements in these metals also confirm the Curie free-ion paramagnetic
behaviour at temperatures well above the crystal field splitting:

$$\chi_0 = (g_J \mu_B)^2 \frac{J(J+1)}{3 k_B T} \tag{10.4}$$

supporting the view that the $4f$ spins are localized, and behaving as free ions
in the paramagnetic phase $(T \gg T_c)$ (Compare with Eq. (1.6)).

As a consequence of the negligible overlap among $4f$ orbitals on different
sites, magnetic cooperative effects in these metals depend upon the coupling of
spins through the agency of the conduction electrons.

We have already seen that within the Heitler-London approximation, the
main factor leading to spin dependent interactions among electrons on different
atoms is the exchange integral j (Eq. (2.28)), involving the Coulomb electrostatic interaction potential. In the problem considered in Chap. 2, we were con-
cerned with two atoms, a and b, which could exchange electrons. In the present
case we shall consider exchange processes involving two types of states:localized
states, either d or f, on the one hand; on the other hand, itinerant delocalized
Bloch states, originating in predominantly s and p atomic states with appreciable overlap belonging to one or more conduction bands, which are reasonably
wide (several eV).

10.2   Effective *s-d* exchange interaction

We construct Slater determinants as a convenient basis for the Hilbert space
of the many electron system from anti-symmetrized products of wide band s
and p states and narrow band d or f localized states. In the simplest case we
consider only two kinds of states. One is fairly localized within a Wigner-Seitz
cell around a given site, which we shall call a "d" state and which we take
as a singlet for simplicity. A second set of states, which we shall call "s" do
overlap considerably, and form a conduction band. We expand the matter-wave
operators $\Psi_\sigma(\mathbf{x})$, $\Psi_\sigma^\dagger(\mathbf{x})$ in the basis of the d and s orbitals (we may assume that
these orbitals are mutually orthogonal), so that we write

$$\Psi_\sigma(\mathbf{x}) \simeq \sum_{n\sigma} \phi_n(\mathbf{x}) d_{n\sigma} + \sum_{k\sigma} a_{k\sigma} \psi_{k\sigma}(\mathbf{x}) \tag{10.5}$$

The Coulomb interaction in this representation is

$$V_C = \frac{1}{2} \int \int d\mathbf{r}\, d\mathbf{r}' \sum_{\sigma\sigma'} \Psi_\sigma^\dagger(\mathbf{r}) \Psi_{\sigma'}^\dagger(\mathbf{r}') v(\mathbf{r} - \mathbf{r}') \Psi_{\sigma'}(\mathbf{r}') \Psi_\sigma(\mathbf{r}) \tag{10.6}$$

where $v(\mathbf{r} - \mathbf{r}') = e^2 / \mid \mathbf{r} - \mathbf{r}' \mid$. Upon substitution of (10.5) into (10.6) there will be 8 types of terms, which can be schematically represented as

1. $d^\dagger d^\dagger dd$

2. $c^\dagger c^\dagger cc$

3. $d^\dagger d^\dagger cc$

4. $c^\dagger c^\dagger dd$

5. $d^\dagger c^\dagger cd$

6. $d^\dagger c^\dagger dc$

7. $c^\dagger d^\dagger cd$

8. $c^\dagger d^\dagger dc$

The terms involving exchange are (3), (4), (6) and (7). Among these (3) and (4) involve two electrons on a d orbital. If the orbitals are on the same site, this integral is of the order of the intra-atomic U for the d orbitals, by assumption a large quantity, so that the corresponding energies are outside the range of the excitations we want to describe. In the perturbation expansion of the energy that we shall perform below, these states will contribute terms with very large denominators, so that we shall neglect them altogether. If both d orbitals are on different sites, they are orthogonal by hypothesis, and we also neglect these terms.

We are then left with exchange terms (6) and (7) which involve the exchange of an electron between a d and an s state. It is easy to see that both terms are equal, since we can interchange them by renaming the integration variables in the expression for the matrix element. We shall consider only the on-site terms $n = m$ among all products involving two $d, d^\dagger$ operators , since by assumption the d orbitals are well localized.

The conduction band s wave functions are Bloch states

$$\psi_k(\mathbf{r}) = \frac{1}{\sqrt{N}} u_k(\mathbf{r}) e^{i\mathbf{k}\cdot\mathbf{r}} \tag{10.7}$$

where $u_k(\mathbf{r}) e^{i\mathbf{k}\cdot\mathbf{r}}$ has the periodicity of the lattice and is normalized within a crystal cell, so that $\psi_k(\mathbf{r})$ is normalized over the whole crystal. If we add together terms (6) and (7) we get

$$H_{sd} = -\frac{1}{N} \sum_{n\sigma\sigma'} \sum_{kk'} J_{kk'} e^{i(\mathbf{k}-\mathbf{k}')\cdot\mathbf{R_n}} \, d_{n\sigma}^\dagger d_{n\sigma'} a_{k\sigma'}^\dagger a_{k'\sigma} \tag{10.8}$$

where the exchange integral $J_{kk'}$ is

$$J_{kk'} = \int d\mathbf{r} \int d\mathbf{r}' \phi^*(r) u_k^*(r') e^{-i\mathbf{k}\cdot\mathbf{r}'} \phi(r') u_{k'}(r) e^{i\mathbf{k}'\cdot\mathbf{r}} v(r - r') \tag{10.9}$$

Exercise 10.1
Show that leaving aside a term which is spin-independent the other terms in
(10.11) can be cast in the form [5]

$$
\begin{aligned}
H_{sd} & = -\frac{1}{N} \sum_n \sum_{kk'} J_{kk'} e^{i(\mathbf{k}-\mathbf{k}')\cdot\mathbf{R_n}} \Big\{ S_n^z \big(a_{k'\uparrow}^\dagger a_{k'\uparrow} \\
& \quad - a_{k\downarrow}^\dagger a_{k'\downarrow} \big) + S_n^+ a_{k\downarrow}^\dagger a_{k'\uparrow} + S_n^- a_{k\uparrow}^\dagger a_{k'\downarrow} \Big\}
\end{aligned}
\tag{10.10}
$$

If we substitute the Bloch states by plane waves, and write explicitly the
Coulomb potential, the expression for the exchange integral above simplifies
(we call $\mathbf{k}' = \mathbf{k} + \mathbf{q}$):

$$
J(\mathbf{k},\mathbf{q}) = e^2 \int d\mathbf{r}_1 d\mathbf{r}_2 \frac{e^{i\mathbf{k}\cdot\mathbf{r}_1} e^{-i(\mathbf{k}+\mathbf{q})\cdot\mathbf{r}_2} \phi(\mathbf{r}_1)\phi(\mathbf{r}_2))}{\mid \mathbf{r}_1 - \mathbf{r}_2 \mid}
\tag{10.11}
$$

More complicated (and realistic) cases in which the orbital degeneracy of the
localized orbitals is considered, as well as the possibility of core states with
several electrons, are also considered in the literature [6]. We shall restrict
ourselves to the simple form above for the $s-d$ interaction We consider now the
conduction electrons as independent particles and proceed, following Yosida [5],
to calculate in perturbation theory the effect of H_{sd} on the total energy of the
conduction band electrons, considered as a Fermi gas. The first order correction
to the energy of the s electrons is given by the expectation value of H_{sd} in the
Fermi sea ground state:

$$
\Delta E^{(1)} = -\frac{1}{N} J(0) \left(\langle n_+ \rangle - \langle n_- \rangle \right) \sum_n < S_n^z >
\tag{10.12}
$$

where z is the spin quantization axis and $\langle n_\pm \rangle$ are the total numbers of conduction electrons of spin $+$ and $-$. We assume that $J(k,q)$ above can be approximated by a constant J_0. Assuming $J_0 > 0$, (10.12) implies, if $\langle S_n^z \rangle > 0$, that to
minimize the energy to this order one would obtain a net uniform polarization
of the conduction electrons, with $\langle n_+ \rangle - \langle n_- \rangle > 0$. We denote with $n_\pm$ the total
number of electrons with the corresponding z component of spin.
If the resultant net polarization in (10.12) is $\uparrow$, then for an isotropic band
$k_{F\uparrow} > k_{F\downarrow}$. The expectation value of the total energy is:

$$
E^{kin} + \Delta E^{(1)} = \sum_\sigma \sum_k^{k_{F\sigma}} \left[\epsilon_{k\sigma} - \frac{\sigma J(0)}{N} \sum_n < S_n^z > \right] n_{k\sigma}
\tag{10.13}
$$

with an obvious notation. In this chapter the equivalent notations $\pm$ or $\uparrow, \downarrow$ will
be used for spin, according to convenience.

Exercise 10.2
Prove (10.13)

The Fermi level μ in the presence of the perturbation must be the same for both spins, so that

$$\mu = \epsilon_{k_{F\uparrow}} - \frac{\sigma J(0)}{N} \sum_n < S_n^z > = \epsilon_{k_{F\downarrow}} + \frac{\sigma J(0)}{N} \sum_n < S_n^z > \qquad (10.14)$$

The total number of conduction electrons of a given spin satisfies the relation

$$n_\sigma = \left(\frac{V}{6\pi^2} \right) k_{F\sigma}^3 \qquad (10.15)$$

Consider the perturbation of k_F of the paramagnetic state when the perturbation is present. From (10.15) we see that, to first order, we can write

$$k_{F\sigma} = k_F \pm \Delta k_F \qquad (10.16)$$

with

$$k_F = \frac{1}{2} (k_{F\uparrow} + k_{F\downarrow})$$

$$\Delta k_F = \frac{1}{2} (k_{F\uparrow} - k_{F\downarrow}) \qquad (10.17)$$

To the same order, we have

$$\mu = E_F \pm 2E_F \frac{\Delta k_F}{k_F} \qquad (10.18)$$

where E_F is the unperturbed Fermi level.
The total number of conduction electrons, which we call $2n$ for convenience, is

$$2n = n_+ + n_- \qquad (10.19)$$

so that

$$n = \frac{V}{6\pi^2} k_F^3 + O\left((\Delta k_F)^2 \right) \qquad (10.20)$$

Exercise 10.3
Prove (10.20)

Eq.(10.20) the conservation of the number of electrons (or charge) to first order.

Exercise 10.4
Prove the following relations to first order:

$$\frac{2\Delta k_F E_F}{k_F} = J(0)M \qquad (10.21)$$

and

$$n_\sigma = n + 3\sigma \frac{J(0)Mn}{2E_F} \qquad (10.22)$$

where

$$M \equiv \frac{1}{N} \sum_n < S_n^z > \qquad (10.23)$$

and $< S_n^z >$ is the expectation value of the d state spin polarization at site n.

It is convenient to define the parameter

$$\delta n \equiv n^\uparrow - n^\downarrow \qquad (10.24)$$

which can be expressed as:

$$\delta n = \frac{3nJ(0)M}{E_F} \qquad (10.25)$$

One can calculate the change in the kinetic energy to lowest order (which turns out to be *second order*) in this parameter:

Exercise 10.5
Prove that

$$\delta E^{kin} = \frac{(\delta n)^2}{6n} E_F \qquad (10.26)$$

We can write the total change in energy due to the interaction as:

$$\Delta E^{tot} = -\frac{3}{2} \frac{(J(0)M)^2}{E_F} \qquad (10.27)$$

Exercise 10.6
Express the change in kinetic energy and the first order perturbation $\Delta E^{(1)}$ in Eq. (10.12) in terms of δn, and verify that the value (10.25) for δn is obtained upon minimizing the total energy change with respect to δn.

We just performed averages over the whole system, so that we can consider the equations above to be the result of a uniform mean-field treatment. However we can also ask ourselves about the local non-uniform spin polarization of the conduction electrons, from the first order change in the wave functions of the conduction electrons.

We shall write the correction to each conduction electron ket -whose expression contains local spin *operators* - due to the perturbation Hamiltonian (10.10). The conduction electrons are treated as non-interacting particles. Then we write the wave function to first order in the perturbation expansion as

$$\overline{\mid k\sigma\rangle} = \mid k\sigma\rangle + \sum_{k'\sigma'} \frac{(H_{sd})_{k'\sigma',k\sigma}}{\epsilon_k - \epsilon_{k'}} \mid k'\sigma'\rangle \qquad (10.28)$$

Substitution of (10.10) into (10.28) leads to

$$\overline{| \, k\sigma \rangle} \;=\; | \, k\sigma \rangle - \sum_{k'} \frac{J(\mathbf{k}-\mathbf{k}')}{\epsilon_k - \epsilon_{k'}}$$

$$\times \frac{1}{N} \sum_n e^{i(\mathbf{k}-\mathbf{k}')\cdot \mathbf{R_n}} \left(\sigma S_n^z \, | \, k'\sigma \rangle + S_n^\sigma \, | \, k', -\sigma \rangle \right) \qquad (10.29)$$

where as usual the unperturbed Bloch functions were substituted by plane waves.

We can now obtain the spin-up and spin-down electron density of the conduction band at low T, as

$$\rho_\sigma(\mathbf{r}) = \sum_k^{k_{F\sigma}} < \mathbf{r} | \, \mathbf{k}, \sigma > < \mathbf{k}, \sigma \, | \mathbf{r} > \qquad (10.30)$$

Since the local and conduction electron spins act independently as operators on the corresponding kets, only diagonal terms in spin will be left upon calculating the average on the spin states, so we retain only these:

$$\rho_\sigma(\mathbf{r}) \;=\; \frac{1}{V} \sum_k^{k_{F\sigma}} (\, 1 - \sigma \frac{2m}{\hbar^2 N} \sum_{k \neq k'} \frac{J(\mathbf{k}-\mathbf{k}')}{k^2 - (k')^2}$$

$$\times \sum_n \left\{ e^{i(\mathbf{k}-\mathbf{k}')\cdot(\mathbf{r}-\mathbf{R_n})} + e^{-i(\mathbf{k}-\mathbf{k}')\cdot(\mathbf{r}-\mathbf{R_n})} \right\} \langle S_n^z \rangle \,) \qquad (10.31)$$

With the aid of (10.22) one can rewrite the first term of (10.31) as

$$\frac{1}{V} \sum_k^{k_{F\sigma}} = \frac{1}{V} \left[n + 3\sigma \frac{J(0) \sum_n \langle S_n^z \rangle}{2 E_F N} \right] \qquad (10.32)$$

Now call $\mathbf{k} - \mathbf{k}' = \mathbf{q}$, and perform the sum over $\mathbf{k}$ in the second term of (10.31). Let us define

$$f(q) = \frac{k_F^2}{N} \sum_{k \leq k_F} \frac{1}{(\mathbf{k}+\mathbf{q})^2 - k^2} \qquad (10.33)$$

where we neglect the spin dependence of $k_{F,\sigma}$, since it introduces a second order correction. The second term of (10.31) then becomes

$$\sigma \frac{2m}{\hbar^2 V N} \sum_{q \neq 0} J(\mathbf{q}) f(q) \sum_n \left\{ e^{i\mathbf{q}\cdot(\mathbf{r}-\mathbf{R_n})} + e^{-i\mathbf{q}\cdot(\mathbf{r}-\mathbf{R_n})} \right\} \langle S_n^z \rangle \qquad (10.34)$$

By use of (10.20) we can write then the total densities as

$$\rho_\sigma(\mathbf{r}) \;=\; \frac{1}{V} \left(n + 3n\sigma \frac{J(0) \sum_n \langle S_n^z \rangle}{2 E_F N} \right)$$

$$+ \sigma \frac{3n}{8 E_F V N} \sum_{q \neq 0} J(\mathbf{q}) f(q)$$

$$\times \sum_n \left\{ e^{i(\mathbf{k}-\mathbf{k}')\cdot(\mathbf{r}-\mathbf{R_n})} + e^{-i(\mathbf{k}-\mathbf{k}')\cdot(\mathbf{r}-\mathbf{R_n})} \right\} \langle S_n^z \rangle \qquad (10.35)$$

We rewrite (10.33) as

$$f(q) = \frac{V}{8\pi^3 N} \int_{k \le k_{F\sigma}} d^3k \frac{1}{q^2 + 2qk \cos\theta} \tag{10.36}$$

which yields the result:

$$f(q) = 1 + \frac{4k_F^2 - q^2}{4qk_F} \ln \left| \frac{2k_F + q}{2k_F - q} \right| \tag{10.37}$$

Exercise 10.7

- *Prove (10.37);*

- *Show that $\lim_{q \to 0} f(q) = 2$;*

- *Show that $\frac{\partial f(q)}{\partial q}$ is infinite at $q = 2k_F$.*

We see that the value of the second term in (10.35) is exactly the same of the third term for $q = 0$, so one can just omit the second one and include $q = 0$ in the third term.

The final result is therefore

$$\rho_\sigma(\mathbf{r}) = \frac{n}{V} + \sigma \frac{3n}{8E_F V N} \sum_q J(\mathbf{q}) f(q)$$

$$\times \sum_n \{ e^{i\mathbf{q}\cdot(\mathbf{r} - \mathbf{R_n})} + e^{-i\mathbf{q}\cdot(\mathbf{r} - \mathbf{R_n})} \} S_n^z \tag{10.38}$$

We note at this point that $f(q)$ is, apart from a constant factor, the free electron susceptibility for $\omega = 0$, as can be verified by comparing it with Eqs. (7.76) to (7.78). If we take for simplicity $J(q) \approx constant \simeq J_0$, the spin polarization of the electron gas at $\mathbf{r}$ due to a non-zero average core spin at point $\mathbf{R_n}$ is just proportional to the space-dependent free-electron static susceptibility $\chi_0(\mathbf{r} - \mathbf{R_n})$, which is also the result of linear response theory in the static limit.

Exercise 10.8
Verify that

$$\chi_0(\mathbf{r} - \mathbf{R_n}) = \frac{1}{N} \sum_q e^{-i\mathbf{q}\cdot\mathbf{R}} f(q) = -\frac{24\pi n}{N} \Phi(2k_F R) \tag{10.39}$$

where

$$\Phi(x) = \frac{x \cos x - \sin x}{x^4} \tag{10.40}$$

is Ruderman-Kittel range function.[4]
Hint: you can use the integral representation:

$$\ln \left| \frac{a + q}{a - q} \right| = 2 \int_0^\infty \frac{\sin(ax) \sin(qx)}{x} \, dx \tag{10.41}$$

One obtains finally

$$\rho_\sigma(\mathbf{r}) = \frac{n}{V} - \frac{2\pi(3n^2)\sigma J_0}{N E_F} \times \sum_n \Phi(2k_F \mid \mathbf{r} - \mathbf{R_n} \mid)\langle S_n^z \rangle \tag{10.42}$$

The conclusion is that the perturbation of the spin-dependent density at the point $\mathbf{r}$ is a sum of contributions from all magnetic sites, each of which oscillates in space with period $1/(2k_F)$ and decrease with distance as

$$|\mathbf{r} - \mathbf{R_n}|^{-3} \quad .$$

Yosida [5] observes that if the approximation $J(q) = const.$ is improved by introducing a cutoff in $J(q)$ then the linear perturbation of the electron density is finite at the position of the ions. To follow Yosida's argument we remind ourselves that in Exercise 10.7 we found that $f(q)$ decreases abruptly at $q = 2k_F$, where the derivative is singular. We notice as well that it vanishes for $q \gg 2k_F$ as $\sim (q/2k_F)^{-2}$. Since $J(\mathbf{k}, \mathbf{k} + \mathbf{q})$ certainly vanishes as $q \to \infty$ he makes the following reasonable approximation:

$$f(q)J(q) = \begin{cases} 2J_0 & \text{if } q \leq 2k_F \\ 0 & \text{if } q > 2k_F \end{cases} \tag{10.43}$$

Then the $\mathbf{r}$ dependent term in the r. h. s. of (10.38) is

$$-\frac{4\sigma(3n)^2}{V E_F} \frac{J_0}{N} \sum_n 2k_F |\mathbf{r} - \mathbf{R_n}| \Phi(2k_F |\mathbf{r} - \mathbf{R_n}|) S_n^z \tag{10.44}$$

which is finite at $\mathbf{r} \to \mathbf{R_n}$.

Exercise 10.9
Obtain Eq. (10.44).

10.3 Indirect exchange Hamiltonian

We shall now obtain the second order correction to the energy of the Fermi gas of conduction electrons due to the exchange interaction with localized spins. In the following we shall consider the localized-electron spins as fixed external variables, and we shall obtain an expression for the perturbation of the total energy of the system that will depend on the detailed configuration of the localized spins. The result in the simplest possible form is an effective interaction among the localized spins of the form of the Heisenberg isotropic Hamiltonian. Let us write the expression for the second-order correction to the energy of the electron gas:

$$\Delta E^{(2)} = \langle 0 \left| H_{sd} \frac{P}{E_0 - H_0} H_{sd} \right| 0 \rangle \tag{10.45}$$

where the projection operator $P = 1 - |0\rangle\langle 0|$ eliminates the ground state of the Fermi gas from the sum over intermediate states. Upon substituting the definition of H_{sd} from (10.10), we obtain the isotropic Heisenberg interaction, with an effective exchange integral of the form [7]:

$$H_{eff} = \Delta E^{(2)} = \sum_{n,m} (J_0)^2 \left(\frac{3n}{N}\right)^2 \frac{2\pi}{E_F} \Phi(\mathbf{R_n} - \mathbf{R_m})\mathbf{S}_n \cdot \mathbf{S}_m \qquad (10.46)$$

Exercise 10.10
Verify Eq. (10.46) and show that in the case of free electrons and within the approximation $J(q) = constant = J_0$, $\Phi(x)$ is the Ruderman-Kittel range function of Eq. (10.40).

We remark that the particular form (10.46) is obtained in the case of free electrons in a box and under the simplifying assumptions on $J(q)$ that [8]:
1) $J(\mathbf{k}, \mathbf{k}') \approx J(\mathbf{k} - \mathbf{k}')$;
2) $J(q) \approx J_0$.
   We shall see in the following section that if the periodic lattice potential is considered, the resulting band structure, and as a consequence, the usually complex geometry of the Fermi surface, demand a non-trivial generalization of the RKKY formulation.

10.4   Range function and band structure

Let us now extend the theory by including more than one conduction band, and also by representing the one electron band states by Bloch waves. Instead of (8.6) then we should write:

$$J_{n\,n'}(\mathbf{k}, \mathbf{k}') = e^2 \int\int d\mathbf{r}_1\, d\mathbf{r}_2 \phi(\mathbf{r}_1)\phi(\mathbf{r}_2) \frac{1}{|\,\mathbf{r}_1 - \mathbf{r}_2\,|} \cdot$$
$$u^*_{\mathbf{k}n}(\mathbf{r}_1)u_{\mathbf{k}'n'}(\mathbf{r}_2)e^{-i\mathbf{k}\cdot\mathbf{r_1}}e^{i\mathbf{k}'\cdot\mathbf{r_2}} \qquad (10.47)$$

Then the range function in the general case is the sum

$$\Phi(\mathbf{R}) = \sum_{n,n'} \Phi_{nn'}(\mathbf{R}) \qquad (10.48)$$

over all relevant bands of the contributions due to each pair n, n' (where $n = n'$ is the previous case and will be naturally included in the sum):

$$\Phi_{nn'}(\mathbf{R}) = -\frac{1}{2V^2}\sum_{\mathbf{k}\,\mathbf{k}'} I_{n,n'}(\mathbf{k}\,\mathbf{k}')\frac{f(\varepsilon_{\mathbf{k}n}) - f(\varepsilon_{\mathbf{k}'n'})}{\varepsilon_{\mathbf{k}n} - \varepsilon_{\mathbf{k}'n'}}e^{-i(\mathbf{k}-\mathbf{k}')\cdot\mathbf{R}} \qquad (10.49)$$

where

$$I_{n,n'}(\mathbf{k}\,\mathbf{k}') = |\,J_{n\,n'}(\mathbf{k}, \mathbf{k}')\,|^2 \qquad (10.50)$$

Let us assume, as before, that I is slowly varying in k and k'. The sums over the reciprocal space in (10.49) are to be performed within the first BZ. By definition, the function $I_{n,n'}(\mathbf{k}\,\mathbf{k}')$ is invariant under time-reversal:

$$I_{n,n'}(-\mathbf{k} - \mathbf{k}') = I_{n,n'}(\mathbf{k}\,\mathbf{k}') \qquad (10.51)$$

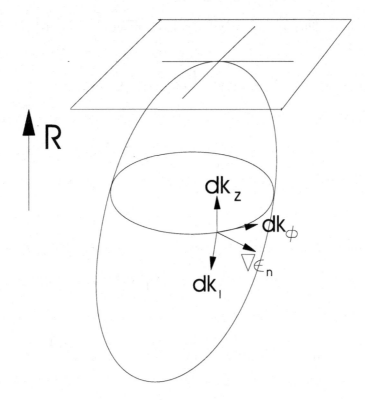

Figure 10.1: *Orthogonal coordinates for integration in k space*

We perform as usual the thermodynamic limit $V \to \infty$ and substitute sums by integrals over the first BZ.

To start with, let us define a system of coordinates in k-space appropriate to describe arbitrary surfaces of constant energy. We choose for convenience the k_z axis paralel to the vector R separating the two interacting localized spins. For the n-th band we perform the transformation

$$(k_x, k_y, k_z) \implies (\varepsilon_{\mathbf{k}n}, k_l, k_\phi) \tag{10.52}$$

from the original cartesian coordinates to a new system of orthogonal coordinates which are defined in the following way: [9, 10]

Let us consider the family of surfaces $\varepsilon_{\mathbf{k}n} =$ constant. The intersection of the surface of constant energy with the plane $k_z = const.$ defines a curve on the plane. At any point P on such a curve we can define an orthogonal basis triad, with versors along, respectively, the gradient of the energy $\nabla_k \varepsilon_{\mathbf{k}n}$ (which is perpendicular to the surface), the tangent to the curve and the tangent to the constant energy surface which is perpendicular to both previous versors, as shown in Fig. 10.1. We call the differentials of $\mathbf{k}$ along those directions dk_ε, dk_ϕ and dk_l respectively.

The volume differential in k space in a neighbourhood of P is:

$$dV = dk_\phi dk_z \left(\frac{\partial k_l}{\partial k_z} \right)_{\varepsilon, k_\phi} \frac{d\varepsilon}{|\nabla \varepsilon|} \tag{10.53}$$

Let us choose the sense of k_l such that $\left(\frac{\partial k_l}{\partial k_z} \right)_{\varepsilon, k_\phi} > 0$. The double volume integral in k space of (10.49) can now be performed in the new coordinate system. We perform first the curvilinear integrals for given energies ε and ε' and fixed k_z, k_z' which yield as a result a function γ of all these four variables:

$$\gamma(\varepsilon_n', k_z' ; \varepsilon_n, k_z) = \frac{1}{(2\pi)^6} \oint_{\varepsilon, \, k_z} \oint_{\varepsilon', \, k_z'} \frac{dk_\phi}{|\nabla_k \varepsilon_n(\mathbf{k})|} \left(\frac{\partial k_l}{\partial k_z} \right)$$
$$\times \frac{dk_\phi'}{|\nabla_{k'} \varepsilon_n'(\mathbf{k}')|} \left(\frac{\partial k_l'}{\partial k_z'} \right) I_{nn'}(\mathbf{k}, \mathbf{k}') \tag{10.54}$$

We assume that $I_{nn'}(\mathbf{k}, \mathbf{k}')$ and $\nabla_k \varepsilon_n(\mathbf{k})$ are slowly varying functions of k_ϕ, so that we can perform the curvilinear integral over k_ϕ and use

$$dS = dk_l \oint dk_\phi \tag{10.55}$$

as element of surface area on the constant energy surface. Then we can rewrite Eq. (10.54) as the second derivative of a double surface integral

$$\gamma(\varepsilon_{n'}, k_z' ; \varepsilon_n, k_z) = \frac{1}{(2\pi)^6} \frac{\partial}{\partial k_z} \frac{\partial}{\partial k_z'} \oint_{\varepsilon, \, k_z} \oint_{\varepsilon', \, k_z'}$$
$$\times \frac{dS_n}{|\nabla_k \varepsilon_n(\mathbf{k})|} \frac{dS_{n'}}{|\nabla_{k'} \varepsilon_{n'}(\mathbf{k}')|} I_{nn'}(\mathbf{k}, \mathbf{k}') \tag{10.56}$$

Notice that integrations in (10.56) are performed over the curves defined by $\varepsilon_n = const., k_z = const.$, and the corresponding conditions for $\varepsilon_{n'}$ and k_z'.

Exercise 10.11
Prove the equivalence of (10.54) and (10.56) under the present assumptions.

Finally, the expression for $\Phi_{nn'}$ is obtained by integrating with respect to k_z, k_z' and the energy variables:

$$\Phi_{nn'} = -\frac{1}{2} \int \int \int \int d\varepsilon_n d\varepsilon_{n'} dk_z dk_z' \gamma(\varepsilon_{n'}, k_z'; \varepsilon_n, k_z)$$

$$\times \left[\frac{f(\varepsilon_n) - f(\varepsilon_{n'})}{\varepsilon_n - \varepsilon_{n'}} \right] e^{iz(k_z - k_z')} \tag{10.57}$$

Consider a given constant energy surface and assume, for the time being, that it is contained within a finite volume inside the first BZ. Then in the simplest case of a simply connected, smooth surface, of which a sphere is the ideal case, one finds several extreme values of k_z. Since the energy eigenvalues are real and invariant under inversion in k space, for each extreme value of k_z, say k_i, there is another one $k_j = -k_i$. Let us now label all extreme values of k_z for both energy surfaces, as k_i, k_j'. We sketch a simple example in Fig. 10.2.

Let us change notation, and, since z was chosen in the direction of $\mathbf{R}$, call

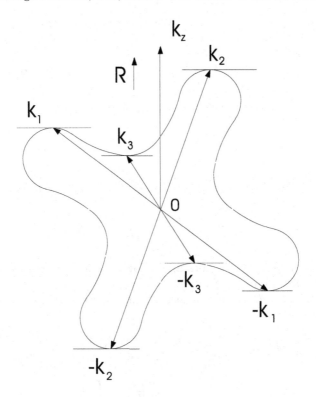

Figure 10.2: *Extreme values of k_z on constant energy surfaces in k space*

$z = R$. The expression (10.57) is particularly adequate for obtaining an asymptotic expansion of the range function for large R. To this end it is convenient to use the identity

$$e^{ikR} \equiv \frac{1}{iR} \frac{\partial}{\partial k}(e^{ikR}) \tag{10.58}$$

and rewrite (10.57) as:

$$\Phi_{nn'} = \frac{1}{2} \int \int \int \int d\varepsilon_n d\,\varepsilon_{n'} dk_z dk'_z \gamma(\varepsilon_{n'}, k'_z; \varepsilon_n, k_z)$$

$$\times \left[\frac{f(\varepsilon_n) - f(\varepsilon'_n)}{\varepsilon_n - \varepsilon_{n'}} \right] \frac{\partial}{\partial k_z} \frac{\partial}{\partial k'_z} \left(e^{iR(k_z - k'_z)} \right) \qquad (10.59)$$

The double integral over k_z, k'_z above can be performed by parts. The end points of the interval for the k_z and k'_z integrations are the extreme values. One can use repeatedly transformation (10.58) and integration by parts to obtain an expansion of Φ in inverse powers of R, which one expects will converge for large R. We write it down below, simplifying a bit the notation, by using k, k' for k_z etc. and defining

$$\frac{\Delta f}{\Delta \varepsilon} = \frac{f(\varepsilon_n) - f(\varepsilon'_n)}{\varepsilon_n - \varepsilon_{n'}} \quad . \qquad (10.60)$$

We find

$$\Phi_{nn'} = -\frac{1}{2R^2} \sum_{j,l} \int \int d\varepsilon_n d\varepsilon_{n'} \Gamma_{j,l}(; \varepsilon_n, \varepsilon_{n'}, R)$$

$$\times \frac{\Delta f}{\Delta \varepsilon} e^{iR(k_j - k'_l)} e^{i(\phi_j - \phi_l)} \qquad (10.61)$$

where

$$\Gamma_{j,l}(\varepsilon, \varepsilon', R) = \gamma(\varepsilon, k_j; \varepsilon', k'_l) - \frac{1}{iR} \left(\frac{\partial \gamma}{\partial k} - \frac{\partial \gamma}{\partial k'} \right)_{k_j, k'_l}$$

$$+ \frac{1}{(iR)^2} \left(\frac{\partial^2 \gamma}{\partial k^2} - 2 \frac{\partial^2 \gamma}{\partial k \partial k'} + \frac{\partial^2 \gamma}{\partial k'^2} \right)_{k_j, k'_l} + \cdots \qquad (10.62)$$

The phase factors $e^{i\phi_i}$, $e^{i\phi_j}$ from upper/lower limits are respectively $+1$ for a maximum, -1 for a minimum.

For large enough R we can neglect all but the first term in (10.62). In this case, Γ reduces to

$$\Gamma_{j,l}(\varepsilon, \varepsilon', R) \simeq \frac{I_{nn'}(\mathbf{k_j}, \mathbf{k'_l})}{(2\pi)^6 \mid \nabla_{\mathbf{k_j}} \varepsilon_n \mid\mid \nabla_{k_l} \varepsilon_{n'} \mid}$$

$$\times \left(\frac{\partial S_n}{\partial k_z} \right)_{k_j} \left(\frac{\partial S_{n'}}{\partial k_{z'}} \right)_{k'_l} \qquad (10.63)$$

In the cases of ellipsoidal or spherical surfaces, γ is independent on k_z and k'_z, so that (10.63) is in fact exact [10]. The special cases of a sphere and an ellipsoid with axial symmetry around the direction of $\mathbf{R}$ can be easily worked out.

Exercise 10.12

Prove that if I is constant and if the constant-energy surface is an ellipsoid of revolution with symmetry axis parallel to $\mathbf{R}$, so that

$$\varepsilon = \frac{\hbar^2}{2}\left[\frac{(k_x^2 + k_y^2)}{m_\parallel^*} + \frac{k_z^2}{m_\perp^*}\right] \tag{10.64}$$

one gets

$$\Gamma = \frac{4\pi^2 I}{\hbar^2} m_\perp^* m_\perp^{*\prime} \tag{10.65}$$

The case of saddle points can be analyzed by similar methods. Following Roth *et al.* [10] we can write an unique expression for all kinds of extreme points, namely bulges, dimples and saddle points, as

$$\Phi(R)_{nn'} = -\sum_{jl}$$

$$\frac{e^{i(\phi_j - \phi_l)}}{2R^2} \int\int \mathrm{d}\varepsilon_n \mathrm{d}\varepsilon_{n'} \Gamma_{jl} e^{iR(k_z^j - k_z^l)} \frac{f(\varepsilon_n) - f(\varepsilon_{n'})}{\varepsilon_n - \varepsilon_{n'}} \tag{10.66}$$

Here the expression for Γ has the general form

$$\Gamma_{jl}(R, \varepsilon_n, \varepsilon_{n'}) = \gamma_{jl}(\varepsilon_n, \varepsilon_{n'}) - \frac{\gamma_{jl}^{(1)}(\varepsilon_n, \varepsilon_{n'})}{iR}$$

$$+ \frac{\gamma_{jl}^{(2)}(\varepsilon_n, \varepsilon_{n'})}{(iR)^2} + \cdots \tag{10.67}$$

When the surface of constant energy has bulges, dimples and saddle points, as in the one shown in Fig. 10.2, the integral by parts we must perform to obtain γ has to be broken into the contributions from the different extrema, and the sum over k_i, k_j' contains all these separate contributions [10]. The last factor in (10.66) involves the Fermi functions in an expression which has a maximum when both band energies are near the Fermi level, so let us assume that most of the contribution to the double integral comes from that region. Then we extend the energy integrals to $\pm\infty$. This approximation is quite reasonable even at high temperatures, because the bandwidths are typically several eV, while the energy interval where the Fermi function varies appreciably is of the order of $k_B T$ around the Fermi energy E_F. The integrals can be performed by closing the contours in the complex plane. This requires the analytic continuation of k_j and k_l as functions of ε or ε' to complex values of the energies. In practice, though formally extending the integrals to infinity, one only needs the properties of these functions near the real axis. Let us assume that $k(\varepsilon)$ is analytic, so that we expand it in a series around the real value ε_n^r for complex $\varepsilon_n = \varepsilon_n^r + i\varepsilon_n^i$:

$$k_z^j(\varepsilon_n) = k_z^j(\varepsilon_n^r) + \left(\frac{\mathrm{d}k_z^j(\varepsilon_n)}{\mathrm{d}\varepsilon_n}\right)_{\varepsilon_n^r} \cdot i\varepsilon_n^i + \cdots \tag{10.68}$$

The sign of the z component of the electron group velocity

$$v_z = \left(\frac{d\varepsilon_n}{dk_z^{\bar{\jmath}}}\right)_{\varepsilon_n^r}$$

at the extremum, determines whether the contour must be closed on the upper or lower half- plane. Analytic continuation of the dispersion relation of elementary excitations is a standard problem in solid state physics, as in the calculation of surface or interface states [11], the localization properties of Wannier functions [12], *etc.* Let us consider the integral over ε_n. The function

$$g(\varepsilon_n) = \frac{f(\varepsilon_n) - f(\varepsilon_{n'})}{\varepsilon_n - \varepsilon_{n'}} \tag{10.69}$$

is regular, at finite temperature, for $\varepsilon_n \to \varepsilon_{n'}$. We can close the contour in the half plane in which the integrand converges exponentially at infinity, that is, for the ε integral, that where

$$v_z^{nj} \cdot \mathcal{I}m(\varepsilon_n) > 0 \ .$$

The poles which contribute with finite residues to the Cauchy integral are $\varepsilon_n = E_F + i\omega_{m_j}$, where, if n is the band index , j denotes one particular extremum of that band, and $\sigma_{nj} = \text{sign}(v_z^{nj})$, we have

$$\omega_{m_j} = (2m_j + 1)\pi k_B T \sigma_{nj} \ , \tag{10.70}$$

Therefore, calling $\beta = 1/k_B T$, we obtain the following expression for (10.66):

$$\Phi(R)_{nn'} = -\frac{2\pi i}{2R^2\beta} \sum_{lj} \sum_{m_j=0}^{\infty} \sigma_{nj} e^{i(\phi_j - \phi_l)} \times$$

$$\oint \frac{d\varepsilon_{n'} e^{-iRk_z''}}{E_F + i\omega_{m_j} - \varepsilon_{n'}} \left[\Gamma_{jl}(R, \varepsilon_n, \varepsilon_{n'}) e^{iRk_z^j} \right]_{\varepsilon_n = E_F + i\omega_{m_j}} \tag{10.71}$$

Due to the minus sign of the exponent in the factor $e^{-iRk_z''}$ the integral over $\varepsilon_{n'}$ converges on a contour for which $v_z v_z' < 0$ Therefore, there will be no contribution from residues to the integral unless the condition

$$\frac{\partial k_z^j}{\partial \varepsilon} \cdot \frac{\partial k_z''}{\partial \varepsilon'} < 0 \tag{10.72}$$

is satisfied. The conclusion is that a given pair (j, l) of extrema in k space contributes to the range function only if the z components of their group velocities have opposite sign.

The resultant expression for the range function is:

$$\Phi(R))_{nn'} = \frac{2\pi i}{2R^2\beta} \sum_{jl}{}' \sum_{m=0}^{\infty}$$

$$\left[e^{i(\phi_j - \phi_l)} e^{iR(k_z^j - k_z'')} \Gamma_{jl}(R, \varepsilon_n, \varepsilon_{n'}) \right]_{\varepsilon_n = \varepsilon_{n'} = E_F + i\omega_{m_j}} \tag{10.73}$$

where the prime in the summation means that it is limited to the case $v_z^j \cdot v_z^l < 0$. We expand now the term in square brackets in (10.73) around E_F, assuming that Γ can be approximated by a constant. The expansion of the exponent yields:

$$
\begin{aligned}
\left(k_z^j - k_z^l\right)_{\varepsilon_n = \varepsilon_{n'} = E_F + i\omega_{mj}} = & \left(k_z^j - k_z'^l\right) + i\omega_{mj}\left(\frac{\partial k_{zj}}{\partial \varepsilon_n} - \frac{\partial k_z'^l}{\partial \varepsilon_{n'}}\right)_{E_F} \\
& + \frac{1}{2}(i\omega_{mj})^2\left(\frac{\partial^2 k_z^j}{\partial \varepsilon_n^2} - \frac{\partial^2 k_z'^l}{\partial \varepsilon_{n'}^2}\right)_{E_F} + \cdots \quad (10.74)
\end{aligned}
$$

We assume that terms involving the second derivatives of k_z above are small, and neglect higher order derivatives. Note that the pairs $\mathbf{k_j}$, $-\mathbf{k_j}$ as well as $\phi_j, -\phi_j$ etc. appear always associated in the sum over j, l in (10.72), so that the result of (10.73) is real. Pairs of points like $(\mathbf{k_j}, -\mathbf{k_j})$ are called *calipers* because $2 \mid k_j^z \mid$ is a diameter of the Fermi surface in the direction of $\mathbf{R}$. After performing the summation over m in (10.73), we arrive at [9]:

$$
\begin{aligned}
\Phi_{n,n'} = & \sum_{j,l}' \frac{I_{n,n'}\left(k_j(\varepsilon_n), k_l(\varepsilon_{n'})_{E_F}\right) m_j^*(E_F) m_l^*(E_F)}{16\pi^3 \hbar^4 R^3} \\
& \times \left(\cos\psi(j,l) + \frac{b_{jl}}{R}\sin\psi(j,l)\frac{\partial^2}{\partial a_{jl}^2}\right)\frac{\pi R k_B T}{\sinh\left(a_{jl}\pi R k_B T\right)} \quad (10.75)
\end{aligned}
$$

where

$$
\psi(j, i) \equiv R \cdot \{(k_j(\varepsilon_n t)_{E_F} - k_l(\varepsilon_{n'})_{E_F}\} + \phi_j - \phi_l \quad (10.76)
$$

$$
a_{jl} \equiv \left|\left(\frac{\partial k_z^j(\varepsilon_n)}{\partial \varepsilon_n}\right)_{E_F} - \left(\frac{\partial k_z^l(\varepsilon_{n'})}{\partial \varepsilon_{n'}}\right)_{E_F}\right| \quad (10.77)
$$

and

$$
b_{jl} \equiv \frac{1}{2}\left|\left(\frac{\partial^2 k_z^j(\varepsilon_n)}{\partial \varepsilon_n^2}\right)_{E_F} - \left(\frac{\partial^2 k_z^l(\varepsilon_{n'})}{\partial \varepsilon_{n'}^2}\right)_{E_F}\right| \quad (10.78)
$$

The summation in (10.75) extends over all *calipering* pairs of points on the Fermi surface, and also on spin.

Exercise 10.13
Prove (10.75).

The temperature dependent term on the r.h.s. of (10.75) yields an exponentially decreasing amplitude at large R. At low temperatures the range cutoff

$$
R_c = \frac{\hbar v_F^z}{k_B T}
$$

can be very much larger than the mean free path due to electron scattering off impurities or defects [1]. In concrete cases however the geometry of the Fermi surface can produce small values of the z component of the Fermi velocity at some particular extrema, yielding shorter cutoff lengths.

For certain directions of **R** there might be no calipering points contributing to (10.75). This happens in particular when along that direction the Fermi surface touches the boundary of the Brillouin zone. In this case, due to the simultaneous effect of time reversal and translation symmetry, the constant energy surface must be orthogonal to the boundary and the present method of evaluation of the range function does not apply [10].

We remark that the range function is in general multi-periodic in R, with wavelengths determined by the various calipers of the Fermi surface for that direction.

We can verify that Eq. (10.75) reduces to the simple RKKY result at low T for a spherical FS. In this case

$$\varepsilon(k) = \frac{\hbar^2 k^2}{2m^*} \ .$$

In the $T \to 0$ limit we find the original RKKY expression:

$$\Phi(R) = -\frac{m^* k_F^4 I(\mathbf{k_F}, -\mathbf{k_F})}{\pi^3 \hbar^2} \left(\frac{x \cos x - \sin x}{x^4} \right)$$

where we included a factor 2 for spin and a second factor 2 due to the fact that the two equivalent caliper pairs $(\mathbf{k}_F, -\mathbf{k}_F)$, $(-\mathbf{k}_F, \mathbf{k}_F)$ must be considered.

Exercise 10.14
Prove the above result and verify that the higher order terms in powers of $1/R$ exactly cancel each other in the limit $T \to 0$.

10.5   Semiconductors

There are several different cases to consider, depending on the temperature and the doping level. In the case of intrinsic semiconductors, that is those with a negligible concentration of electronically active impurities, we must consider separately the limits of high and low temperature. Here, "high" or "low' refers to the value of the ratio $k_B T/E_g$ where E_g is the gap at $T = 0$ between the uppermost occupied (valence) band and the lowest unoccupied (conduction) band.

10.5.1   Intrinsic semiconductors, high T

At $T > 0$ there is a finite concentration of electrons in the conduction band and holes in the valence band. The carriers, within the effective mass approximation, can be described as two Maxwell-Boltzmann gases, one of electrons and one of holes. We simplify the problem assuming that the dominant contribution is due to the conduction band. This would be the case if $m_e^* \ll m_h^*$. Let us then calculate the contribution to $\Phi(R)$ from a spherical band with effective mass m^* at high T. The high T form of the Fermi-Dirac distribution for electrons in the conduction band is [13]

$$f(\varepsilon - E_c) \simeq \frac{nh^3}{2(2\pi m^* k_B T)^{3/2}} e^{-(\varepsilon - E_c)/k_B T} \tag{10.79}$$

where E_c is the bottom of the conduction band. We can apply equation (10.66).

Let us define the new variables:
$kR = x$, $k'R = y$, $b^2 = \hbar^2/(2m^* k_B T)$. The exponential Boltzmann factor allows us to extend the integral over energy in the conduction band up to infinity. Then (10.66) becomes

$$\Phi(R) = -\frac{n\hbar I}{4\pi^4 (2\pi m^* k_B T)^{3/2}} \times$$

$$\int_{-\infty}^{\infty} \int_{-\infty}^{\infty} \frac{\left(e^{-b^2 x^2} - e^{-b^2 y^2}\right)}{x^2 - y^2} e^{i(x-y)} xy \, dx \, dy \tag{10.80}$$

The integral in (10.80) may be performed by contour integration followed by the use of the definite integral

$$\int_0^{\infty} e^{-b^2 t^2} t \sin t \, dt = \frac{\sqrt{\pi}}{2b^3} e^{-1/b^2} \tag{10.81}$$

Exercise 10.15
Prove that the final result is

$$\Phi(R) = \frac{nm^* I}{8\pi \hbar^2 R} e^{-(R/R_0)^2} \tag{10.82}$$

where the screening length R_0 is

$$R_0 = \sqrt{\left(\frac{\hbar^2}{2m^* k_B T}\right)}. \tag{10.83}$$

10.5.2   Intrinsic semiconductors, low T

In this case the indirect interactions are due to electron excitation from a filled spherical valence band of effective mass m_h^* to an empty spherical conduction band with effective mass m_e^*, separated by the energy gap E_g. We start from (10.66), assume as usual that I is independent on $\mathbf{k}, \mathbf{k}'$, and define new variables

$$x = k_e \sqrt{\frac{\hbar^2}{2m_e^* E_g}} \ , \quad y = k_e \sqrt{\frac{\hbar^2}{2m_h^* E_g}} \ .$$

and the parameters

$$a_1 = R\sqrt{\frac{2m_e^* E_g}{\hbar^2}} \ , \ a_2 = R\sqrt{\frac{2m_h^* E_g}{\hbar^2}} \ .$$

Now the expression for Φ becomes:

$$\Phi = -C \int_{-\infty}^{\infty} \int_{-\infty}^{\infty} \frac{e^{i(a_1 x - a_2 y)}}{1 + x^2 + y^2} xy \, dx \, dy \tag{10.84}$$

where

$$C = \frac{2I E_g m_e^* m_h^*}{h^4 R^2}$$

Exercise 10.16
Verify (10.84).

We perform an orthogonal coordinate transformation in the plane x, y, such that the new axis x' is in the direction of $a_1 x - a_2 y$, by defining:

$$(a_1^2 + a_2^2)^{1/2} x' = a_1 x - a_2 y$$
$$(a_1^2 + a_2^2)^{1/2} y' = a_1 x + a_2 y \tag{10.85}$$

Now one can perform the integral over x' by contour integration and obtain:

$$\Phi = \frac{4\pi a_1 a_2 C}{\eta^2} \int_0^{\infty} \frac{e^{-\eta\sqrt{1+y'^2}}}{\sqrt{1+y'^2}} \cdot (1 + 2y'^2) \, dy' \tag{10.86}$$

where $\eta = (a_1^2 + a_2^2)^{1/2}$ is proportional to R, so that in the large R limit we can evaluate the integral approximately, obtaining:

$$\Phi = -\frac{2I E_g^2}{\hbar^6} \frac{(m_e^* m_h^*)^{3/2}}{2\pi^{5/2}} \frac{1}{(k_0 R)^{5/2}} e^{-k_0 R} \tag{10.87}$$

where the inverse screening length is

$$k_0 = \sqrt{\frac{2E_g(m_e^* + m_h^*)}{\hbar^2}} .$$

Exercise 10.17
Prove (10.87).

Semiconductors are therefore capable of sustaining indirect exchange interactions among magnetic impurities. However, in distinction with the case of metals, the results above predict an effective interaction which does not oscillate in space. Detailed calculations by Narita [14] -which also include the case of metallic systems with a saddle point Fermi surface- generalize previous calculations [15] and show that the sign of the interaction depends on the band parameters. Experimental results for systems called magnetic "trilayers" (see next section), [17] in which two ferromagnetic layers are deposited on a convenient substrate with a non-magnetic layer in between, called "spacer", reveal an effective coupling among the magnetic layers whose sign depends upon the spacer width, even in cases where the latter is a semiconductor. Within the simple model discussed above this is not possible when the Fermi level in the semiconductor lies inside the gap.

10.5.3 Degenerate semiconductors

For very highly doped semiconductors at low temperatures one may find the Fermi level inside a partially occupied band and, at low enough temperatures the semiconductor transforms into a low conductivity metal with either kind of carriers. This happens in covalent semiconductors, typically Ge and Si, when the doping impurity concentration is large enough that a reasonable statistical fraction of their wave fuctions start to overlap and generate a band of states with energies contained in the gap [18]. At low enough temperatures (in general about a couple of K) the electrons (holes) for $n(p)$ type systems become a degenerate Fermi liquid.These systems are completely disordered, beause the impurities occupy random substitutional sites in the host lattice. Therefore, the electronic impurity states are solutions of Schrödinger's equation with a random potential, and the absence of translational symmetry precludes the application of Bloch's theorem. The k vector is no longer a "good quantum number" and accordingly the concept of Fermi surface simply does not occur in this context. Neglecting Coulomb interactions between the carriers it is possible to obtain a theoretical description of the impurity bands which leads to a range function for the indirect exchange interaction among eventual diluted magnetic impurities that oscillates in space [19], in spite of the absence of a Fermi surface. One also finds an oscillating range funcion in a metallic disordered alloy [20, 21].

Finally, let us mention that in mixed finite clusters of a transition metal surrounded by noble metal atoms, long-range magnetic polarization oscillations are also found, both theoretically [27] and experimentally [28].

10.6 Magnetic multilayer systems

These are systems produced artificially in the laboratory in which ferromagnetic and non-magnetic layers alternate along the direction perpendicular to the layer planes. They have been the object of intense research since the observation by Grünberg et al. [17] that Fe films separated by a Cr spacer coupled antiferromagnetically. Upon varying the width of the spacer, Parkin [22] et al. discovered that the interlayer effective interaction in $Fe/Cr/Fe$ and $Co/Ru/Co$ multilayers changes alternatively from FM to AFM when the spacer thickness varies. The layers, either of magnetic or non-magnetic elements, can contain from just a few to a hundred atomic layers. The oscillations are found with any non-magnetic transition metal as spacer. The oscillations, which in general are multi-periodic, are in many cases consistent with the generalized RKKY theory presented above, and in principle depend mainly on the properties of the band structure of the spacer metal. Further study has shown that there is an alternative theory which describes these oscillations as originating in the confining effect of the spin-dependent potential barriers at the interface region between both materials. In the FM alignment in which all majority-bands are below the Fermi surface (complete ferromagnetism), electrons of this spin with energy E_F are confined to the non-magnetic spacer. This is accordingly known as the "quantum-well" theory of interlayer coupling [23]. It has been found that it is possible to obtain both theories as limiting cases of a general approach in which a total energy calculation of a multilayer is performed based on Green's functions methods [24]. We refer the interested reader to some excellent review articles [25, 26].

References

1. De Gennes, P. G. (1962) *J. Phys. Radium* **23**, 510; 630.

2. Elliott, R. J. (1965) *Magnetism*, eds. Rado, G. T. and Suhl, H. H. **2a**, Academic Press, New York, p. 355.

3. Kittel, C. (1971) *" Introduction to Solid State Physics"*, *Chap. 15*, John Wiley & Sons, Inc., New York.

4. Ruderman, M. A. and Kittel, C. (1954) *Phys. Rev.* **96**, 99.

5. Yosida, K. (1957) *Phys. Rev.* **106**, 893.

6. Specht, Frederick (1967) *Phys. Rev.* **162**, 389.

7. Kasuya, T. (1956) *Prog. Theoret. Phys.* **16**), 45.

8. Kaplan, T. A. (1961) *Phys. Rev.* **124**, 329.

9. Zeiger, H. J. and Pratt, G. W. (1973) *"Magnetic interactions in solids"*, Clarendon, Oxford.

10. Roth, Laura M., Zeiger, H. J. and Kaplan, T. A. (1966) *Phys. Rev.* **149**, 519.

11. Chaves, C. M., Majlis, N. and Cardona, M. (1966) *Sol. State. Comm.* **4**, 631.

12. Blount, E. I. (1962) *"Solid State Physics, Adv. in Res. and Appl."* eds. F. Seitz and D. Turnbull, Vol. 13, pp. 306-70.

13. Ziman, J. M., (1964) *"Principles of the Theory of Solids"*, Cambridge at the University Press.

14. Narita, Akira (1986) *J. Phys.* **C19**, 4797.

15. Baltensperger, W. and de Graaf, A. M. (1960) *Helv. Phys. Acta* **33**,881.

16. Bloembergen, N. and Rowland, T. J. (1955) *Phys. Rev.* **97**, 1679.

17. Grünberg, P., schreiber, R., Pang, Y., Brodsky, M. B. and Sower, H., (1986) *Phys. Rev. Lett.* **57**, 2442.

18. Majlis, N. (1967) *Proc. Phys. Soc. London* **90**, 811.

19. Ochi, Carmen L. C.. and Majlis, N. (1995) *Phys. Rev.* **B51**, 14221.

20. Bulaevskii, L. N. and Pannyukov, S. V. (1986) *JETP Lett.* **43**, 190.

21. Bergmann, Gerd (1987) *Phys. Rev.* **B36**, 2649.

22. Parkin, S. S. P., More, N. and Roche, K. P. (1990) *Phys. Rev. Lett.* **64**, 2304.

23. Edwards, D. M., Mathon, J., Phan, M. S. and Muniz, R. B. (1991) *Phys. Rev. Lett.* **67**, 493.

24. d' Albuquerque e Castro, J., Ferreira, M. S. and Muniz, R. B. (1994) *Phys. Rev.* **B 49**, 16062.

25. Bruno, P. (1995) *Phys. Rev.* **B52**, 411.

26. Fert, A. and Bruno, P. (1992) *"Ultrathin Magnetic Structures"*, eds. Heinrich, B. and Bland, A., Springer–Verlag.

27. Guevara, Javier, Llois, Ana Maria and Weissmann, Mariana (1998) *Phys. Rev. Lett.* **81**, 5306.

28. Guevara, Javier and Eastham, D. A. (1997) *J. Phys.-Condens. Matter* **9**, L497.

Chapter 11

Local Moments

11.1 The s-d and Anderson Hamiltonians

In this chapter we shall study the process of formation of local magnetic moments and the effect of substitutional magnetic impurities in a host non-magnetic metal.

The basic models which have been the basis of the theoretical study of these problems are the *s-d* and the *Anderson model*. The *s-d* exchange interaction was already introduced in Chap. 10, where we derived the form of the long-range RKKY interaction among spins localized on different sites in a non-magnetic metal, like Mn in Cu.

However, when iron-group elements are introduced into a non-magnetic metal, it is not always the case that they display a permanent magnetic moment. Fe and Mn in Cu maintain their spins, while Mn in Al does not. For the first case, we may use the *s-d* exchange interaction model of Chap. 10. This interaction is the main mechanism of the Kondo effect, that we shall discuss further on in the second part of this chapter.

In the second case, one must understand why there is no magnetic moment on the transition ion impurity. In order to study the conditions for an impurity to sustain a permanent magnetic moment we require an approach departing from a more basic standpoint, such that both cases can be explained with the same model.

11.2 Anderson model

To this end *Anderson* proposed the following Hamiltonian:[1]

$$H = \sum_{\mathbf{k}\sigma} \epsilon_{\mathbf{k}} a^{\dagger}_{\mathbf{k}\sigma} a_{\mathbf{k}\sigma} + E_d \sum_{\sigma} a^{\dagger}_{d\sigma} a_{d\sigma}$$

$$+ \sum_{\mathbf{k}\sigma} V_{\mathbf{k}d} a_{\mathbf{k}\sigma}^{\dagger} a_{d\sigma} + V_{d\mathbf{k}} a_{d\sigma}^{\dagger} a_{\mathbf{k}\sigma} + U a_{d\uparrow}^{\dagger} a_{d\uparrow} a_{d\downarrow}^{\dagger} a_{d\downarrow} \qquad (11.1)$$

where E_d, assumed to lie below E_F, is the energy level of the electron bound to the transition metal ion whose location one may choose as the origin of the coordinate system. The conduction electrons are taken as mutually independent, while Hubbard repulsion acts on the impurity bound electrons. Degeneracy of the d states is also neglected. The parameters above are:

$$U = \int d\,\mathbf{r}_1 d\,\mathbf{r}_2 \, |\phi_d(\mathbf{r}_1)|^2 \, |\phi_d(\mathbf{r}_2)|^2 \, \frac{e^2}{r_{12}} \qquad (11.2)$$

$$V_{\mathbf{k}d} = \frac{1}{\sqrt{N}} \int d^3\mathbf{r} \sum_{\mathbf{n}} \phi_d(\mathbf{r}) V_{ion}(\mathbf{r}) W_{\mathbf{k}}(\mathbf{r} - \mathbf{R_n}) e^{i\mathbf{k}\cdot\mathbf{R_n}} \qquad (11.3)$$

with $W_{\mathbf{k}}(\mathbf{r} - \mathbf{R_n})$ = Wannier function, centered at site $\mathbf{n}$, for a conduction electron.

11.3   Hartree–Fock solution of Anderson Hamiltonian

We can define four retarded Green's functions for this problem:

$$G_{k\,k'}^{\sigma} \equiv \langle\langle a_{k\sigma}(t) \,|\, a_{k'\sigma}^{\dagger}(0) \rangle\rangle$$
$$G_{k\,d}^{\sigma} \equiv \langle\langle a_{k\sigma}(t) \,|\, a_{d}^{\dagger}(0) \rangle\rangle$$
$$G_{d\,d}^{\sigma} \equiv \langle\langle a_{d\sigma}(t) \,|\, a_{d\sigma}^{\dagger}(0) \rangle\rangle$$
$$G_{d\,k}^{\sigma} \equiv \langle\langle a_{d\sigma}(t) \,|\, a_{k\sigma}^{\dagger}(0) \rangle\rangle \qquad (11.4)$$

Observe that $G_{k,k'}$ is *not* diagonal in the quasi-momentum variables, because the presence of an impurity at a given site breaks the translation symmetry of the Hamiltonian.

We repeat now for the Hamiltonian defined in Eq. (11.1) the procedure followed in Chapters 6 and 9 to calculate the retarded Green's functions:

we write down the equations of motion for all Green's functions, perform the time Fourier transform to the frequency domain, and whenever a product of 4-fermion operators appears we substitute in all possible ways binary products of a creation times an annihilation operator by their statistical average, to be calculated self-consistently afterwards.

Exercise 11.1
Verify that the result is the following system of algebraic linear equations:

$$(\omega - E_{\sigma}) G_{d\,d}^{\sigma} - \sum_{k} V_{d\,k} G_{k\,d}^{\sigma} \; = \; 1$$
$$(\omega - \epsilon_k) G_{k,d}^{\sigma} - V_{k\,d} G_{d\,d}^{\sigma} \; = \; 1$$

$$(\omega - E_\sigma)G^\sigma_{dk} - \sum_{k'} V_{dk'}\, G^\sigma_d \;=\; 1$$

$$(\omega - E_\sigma)G^\sigma_{kk'} - V_{kd}\, G^\sigma_{dk'} \;=\; 1 \tag{11.5}$$

where we defined

$$E_\sigma = E_d + U\langle n_{d-\sigma}\rangle \tag{11.6}$$

We introduce now the auxiliary function

$$S_k(\omega) = \sum_{k''} V_{dk''}\, G_{k''k}(\omega + i\eta) \tag{11.7}$$

and we find:

$$S_k(\omega) = \frac{V_{dk}\, g_k(\omega)}{1 - g_d(\omega) \sum_{k''} g_{k''}(\omega)\, V_{k''d}} \tag{11.8}$$

where the unperturbed Green's functions are

$$g_k(\omega) \;=\; \frac{1}{\omega - \epsilon_k}$$

$$g_d(\omega) \;=\; \frac{1}{\omega - E_d} \tag{11.9}$$

Finally, we obtain:

$$G^\sigma_{dd} = (\omega - E_\sigma - \Sigma(\omega + i\eta))^{-1}$$

$$G^\sigma_{kk'}(\omega) = g_k(\omega)\,\delta_{kk'} + \frac{g_k(\omega)\, V_{kd}\, V_{dk'}\, g_{k'}(\omega)}{1 - g_d(\omega)\,\Sigma(\omega)} \tag{11.10}$$

where $\Sigma(\omega)$, the local state self-energy, is

$$\Sigma(\omega) = \sum_{k''} V_{dk''} g_{k''}(\omega) V_{k''d} \equiv \Delta(\omega) + i\Gamma(\omega) \tag{11.11}$$

The imaginary part of Σ is

$$\Gamma(\omega) \;\equiv\; \sum_k \frac{|V_{dk}|^2}{\omega - \epsilon_k + i\eta}$$

$$= -i\pi \int d\epsilon_k \oint_{\epsilon_k = \omega} \frac{dS_k\, |V_{dk}|^2}{|\nabla_{\mathbf{k}}\, \epsilon_k|}$$

$$\equiv \rho(\epsilon_k)\overline{V^2}(\epsilon_k) \tag{11.12}$$

where $\overline{V^2}(\epsilon_k)$ is the average of $|V_{dk}|^2$ over the surface $\epsilon_k = \omega$ and $\rho(\epsilon_k)$ is the unperturbed density of states of the conduction electrons.
The real part of the local self-energy is

$$\Delta(\omega) = P \int d\epsilon_k \rho(\epsilon_k) \frac{\overline{V^2(\epsilon_k)}}{\omega - \epsilon_k} \tag{11.13}$$

Let us suppose that $\overline{V^2(\epsilon_k)}$ is only appreciable in an interval of energies over which $\rho(E)$ is almost constant. Then the principal value integral is very small, and we can neglect the displacement Δ of the real part of the energy. As a further approximation we take $\rho(E)$ at the fermi energy. Now we get the local density of states as

$$\rho_{d\sigma}(\omega) = -\frac{1}{\pi} \mathcal{I}m \, G_{dd}^{\sigma} = \frac{1}{\pi} \frac{\Gamma}{(\omega - E_\sigma)^2 + \Gamma^2} \tag{11.14}$$

and we can immediately calculate the average number of d electrons at $T = 0$ for a given spin, which enters into the definition (11.6) of the effective local state binding energy E_σ. We obtain a pair of self-consistent equations to solve, for $\langle n_{d\uparrow} \rangle$ and $\langle n_{d\downarrow} \rangle$:

$$\langle n_{d\uparrow} \rangle = \int_{-\infty}^{E_F} \rho d\sigma(\omega) d\omega = \frac{1}{\pi} \cot^{-1} \left(\frac{E_d + U\langle n_{d\downarrow} \rangle - E_F}{\Gamma} \right)$$

$$\langle n_{d\downarrow} \rangle = \int_{-\infty}^{E_F} \rho d\sigma(\omega) d\omega = \frac{1}{\pi} \cot^{-1} \left(\frac{E_d + U\langle n_{d\uparrow} \rangle - E_F}{\Gamma} \right) \tag{11.15}$$

Clearly, they have always a non-magnetic solution

$$\langle n_{d\uparrow} \rangle = \langle n_{d\downarrow} \rangle = n_d \, .$$

This is certainly the case for $U/\Gamma = 0$, because for this situation $E_\uparrow = E_\downarrow$ and both equations are identical. Therefore one suspects that magnetic solutions only exist when this ratio is greater than some minimum value. Let us simplify the notation. We define $n_1 \equiv \langle n_{d\uparrow} \rangle$, $n_2 \equiv \langle n_{d\downarrow} \rangle$. It is clear from the structure of the pair of equations (11.15) that the permutation of the spin orientation $n_1 \leftrightarrow n_2$ is a symmetry operation. In the plane (n_1, n_2) this operation is the mirror reflection of the curves $n_2 = F(n_1)$, $n_1 = F(n_2)$ on the line $n_1 = n_2$. In the absence of a magnetic solution both curves coincide, and the common curve must be normal to the line $n_1 = n_2$ at the point where they intersect, or , in other words, the tangent of the curve is a line $n_1 + n_2 = const$. In order that there be a magnetic solution within the physical domain, which is the square $0 \leq n_i \leq 1$, both curves must not coincide, and therefore they must have at least three intersections: one, on the mirror line, one for $n_1 > n_2$ and a symmetric one for $n_2 > n_1$. This implies that at the intersection of both curves with the mirror line, the derivative of, say, $n_1 = F(n_2)$, must be > 1 in absolute value, which implies

$$|F'(n_2)|_{n_1=n_2} > 1 \tag{11.16}$$

that is

$$\frac{U}{\Delta} \geq \frac{\pi}{\sin^2 \pi n_d} \tag{11.17}$$

where we substituted $n_2 = n_d$ to obtain the limiting case in which the magnetic solution just appears. Comparing the r. h. s. of Eq. (11.14) with that of the inequality above and using Eq. (11.15), we can rewrite (11.17) as

$$\rho_d(E_F)U \geq 1 \tag{11.18}$$

where $\rho_d(E_F)$ is the density of states of d electrons at the Fermi level for the non-magnetic solution with the same parameters. We see that this condition has a familiar form.

One suspects that the zero-field static susceptibility would diverge when the equality applies in (11.18). Let us verify this. Under an infinitesimal applied magnetic field δB the Hartree–Fock energy of the local state changes by

$$\delta E_\sigma = -\frac{g}{2}\sigma\mu_B\,\delta B + U\delta\langle n_{d-\sigma}\rangle \tag{11.19}$$

The local susceptibility of the d state is

$$\chi_d = -\frac{g\,\mu_B}{2}\frac{(\delta n_\uparrow - \delta n_\downarrow)}{\delta B} \tag{11.20}$$

Exercise 11.2
Show that

$$\chi_d = \frac{1}{2}\frac{g^2\mu_B^2\rho_d(E_F)}{1 - \rho_d(E_F)U} \tag{11.21}$$

which behaves as expected.

It is possible to approach the study of Anderson's Hamiltonian with perturbation methods. The two cases in which this is reasonable are either the large U, small limit, or the inverse one with large V_{kd}, small U. For the iron group U is estimated to be large. In this limit, assuming (U, E_d) large compared to V_{kd}, second order perturbation in the exchange interaction leads to the *s-d* model hamiltonian, which is the basis of the *Kondo effect* [2] The complete Hamiltonian for this case is the sum of the band Hamiltonian with independent electrons and the exchange H_{sd} Hamiltonian of Eq. (10.10):

$$H = \sum_{k\sigma} \epsilon_{k\sigma}\hat{n}_{k\sigma} + H_{sd} \tag{11.22}$$

which we used to study the case where there are permanent magnetic moments on substitutional impurities in the metal lattice.

11.4 Kondo effect

According to the results of the previous section, a non-zero magnetic moment requires, within the Hartree–Fock approximation, that inequality (11.18) be satisfied. As mentioned before, this case leads to the *Kondo effect*, which was discovered while looking for the explanation of the behaviour of the resistance of an alloy like $Cu\text{-}Mn$ as a function of temperature.

The electrical resistance $R(T)$ of normal metals at large T is dominated by the electron-phonon interaction [3]. At high $T > \Theta_D$ it grows linearly with T. For $T \ll \Theta_D$ it decreases as T^5. Therefore, as $T \to 0$, the phonon contribution to electron scattering becomes negligible. Due to the unavoidable presence of impurities which are responsible for elastic scattering of conduction electrons,

$$\lim_{T \to 0} R(T) \to \text{constant} \times \text{impurity concentration} \equiv R_0 \, ,$$

which is called the *residual resistance*.

Experiments performed in the 30's had already shown that the electrical resistance $R(T)$ of some alloys of a noble metal with a small ($< 0.1\%$) atomic concentration of a magnetic element has a minimum at low $T \approx 10K$. [4, 5]

The explanation of this minimum was found by Kondo in 1964.[6]. He calculated the scattering amplitude for conduction electrons by a localized spin S due to the exchange interaction of Eq. (10.10), and he discovered that in the second order of the Born perturbation series, the scattering amplitude displays a singular behaviour which leads to a $\ln T$ low temperature dependence of the resistance. For an AF exchange ($J < 0$) this term increases as T decreases, while the phonon contribution decreases, as we have just mentioned, so the competition between these opposite behaviours explains the minimum of $R(T)$. At very low T the logarithmic term would however diverge as $T \to 0$, while it was found experimentally that the resistance deviates from its $T = 0$ value by a term in T^2, so that Kondo's calculation had to be modified. A detailed survey, though, of the evolution of the *"Kondo problem"* is outside the scope of this book. I refer the interested reader to the excelent monograph by Hewson [7].

11.4.1   Calculation of resistivity

We consider the simple case in which the conduction band electrons are well described as almost free, with kinetic energy

$$\epsilon_k = \frac{\hbar^2 k^2}{2m^*}$$

where m^*=effective mass. In the following we shall not write the asterisc on the effective mass to simplify the notation. In the steady state, under the application of a constant uniform electric field F,the electron distribution function in $f_\sigma(\mathbf{k})$ satisfies the Boltzmann equation:[8]

$$\left(\frac{\partial f_\sigma(\mathbf{k})}{\partial t}\right)_{field} + \left(\frac{\partial f_\sigma(\mathbf{k})}{\partial t}\right)_{scatt} = 0 \qquad (11.23)$$

The field time-derivative is

$$\left(\frac{\partial f_\sigma(\mathbf{k})}{\partial t}\right)_{field} = \frac{\partial f_\sigma(\mathbf{k})}{\partial \epsilon_k} \nabla_{\mathbf{k}} \epsilon_k \cdot \frac{d\mathbf{k}}{dt} \qquad (11.24)$$

The time-derivative of the quasi-momentum is:[9]

$$\frac{d\mathbf{k}}{dt} = (e/\hbar)\mathbf{F} \qquad (11.25)$$

so that we have:

$$\left(\frac{\partial f_\sigma(\mathbf{k})}{\partial t}\right)_{field} = \frac{\partial f_\sigma(\mathbf{k})}{\partial \epsilon_k} \nabla_{\mathbf{k}}\epsilon_k \cdot (e/\hbar)\mathbf{F} \qquad (11.26)$$

where $e =$ electron charge.

Let us now obtain the scattering term. Collisions with impurities are responsible for a finite life-time of an electron state. We consider isotropic scattering, so that the life-time τ_c depends only on the electron energy. In this case, a deviation from the equilibrium distribution $f_0(\mathbf{k})$ at the fixed temperature decays according to

$$\left(\frac{\partial f_\sigma(\mathbf{k})}{\partial t}\right)_{scatt} = -\frac{f - f_0}{\tau_c} \qquad (11.27)$$

where $\tau_c(\epsilon_k)$ is the *collision time*, and $(\tau_c(\epsilon_k))^{-1}$ is the transition probability per unit time for the electron to be scattered from state $\mathbf{k}$. We obtain therefore from Eq. (11.23):

$$f(\mathbf{k}) - f_0(\mathbf{k}) = -\frac{\partial f_0(\mathbf{k})}{\partial \epsilon_k} \nabla_{\mathbf{k}}\epsilon_k \cdot (e/\hbar)\mathbf{F}\tau_c(\epsilon_k) \qquad (11.28)$$

to first order in the field, and assuming that there is no spin dependence of the collision time. We can now calculate the electric current density in a sample of volume V with the field applied in the x direction as

$$j_x = \frac{2e}{V} \sum_k \frac{\hbar k_x}{m}(f(\mathbf{k}) - f_0(\mathbf{k})) \qquad (11.29)$$

where the factor 2 is included to take care of both spins.

Exercise 11.3
Show that the conductivity, defined as

$$\sigma = \frac{\partial j_x}{\partial F} \qquad (11.30)$$

has the form

$$\sigma = -\frac{4e^2}{V} \int d\,\epsilon_k \rho(\epsilon_k)\tau_c(\epsilon_k)\,\epsilon_k\,\frac{\partial f_0(\epsilon_k)}{\partial \epsilon_k} \qquad (11.31)$$

We remark that the last factor in Eq. (11.31) imposes that $|\epsilon_k - E_F| \leq k_B T$.

11.4.2 Calculation of the collision time

In order to complete the calculation of σ we need to obtain the collision time τ_c. We shall do this within the formal scattering theory approach, and we shall go up to the second Born approximation.

We call $\hat{H}_o$ the unperturbed Hamiltonian of the conduction band electrons in the host lattice potential and V the H_{sd} exchange interaction.

We need to calculate the transition probability per unit time $W(k, \sigma \rightarrow k', \sigma')$ from an incoming state $a_{k\uparrow}^{\dagger}|0\rangle$, where $|0\rangle$ is the sea of conduction electrons, to an outgoing state $a_{k'\uparrow}^{\dagger}|0\rangle$ or $a_{k'\downarrow}^{\dagger}|0\rangle$ which is expressed through Fermi's "*Golden Rule*" in terms of the $\hat{T}$ matrix:

$$W(k, \sigma \rightarrow k', \sigma') = \frac{2\pi}{\hbar} \sum_M w_M \, |T(\mathbf{k'}, \sigma', \, M' \, | \, \mathbf{k}, \sigma M)|^2 \delta(\epsilon_k - \epsilon_{k'}) \quad (11.32)$$

where w_M is the probability that the magnetic impurity initial state be $S^z = M$.

We remind at this point the definition of the $\hat{T}$ matrix.

The outgoing state, denoted $|\mathbf{k}\rangle^+$, satisfies the Schrödinger equation

$$|\mathbf{k}\rangle^+ = |\mathbf{k}\rangle + \frac{1}{\epsilon_k - \hat{H}_0 + i\eta} V |\mathbf{k}\rangle^+ \quad (11.33)$$

where the infinitesimal η defines the outgoing (retarded) behaviour in time. One defines the transition matrix $\hat{T}$ as

$$|\mathbf{k}\rangle^+ = \hat{T}(\epsilon_k + i\eta)|\mathbf{k}\rangle \quad (11.34)$$

We verify that $\hat{T}$ can be formally expanded in the series

$$\hat{T}(z) = V + V G_0(z) V + V G_0(z) V G_0(z) V + \cdots \quad (11.35)$$

where the *resolvent operator* $G_0(z)$ is

$$G_0(z) \equiv \frac{1}{z - \hat{H}_0} \quad (11.36)$$

Due to the scalar character of the exchange interaction the transition matrix must have the general form

$$\hat{T}(\epsilon) = \hat{t}(\epsilon) + \hat{\tau}(\epsilon)\sigma \cdot \mathbf{S} \quad (11.37)$$

where the electronic spin-flip transitions are acccounted for by the scalar product term.

There are two kinds of processes contributing to the elastic scattering off the magnetic impurity: a) spin conserving transitions and b) spin-flip transitions. For the spin-conserving case, we have:

$$W(\mathbf{k}, \uparrow M \rightarrow \mathbf{k'}, \uparrow M) = \frac{2\pi}{\hbar} \sum_M w_M \, |T(k', \uparrow M \, | \, k, \uparrow M)|^2 \delta(\epsilon_k - \epsilon'_k) \quad (11.38)$$

For the spin-flip case in the first Born approximation to T we write:

$$W(\mathbf{k} \uparrow M \rightarrow \mathbf{k'} \downarrow, M+1) = \frac{2\pi}{\hbar} \sum_M w_M \, |T(k' \uparrow M \, | \, k \downarrow \, M+1)|^2 \delta(\epsilon_k - \epsilon'_k)$$

$$(11.39)$$

Exercise 11.4
Show that from (11.37) and the matrix elements of **S** *(see Eqs. (4.6)) one obtains:*

$$\hat{T}_{\uparrow\uparrow} = \hat{t} + \hat{\tau}M$$

$$\hat{T}_{\uparrow\downarrow} = \hat{\tau}\sqrt{S(S+1) - M(M+1)} \qquad (11.40)$$

The first two terms in Eq. (11.35) correspond respectively to the first and the second Born approximation. To calculate the collision time τ_c in the first Born approximation we add the transition probabilities for both independent channels (spin-conserving and spin-flip), average over all initial impurity spin orientations (assumed equally probable), and sum over all final states on the energy shell, under the restriction that total spin be conserved:

$$\frac{1}{\tau_c} = \frac{2\pi}{\hbar}\rho(\epsilon_k) \sum_M w_M \left(|T(k'\uparrow M|k\uparrow M)|^2_{\epsilon_k=\epsilon_{k'}} \right.$$

$$\left. + \ |T(k'\downarrow M+1|k\uparrow M)|^2_{\epsilon_k=\epsilon_{k'}} \right) \qquad (11.41)$$

Using Eq. (11.4.2) we get:

$$\frac{1}{\tau_c} = \frac{2\pi}{\hbar}\rho(\epsilon_k) \left[|t(\epsilon_k)|^2 + |\tau(\epsilon_k)|^2 S(S+1) \right] \qquad (11.42)$$

Exercise 11.5
Prove Eq. (11.42)

Since in the first Born approximation $\hat{T} = \hat{V} = \hat{H}_{sd}$, we see that

$$\tau^{(1)} = \frac{-J}{2N} \quad , \quad t^{(1)} = 0. \qquad (11.43)$$

where we take $J = constant$. We can now substitute for τ_c in Eq. (11.31), and invert σ to find the resistivity:

Exercise 11.6
Show that if the $\hat{T}$ *matrix is calculated to first order in* $\hat{H}_{sd}$ *we get for the resistivity*

$$\rho^{(1)} = \frac{3}{2}\frac{m\pi}{e^2\hbar}\frac{V}{E_F}\frac{J^2}{4N^2} S(S+1)\, c_{imp} \qquad (11.44)$$

We must include in $\rho^{(1)}$ the factor N_{imp}=total number of impurities, so that the resistivity is proportional to their atomic concentration $c_{imp} = N_{imp}/N$.
We turn now to the second order contribution. The initial state is in both cases an incoming electron in state $k\uparrow$ and the impurity in state (S, M). We recognize now that there are two possible intermediate states, which generate two possible transitions to the final state of the electron:

(a) in the first case an incoming electron in state $(k \uparrow)$ interacts with the impurity in state (S, M) and is annihilated, while an intermediate electron in state $(k'' \uparrow (\downarrow)$ is created and the impurity either stays in the same state or makes a transition to the state $(S, M + 1)$ to conserve total spin. The intermediate electron is further annihilated by the exchange interaction, the impurity recovers its initial state, and the outgoing electron is created in state $(k', \uparrow)$.

(b) in the second case, the incoming electron does not interact first with the impurity but continues propagating, while the interaction creates an electron-hole pair, with the electron occuppying the final state $(k' \uparrow)$, while the hole can be in either state $(k'', \uparrow, \downarrow)$. For the spin-conserving case, we first annihilate an electron in state $(k'' \uparrow)$ and create another in state $(k' \uparrow)$, while the impurity state does not change. The interaction acts again to ahhihilate the incoming state $(k' \uparrow)$ and create an intermediate electron state $(k'' \uparrow)$. In the spin-flip case the interaction first annihilates an electron in state $(k'' \downarrow)$ and creates another one in the final state $(k' \uparrow)$, while the second interaction event annhilates the initial electron state and creates the intermediate state $(k'' \downarrow)$, so the total spin change of the electron band is $+1$, which must be compensated by a transition of the impurity from $S^z = M$ to $S^z = M - 1$. The matrix elements of $\hat{H}_{sd}$ for the spin-flip transition are of course those proportional to the transverse components $S^{(+,-)}$.

We calculate now the second order contribution to T. Thanks to the decomposition (11.37) it is enough to calculate $T_{\uparrow\uparrow}$. For the two processes described above we find respectively:

$$T_a^{(2)}(k' \uparrow, k \uparrow) =$$

$$\frac{J^2}{4N^2} \langle M|(S^z)^2|M \rangle \frac{\langle 0\, a_{k'\uparrow}|a_{k'\uparrow}^\dagger a_{k''\uparrow} a_{k''\uparrow}^\dagger a_{k\uparrow}|a_{k\uparrow}^\dagger|0\rangle}{\epsilon_k - \epsilon_{k''} + i\eta}$$

$$+ \quad \frac{J^2}{4N^2} \langle M|S_- S_+|M \rangle \frac{\langle 0\, a_{k'\uparrow}|a_{k'\uparrow}^\dagger a_{k''\downarrow} a_{k''\downarrow}^\dagger a_{k\uparrow}|a_{k\uparrow}^\dagger|0\rangle}{\epsilon_k - \epsilon_{k''} + i\eta} \qquad (11.45)$$

$$T_b^{(2)}(k' \uparrow, k \uparrow) =$$

$$\frac{J^2}{4N^2} \langle M|(S^z)^2|M \rangle \frac{\langle 0\, a_{k'\uparrow}|a_{k''\uparrow} a_{k\uparrow} a_{k'\uparrow}^\dagger a_{k\uparrow}|a_{k\uparrow}^\dagger|0\rangle}{\epsilon_{k''} - \epsilon_k + i\eta}$$

$$+ \quad \frac{J^2}{4N^2} \langle M|S_+ S_-|M \rangle \frac{\langle 0\, a_{k'\uparrow}|a_{k''\downarrow}^\dagger a_{k'\uparrow}^\dagger a_{k\uparrow} a_{k'\uparrow}^\dagger|a_{k''\downarrow}|a_{k\uparrow}|0\rangle}{\epsilon_{k''} - \epsilon_k + i\eta} \qquad (11.46)$$

We substitute above the matrix elements of $S^{(+,-)}$ from Eq. (4.6) and the number operator of the intermediate state by its statistical average at temperature T, so that the matrix elements of $\hat{T}$ for both types of processes are:

$$T_a(\epsilon_k) = \left(\frac{J}{2N}\right)^2 [S(S+1) - M] \int_{-\infty}^{\infty} \rho(E) \frac{1 - f(E)}{\epsilon - E - i\eta}\, dE$$

$$T_b(\epsilon_k) = \left(\frac{J}{2N}\right)^2 [S(S+1) + M] \int_{-\infty}^{\infty} \rho(E) \frac{f(E)}{\epsilon - E - i\eta} \, dE \qquad (11.47)$$

Exercise 11.7
Obtain Eqs. (11.46) and (11.47).

The terms proportional to M are those originating from the $\hat{\tau}$ term in $\hat{T}$, so that we get the real part of the second order contribution to the matrix elements of $\hat{\tau}$ as

$$Re\,\tau(\epsilon + i\eta)^{(2)} = -\left(\frac{J}{2N}\right)^2 \int_{-\infty}^{\infty} \rho(E)\,P\left(\frac{1 - 2f(E)}{\epsilon - E - i\eta}\right) dE \qquad (11.48)$$

where $P\,(.)$ is the principal value operator. We concentrate on the real part of $\tau^{(2)}$ because that will generate the singularity we are looking for. To simplify, we take a constant density of states distribution:

$$\rho(E) = \begin{cases} \rho, & -D < E < D \\ 0, & |E| > D \end{cases} \qquad (11.49)$$

Notice that energies are now measured from the Fermi level.

Call P the principal part integral:

$$P(\epsilon) = \int_{-D}^{D} \frac{1 - 2f(E)}{\epsilon - E} \, dE \qquad (11.50)$$

We integrate by parts:

$$P(\epsilon) = -\left[\ln|\epsilon - E|(1 - 2f(E))\right]_{-D}^{D} - 2\int_{-D}^{D} \ln|\epsilon - E|\,f'(E)\,dE \qquad (11.51)$$

We take $D \gg k_B T$, so that $f(-D) \approx 1$ and $f(D) \approx 0$. The first term of (11.51) is then

$$-2\ln|\epsilon - E| \qquad (11.52)$$

For the second term we must consider separately two different cases, according to whether $|\epsilon| >$ or $< k_B T$.

In the first case, since $f'(E) \approx 0$ if $|E| > k_B T$ we neglect E with respect to ϵ in the argument of the logarithm, so the second term is:

$$-2\ln|\epsilon| \int_{-D}^{D} f'(E)\,dE \approx 2\ln|\epsilon| \qquad (11.53)$$

If we add this to the first term we have that, for $|\epsilon| > k_B T$,

$$P(\epsilon) = 2\ln\left|\frac{E}{D}\right| \qquad (11.54)$$

Let us look, for the second case, that is $|\epsilon| < k_B T$, at the second term in (11.51). In the integrand we can neglect ϵ with respect to E in the argument of the logarithm, so that the integral is:

$$-2 \int_{-D}^{D} \ln|\epsilon - E| \, f'(E) \, dE \approx -2 \int_{-\infty}^{\infty} \left(\ln \frac{|E|}{k_B T} + \ln k_B T \right) f'(E) \, dE$$
$$\approx 2 \ln k_B T - 2 \ln \left(2 e^{\gamma} / \pi \right) \tag{11.55}$$

Exercise 11.8
Verify Eq. (11.55).

We add the first and second order contributions to $Re\,(\tau)$ we have just found:

$$\tau^{(1)} + Re\left(\tau^{(2)}\right) \approx -\frac{J}{2N} \left(1 + \frac{J\rho}{N} \ln \frac{\Delta}{D} \right) \tag{11.56}$$

where

$$\Delta = \begin{cases} |\epsilon| & \text{if } |\epsilon| > k_B T \\ k_B T & \text{if } |\epsilon| < k_B T \end{cases} \tag{11.57}$$

Then we get the collision time, including the first order term and the singular part of the second order term just calculated, with $|\epsilon| \leq k_B T$:

$$\frac{1}{\tau_c(\epsilon)} = \frac{2\pi\rho}{\hbar} N_i S(S+1) \left(\frac{J}{2N} \right)^2 \left(1 + \frac{J\rho}{N} \ln k_B T/D \right)^2 \tag{11.58}$$

where we included the factor N_i = the number of impurities to get the total transition probability. For the resistivity, then, we find:

$$\rho^{(2)}(T) = \frac{3m}{2e^2} \frac{V}{4N} J^2 c_{imp} \frac{\pi}{E_F \hbar} S(S+1) \left(1 + 2\frac{J\rho}{N} \ln \Delta/D \right) \tag{11.59}$$

We must add this contribution to the phonon one and to the residual resistivity arising from elastic scattering processes due to fixed defects and non-magnetic impurities. The contribution to the resistivity from the electron-electron collisions, which is of the order of $(k_B T/E_F)^2$ [10] is negligible at low T, where the Kondo resistance is important. Then we can write the resistivity at low T as

$$R(T) = const. + A \left(\frac{k_B T}{\Theta_D} \right)^5 + B \ln \frac{k_B T}{D} \tag{11.60}$$

Exercise 11.9
Show that in the case $J < 0$ there is a minimum of the resistivity at a finite T.

11.4.3 Higher order contributions

Abrikosov [11] made a diagrammatic expansion of the $\hat{T}$ matrix to all orders with the use of a drone fermion representation of the spin operators due to *Eliashberg* [12]. The sum of the most divergent terms for the spin-dependent part $\hat{\tau}$ yields for the matrix element

$$\tau(\epsilon) = \frac{-J}{2N} \frac{1}{1 - \frac{J\rho}{N} \ln \frac{\Delta}{D}} \tag{11.61}$$

and we can take $\Delta \approx k_B T$ as we mentioned before. The non-flip part has a weaker divergence and we can leave it out.

For $J > 0$ (11.61) makes a negligible contribution to the resistance as $T \to 0$. For $J < 0$, which is the case in $Cu - Mn$, there is a divergence at the Kondo temperature, which in this formulation is defined as

$$k_B T_K = D e^{-N/|J|\rho} \tag{11.62}$$

At this temperature the denominator in (11.61) vanishes. *(It is important to notice that T_K does not depend on the impurity concentration.)*

Since the transition probability diverges, the collision time $\tau_c \to 0$ and accordingly $\sigma \to 0$ and $\rho \to \infty$ as $T \searrow T_K$.

Although there is no experimentally observed divergence of $R(T_K)$, subsequent work supports the existence of a characteristic temperature of the order of the above one. An exact theoretical solution of the Kondo problem was obtained by *Andrei et.al.* [13] and by *Tsvelick and Wiegmann*[14].

The picture which has emerged from a large body of experimental and theoretical work is that for $J < 0$ and $T \ll T_K$ the impurity does not sustain a spin, because under the influence of the AF exchange interaction, the resonant behaviour of the $\hat{T}$ matrix at low temperature localizes an electron cloud of opposite spin around the impurity which compensates the spin of the latter. As the temperature grows above T_K this magnetic screening disappears, and the susceptibility recovers the Curie behaviour characteristic of a local spin.[7]

References

1. Anderson, P. W. (1961) *Phys. Rev.* **124**, 41.

2. Yosida, Kei (1998) *"Theory of Magnetism"*, Springer.

3. Peierls, R. E. , (1955) *"Quantum Theory of Solids"*, chap. 6. Re-edited in 2001 by Clarendon Press, Cambridge.

4. de Haas, W. J. *et al.* (1934) *Physica* **1**, 1115.

5. Sarachik *et al.* (1964) *Phys. Rev.* **A 135**, 1041.

6. Kondo, J. (1964) *Prog. Theor. Phys.* **32**, 37.

7. Hewson, A. C. (1997) *"The Kondo problem to heavy fermions"*, Cambridge University Press.

8. Peierls, R. E., *loc. cit.*, Chap. 6.

9. Peierls, R. E., *loc. cit.*, equation (4.42).

10. Peierls, R. E., *loc. cit.*, Chap. 6.

11. Abrikosov, A. A. (1985) *Physics* **2**, 5.

12. Eliashberg, G. M. (1962) *JETP* **42**, 1658.

13. Andrei, N. *et al.* (1983) *Rev. Mod. Phys.* **55**, 331.

14. Tsvelick, A. M. and Wiegmann, P. B. (1983) *Adv. Phys.* **32**, 953.

Chapter 12

Low Dimensions

12.1 Introduction

We shall start the discussion of the theory of low dimensional systems with a survey of the theorem of Mermin and Wagner [1] on the absence of long range order (LRO) in magnetic systems. A similar theorem was proved by Hohenberg [2] on superfluids and superconductors. The contents of this theorem as applied to systems of localized spins is the following:

Theorem 12.1 Mermin–Wagner theorem.
An infinite d dimensional lattice of localized spins cannot have LRO at any finite temperature for $d < 3$ if the effective exchange interactions among spins are isotropic in spin space and of finite range.

Since the conditions above are precisely satisfied by the Heisenberg model that we have used for most of the theoretical description of magnetic insulators until now, it turns out that we cannot extend those results to lower dimensions. We have already arrived at this conclusion when we found divergences upon the application of the free-spin-wave approximation (FSWA) and the RPA to low dimensional systems. What we present here is a general result which does not depend on any approximation.

Let us make a few remarks at this point:

(i) We cannot discard LRO at zero T in the isotropic Heisenberg model. It turns out that low dimensional ferromagnets exhibit LRO at $T = 0$, but anti-ferromagnets do not;

(ii) When infinite range interactions are considered, *e.g.* the dipolar interactions or the RKKY interactions, the theorem does not apply and we might find finite critical temperatures even in low dimensions;

(iii) When we incorporate anisotropy into the model (non-isotropic exchange

constants, dipole-dipole interactions, single-ion anisotropy, *etc.*) we again expect to find cases when low dimensional systems will exhibit LRO.

Although rigorously speaking there are no realizations of 1 or 2 dimensional systems, in many practical cases one can be very close to this situation,
and the theoretical description of the ideal system is extremely useful to analyze the experimental results in cases of quasi-two or quasi-one-dimensional
systems. In practice, these are lattices in which some family of lines or planes
contain strongly interacting spins which instead couple weakly with spins on
other lines or planes. A paradigm of quasi-two dimensional systems is provided
by the highly anisotropic compounds which are parents of the high-T_c superconducting perovskites. In particular, La_2CuO_4 and $YBa_2Cu_3O_6$ show fairly large
Néel temperatures T_N ($\approx 300 - 400K$) and strongly anisotropic AF correlations.
The latter indicate that the effective exchange coupling within Cu^{++} ions in the
CuO_2 planes is unusually large ($\simeq 0.1\ eV$) while the exchange anisotropy ratio
$\epsilon = J_\perp/J_\parallel \ll 1$ (see ref. [3] and references therein). The parameter ϵ controls
the cross-over from $3d$ ($\epsilon \simeq 1$) to $2d$ as $\epsilon \to 0$ or indeed to $1d$ as $\epsilon \to \infty$.
K_2NiF_4 is another quasi- $2d$ AFM with perovskite structure [4]. Experiments
show long range AFM correlations on the NiF_2 planes at temperatures well
above the $3d$ Néel temperature.
    In $1d$ we mention the compound $(C_6H_{11}NH_3)CuBr$ (CHAB) which behaves
like an ideal $1d$ easy-plane FM with $S = 1/2$ [5] and the $1d$ antiferromagnet
$[(CH_3)_4N]\ [MnCl_3]$ (TMMC), which has chains of M_n^{2+} ions ($S = 5/2$) [6].

12.2   Proof of Mermin–Wagner theorem

Following the original paper [1] we start by proving Bogoliubov inequality [7].

12.2.1   Bogoliubov inequality

Let $\{|\ i\rangle\}$ be a complete orthonormalized set of eigenfunctions of a given Hamil-
tonian H, and let A, B be two operators, of which we only assume that all their
matrix elements in the basis above are well defined and bounded. Then we
define an inner product of that pair of operators as:

Definition:
$$(A, B) \equiv \sum_{i \neq j} \langle i|A \mid j\rangle^* \langle i \mid A \mid j\rangle \frac{W_i - W_j}{E_j - E_i}\ , \qquad (12.1)$$

where E_n is an eigenvalue of H, $W_n = e^{-\beta E_n}/\rho$ and

$$\rho = Tr\left(e^{-\beta H}\right) \qquad (12.2)$$

with $\beta = 1/k_B T$. Notice that we exclude the terms with vanishing denominator
from the sum (12.1).

Exercise 12.1
Prove that this inner product has the following properties:

$$(u, v)\, i \;=\; (v, u)^{*}\; ;$$
$$(u, a_1 v_1 + a_2 v_2) \;=\; a_1\, (u, v_1) + a_2\, (u, v_2)\; ;$$
$$(u, u) \;\geq\; 0\; .$$

As a consequence it satisfies the Cauchy-Schwartz inequality:

$$(A, A)(B, B) \;\geq |\,(A, B)\,|^{2} \tag{12.3}$$

The following inequality holds:

$$0 \leq \frac{W_i - W_j}{E_i - E_j} \leq (W_i + W_j)\,\beta/2 \tag{12.4}$$

and therefore

$$0 \leq (A, A) \leq \frac{\beta}{2}\langle[\,A, A^{\dagger}\,]_{+}\rangle \tag{12.5}$$

where $[\,\cdot\,,\cdot\,]_{+}$ = anticommutator. Now we choose $B = [\,C^{\dagger}, H\,]$ and then

$$(A, B) \;=\; \langle[\,C^{\dagger}, A^{\dagger}\,]\rangle$$
$$(B, B) \;=\; \langle[\,C^{\dagger}, [\,H, C\,]\,]\rangle \tag{12.6}$$

From (12.1) to (12.6) we can prove that:

$$\frac{1}{2}\langle[\,A, A^{\dagger}\,]_{+}\rangle\langle[\,[\,C, H\,], C^{\dagger}\,]\rangle \geq k_B T\,|\,\langle[\,C, A\,]\rangle\,|^{2} \tag{12.7}$$

(Bogoliubov inequality) [7].

Exercise 12.2
Prove (12.7).

12.2.2 Application to the Heisenberg model

Let us consider again the Hamiltonian

$$H = -\sum_{\mathbf{R},\mathbf{R'}} J(\mathbf{R} - \mathbf{R'})\mathbf{S}(\mathbf{R}) \cdot \mathbf{S}(\mathbf{R'}) - h\sum_{\mathbf{R}} S^{z}(\mathbf{R})e^{-i\mathbf{Q}\cdot\mathbf{R}} \tag{12.8}$$

In the FM case we take $\mathbf{Q} = 0$; in an AFM the phase $\phi \equiv \mathbf{Q} \cdot \mathbf{R} = (2n + 1)\pi$ when $\mathbf{R}$ connects points on different sublattices, and $\phi = 2n\pi$ in the opposite case. The translational symmetry, for an infinite crystalline sample, is already implicit in the form of J. We assume J is an even function of its argument. The Fourier transformed spin operators

$$S^{\alpha}(\mathbf{q}) = \sum_{\mathbf{R}} e^{-i\mathbf{q}\cdot\mathbf{R}} S^{\alpha}(\mathbf{R}) \tag{12.9}$$

where $\alpha = +, -$ or z, satisfy the commutation relations

$$\begin{aligned}
[\; S^+(\mathbf{k}), \mathbf{S}^-(\mathbf{k}')\;] &= 2S^z(\mathbf{k} + \mathbf{k}') \\
[\; S^\pm(\mathbf{k}), \mathbf{S^z}(\mathbf{k}')\;] &= \mp S^\pm(\mathbf{k} + \mathbf{k}') \;.
\end{aligned} \tag{12.10}$$

We make now the following choices for the operators A and C in Eq. (12.7):

$$A = S^-(-\mathbf{k} - \mathbf{Q}) \;,\; C = S^+(\mathbf{k}) \tag{12.11}$$

The statistical average of the double commutator in (12.7) is:

$$\begin{aligned}
D_k(Q) &= \langle[\;[\;C, H\;], C^\dagger\;]\rangle \\
&= \frac{1}{N}\sum_{k'}(J_{\mathbf{k}'} - J_{\mathbf{k}'+\mathbf{k}})\langle S^-_{\mathbf{k}'}S^+_{-\mathbf{k}'} + S^+_{\mathbf{k}'}S^-_{-\mathbf{k}'} + 4S^z_{\mathbf{k}'}S^z_{-\mathbf{k}'}\rangle + 2\gamma h s_z(-Q)
\end{aligned} \tag{12.12}$$

where

$$s_z(Q) = \frac{1}{N}\sum_{\mathbf{R}}\langle S^z(\mathbf{R})\rangle e^{i\mathbf{Q}\cdot\mathbf{R}} \tag{12.13}$$

and we have explicitly used the property $J(-\mathbf{q}) = \mathbf{J}(\mathbf{q})$.

We remark that in the FM case ($Q = 0$), $s_z(0) \propto$ average uniform magnetization, while in the AFM case

$$|s_z(Q)| = (1/2)|\langle S^a_z - S^b_z\rangle| \tag{12.14}$$

(a and b are the up- and down-sublattices) which is > 0 in the Néel phase.

As to the applied field, it is a uniform field in the FM case and a staggered field in the AFM one.

With our choice of operators Bogoliubov inequality reads:

$$\frac{1}{2}\langle[\; S^+(\mathbf{k} + \mathbf{Q}) , \mathbf{S}^-(-\mathbf{k} - \mathbf{Q})\;]_+\rangle \geq \frac{4k_B T N^2 s_z(Q)^2}{D_k(Q)} \tag{12.15}$$

According to Eq. (9.6) $D_k(Q)$ is a norm squared, so that $|\;D_k(Q)\;| = D_k(Q)$ and we can use the inequality

$$|\sum_n a_n| \leq \sum_n |\;a_n\;| \;. \tag{12.16}$$

to find an upper bound for D.

We define the quantity

$$\Delta = S(S + 1)\sum_R R^2\;|\;J(R)\;| \tag{12.17}$$

Exercise 12.3
Verify that

$$D_k(Q) \leq Nk^2\Delta + N\;|\;hs_z(Q)\;| \tag{12.18}$$

We now substitute D in (12.15) by the upper bound above, and then sum both sides of the resultant inequality over $\mathbf{k}$. Upon use of the identity

$$\sum_{\mathbf{k}'} \mathbf{S}(\mathbf{k}') \cdot \mathbf{S}(-\mathbf{k}') = N^2 S(S+1) \tag{12.19}$$

we verify easily that (12.19) is an upper bound for the sum of the left hand side of Eq. (12.15). Then we obtain:

$$s_z(Q)^2 \le \frac{2\beta S(S+1)}{\Phi_Q} \tag{12.20}$$

where

$$\Phi_Q = \frac{1}{N} \sum_k \frac{1}{\Delta k^2 + \mid hs_z(Q) \mid} \tag{12.21}$$

For $V \to \infty$ and a lattice of ν dimensions Eq. (12.21) reads:

$$\Phi_Q = \frac{1}{\rho(2\pi)^\nu} \int d^\nu \mathbf{k} \left\{ \frac{1}{\Delta + \mid hs_z(Q) \mid} \right\} \tag{12.22}$$

where $\rho = N/V$ and we integrate inside the first BZ.

Since $s_z(Q)^2$ is inversely proportional to Φ we reinforce inequality (12.20) by integrating in Eq. (12.22) over a smaller domain, since the integrand is positive definite. This does not alter any conclusion we may reach on the convergence of the integral leading to Φ, since it is clear that the critical region in the limit $h \to 0$ is the neighbourhood of the origin in k space. Let us then integrate over a spherical region with center at the origin and radius k_0 equal to the minimum distance from the origin to the boundary of the first BZ.

Two dimensions:

$$s_z^2 \le 2 \frac{\pi S(S+1)\rho}{k_B T} \frac{\Delta}{\ln(1 + \Delta/\mid hs_z \mid)} \tag{12.23}$$

which in the limit $h \to 0$ yields

$$\mid s_z \mid \le \frac{Const}{T^{1/2}} \frac{1}{\mid ln \mid h \mid\mid^{1/2}}. \tag{12.24}$$

One dimension:

$$s_z^2 \le \frac{4S(S+1)\pi\rho\Delta^{1/2} \mid hs_z \mid^{1/2}}{k_B T \arctan\left(\Delta^{1/2} k_0 / \mid hs_z \mid^{1/2}\right)} \tag{12.25}$$

and in the limit $h \to 0$ we get:

$$s_z \le \frac{Const}{T^{2/3}} \mid h \mid^{1/3} \tag{12.26}$$

Exercise 12.4
Verify Eqs. (12.22) to (12.26).

Equations (12.24) and (12.26) are the main results of Mermin and Wagner. They imply that in the $h \to 0$ limit the LRO parameter $s_z \to 0$ for $\nu < 3$ under the assumptions of *short range, isotropic* exchange interactions.

As regards the first assumption (short range), if the parameter Δ defined in (12.17) diverges we arrive at no conclusion. If $J(R) \approx R^{-\alpha}$ for large R the condition for convergence of the integral in Eq. (12.17) is $\alpha > \nu + 2$. Consequently we cannot exclude the existence of LRO in low dimensions for dipolar or RKKY interactions. The divergence of Φ, according to the definition (12.21) is due to the behaviour of the denominator for small k. The above condition is the present criterion for what we shall consider *long range* effective exchange interactions. The same condition was found by Dyson [8] for the existence of a phase transition in the one dimensional Ising model, as we shall discuss further on in this chapter.

As to the second condition (isotropy), we remark that the denominator is the small k expansion of the dispersion relation for an isotropic FM in the spin wave approximation. We know that if an anisotropy of the right kind is present the dispersion relation acquires a gap for $k = 0$, and this automatically invalidates MW theorem, since Φ is finite in the $h \to 0$ limit.

12.3 Dipolar interactions in low dimensions

We address now the exchange-dipolar low d magnetic insulator as a particularly interesting system because dipolar forces exhibit both long range and anisotropy and they are always present when there are magnetic moments, while other kinds of anisotropic forces depend on special characteristics of the system, and in many cases are absent.

Let us consider a ferromagnetic chain. If we choose the z axis along the chain, then $x_n = y_n = 0$, $z_n = na$.

The dipolar tensor

$$\mathcal{D}^{\alpha\beta}(\mathbf{R}) \equiv \frac{1}{R^3} \left(\delta_{\alpha\beta} - \frac{3 R_\alpha R_\beta}{R^2} \right) \tag{12.27}$$

is diagonal for the chain, $\mathcal{D}$, and

$$\mathcal{D}^{xx}(n,m) = \mathcal{D}^{yy}(n,m) = \frac{1}{a^3 \, | \, n - m \, |^3}$$

$$\mathcal{D}^{zz}(n,m) = -\frac{2}{a^3 \, | \, n - m \, |^3} \tag{12.28}$$

The classical system has energy

$$U_{class} = \frac{\gamma^2}{2a^3} \sum_{n,m}{}' \mathcal{D}^{\alpha\beta}(n,m) S_n^\alpha S_m^\beta \tag{12.29}$$

Consider first a state with all spins aligned along the chain. Performing the sums we get, for N spins

$$\frac{U_{class}^{\parallel}}{N} = \frac{\gamma^2 \zeta(3)}{a^3} [S(S+1) - 3S_z^2] \tag{12.30}$$

where

$$\zeta(p) = \sum_1^\infty \frac{1}{n^p} \tag{12.31}$$

is Rieman n's ζ function of order p.

For spins perpendicular to the chain we choose z as axis of quantization and we take the chain along the x axis. Therefore, $x_n = na$, $y_n = z_n = 0$ and

$$\frac{U_{class}^{\perp}}{N} = \frac{\gamma^2 \zeta(3)}{a^3} [S(S+1) - 3S_x^2] \tag{12.32}$$

We calculate only the dipolar part of the energy since the exchange part is the same in both cases. Eqs. (12.29) and (12.32) are special cases of

$$\frac{U_{class}^{dip}}{N} = \frac{\gamma^2 \zeta(3) S(S+1)}{a^3} \left[1 - \frac{3S}{S+1} \cos^2 \phi \right] \tag{12.33}$$

valid for any uniform orientation of the spins, and where ϕ is the angle the magnetization makes with the chain. $\phi = 0$ minimizes (12.33), so that the Weiss state is aligned along the chain.

Consider now the quantum case. From Chap. 6 we recall the commutator

$$
\begin{aligned}
[S_a^+, H_{dip}] &= -\frac{\gamma^2}{2} S_a^+ \sum_m {}' S_m^z \mathcal{D}_{am}^{zz} \\
&\quad - \frac{\gamma^2}{4} S_a^z \sum_m {}' S_m^+ \mathcal{D}_{am}^{zz} - \frac{3\gamma^2}{4} S_a^z \sum_m {}' S_m^- B_{am}^* \\
&\quad + \frac{3\gamma^2}{2} S_a^+ \sum_m {}' S_m^+ F_{am} - \frac{3\gamma^2}{2} S_a^z \sum_m {}' S_m^z F_{am}^* \\
&\quad + \frac{3\gamma^2}{4} S_a^+ \sum_m {}' S_m^- F_{am}^* \tag{12.34}
\end{aligned}
$$

The coefficients B and F above vanish if spins are polarized along the chain. Accordingly, in this case there are no zero point fluctuations of the transverse components of the total spin and the z component of the total spin is conserved at $T = 0$.

For spin sites a, b, the double-time retarded Zubarev Green's function satisfies the RPA equation

$$\omega G_{ab}^{+-}(\omega) = \frac{\langle S_a^z \rangle \delta_{ab}}{\pi} - 2 \sum_{m \neq a} J_{ma} \left\{ \langle S_a^z \rangle G_{mb}^{+-}(\omega) \right.$$

$$- \langle S_m^z \rangle G_{ab}^{+-}(\omega) \} - \gamma^2 G_{ab}^{+-}(\omega) \sum_{m \neq a} \mathcal{D}_{ma}^{zz} \langle S_m^z \rangle$$

$$- \frac{\gamma^2}{2} \langle S_a^z \rangle \sum_{m \neq a} \mathcal{D}_{ma}^{zz} G_{mb}^{+-}(\omega) + \gamma B G_{ab}^{+-}(\omega) \qquad (12.35)$$

We see that for spins polarized along the chain there is no coupling of G^{+-} with G^{--}.

Fourier transforming Eq. (12.35) to k space we obtain

$$G_k^{+-}(\omega) = \frac{\sigma/\pi}{\omega - E_k} \qquad (12.36)$$

where the dispersion relation is

$$E_k = 2\sigma \left(\tilde{J}(0) - \tilde{J}(k) \right) + \gamma B - \gamma^2 \sigma \left(\tilde{D}^{zz}(0) + \frac{\tilde{D}^{zz}(k)}{2} \right) \qquad (12.37)$$

with the same notation as in Chap. 6. In order to obtain an explicit expression for E_k we need the Fourier transform

$$\tilde{\mathcal{D}}^{zz}(k) = -\frac{2}{a^3} \sum_n {}' \frac{e^{ikna}}{|n|^3} \equiv -\frac{2}{a^3} A(q) \qquad (12.38)$$

with $q = ka$.

Let us describe now Ewald's method [9, 10] for the calculation of lattice sums like $A(q)$.

As a first step, we make use of a representation of $|n|^{-3}$ based on the definition of the Γ function:

$$x^{-\alpha} = \frac{1}{\Gamma(\alpha/2)} \int_0^\infty t^{(\alpha/2-1)} e^{-tx^2} dt \qquad (12.39)$$

It turns out convenient to include a small shift in the coordinate n: $n \to n+r$, and to take the limit $r \to 0$ at the end of the calculation. Then

$$A(q) = \lim_{r \to 0} \frac{1}{\Gamma(3/2)} \int_0^\infty dt \sum_{n \neq 0} e^{iq(n+r)} e^{-t(n+r)^2} t^{1/2} \qquad (12.40)$$

We now add and subtract the $n = 0$ term in the sum above:

$$A(q) = \lim_{r \to 0} \left[\frac{2}{\sqrt{\pi}} \int_0^\infty dt \sum_n e^{iq(n+r)} e^{-t(n+r)^2} t^{1/2} - \frac{e^{iqr}}{r^3} \right] \qquad (12.41)$$

The integrand above is a periodic function and it can be expanded in a Fourier series in reciprocal $1d$ space. Let us write

$$F(r) = \sum_{n=-\infty}^{\infty} e^{iq(n+r)} e^{-t(n+r)^2} = \sum_p e^{-2i\pi pr} F_p \qquad (12.42)$$

where

$$F_p = \int_0^1 \mathrm{d}x \, F(x) e^{2i\pi px} = \int_{-\infty}^{\infty} \mathrm{d}x \, e^{iGx} e^{-tx^2} = \sqrt{\pi} \, t^{-1/2} e^{-G^2/4t} \qquad (12.43)$$

where $G = 2\pi p + q$. We finally arrive at

$$\sum_n e^{iq(n+r)} e^{-t(n+r)^2} = \sum_p e^{-2\pi pr} \sqrt{\pi} \, t^{-1/2} e^{-(2\pi p+q)^2/4t} \qquad (12.44)$$

This identity is known as the *Jacobi imaginary transformation of the ζ function* [10]. It can be readily extended to multiple summations, *i.e.* higher dimensions. In order to get a final expression for $A(q)$ (Eq. (12.41)) we must perform the integration over t and the limit $r \to 0$. Both sides of Eq. (12.44) are different representations of the same function of t. For small t the convergence of the integral (12.41) near $t = 0$ is better if we use the representation of the r. h. s. of Eq. (12.44), while the opposite is true for large values of t. We introduce therefore Ewald's separation parameter ξ, and we write

$$A(q) = \lim_{r \to 0} \left\{ 2 \sum_{p=-\infty}^{\infty} \int_0^{\xi} e^{-(2\pi p+q)^2/4t} e^{-2\pi pr} \mathrm{d}t + \right.$$

$$\left. \frac{2}{\sqrt{\pi}} \sum_{n=-\infty}^{\infty} \int_{\xi}^{\infty} e^{iq(n+r)} e^{-t(n+r)^2} t^{1/2} \mathrm{d}t - \frac{e^{iqr}}{r^3} \right\} \qquad (12.45)$$

The $n = 0$ term in Eq. (12.45) is

$$e^{iqr} \left\{ \int_0^{\infty} - \int_0^{\xi} \right\} e^{-tr^2} t^{1/2} \mathrm{d}t \qquad (12.46)$$

The first integral cancels exactly the subtraction term, and we are left with the finite correction term

$$-\frac{4}{3} \frac{\xi^3}{\sqrt{\pi}}$$

The integral in the reciprocal lattice sum of Eq. (12.45) (first term) ccan be transformed by the change of variables $t = 1/z$ into the exponential integral of order 2, E_2 [11] of the corresponding argument. The integral in the direct lattice sum (second term) is transformed by the change $t = z^2$ into expressions involving Gaussians and the error function:

$$A(q) = A^{latt}(q) + A^{rec}(q) \,, \quad \text{where}$$

$$A^{latt}(q) = \frac{2}{\sqrt{\pi}} \sum_{n \neq 0} e^{iqn} \left(\frac{\xi^{1/2}}{n^2} e^{-\xi n^2} + \frac{\sqrt{\pi}}{2n^3} \mathrm{erfc}(\xi^{1/2} \,|\, n \,|) \right) - \frac{4}{3} \frac{\xi^3}{\sqrt{\pi}}$$

$$A^{rec}(q) = 2\xi \sum_{p=-\infty}^{\infty} E_2 \left(\frac{(2\pi p + q)^2}{4\xi} \right) \qquad (12.47)$$

The $p = 0$ term in $A^{rec}(q)$ has a logarithmic sigularity when $q \to 0$. In order to display the form of this singularity we use the series expansion of E_2 for small argument [11] and find:

$$2\xi E_2(q^2/4\xi) \; = \; 2\xi \left\{ (q^2/4\xi)(\; \log{(q^2/4\xi)} - \psi(2)\;) - \sum_{m=0,\neq 1}^{\infty} \frac{(q^2/4\xi)^m}{m!(m-1)} \right\}$$

(12.48)

where $\psi(x)$ is the digamma function [11]. One verifies that as $\xi \to 0$ the reciprocal space sum vanishes and the direct lattice sum goes back to the original expression for $A(k)$. Ewald's method yields two fast convergent series. The sum of the two series in (12.47) is identical to the original expression (12.45) and accordingly it is independent on the value of ξ , which can be chosen to optimize the series convergence or simplify the final expressions. In particular, note that the singular term in Eq. (12.48) is independent on ξ. At $q = 0$ we have from Eq. 12.37:

$$E_0 = \gamma B + 6\gamma^2 \sigma \zeta(3)/a^3 \;\;,$$

(12.49)

so that even in the absence of an external field $(B = 0)$ there is a finite gap in the spectrum which warrants LRO. The presence of this gap implies that Goldstone's theorem does not apply in this case. This is in agreement with the fact that the symmetry group of the ground state, for spins along the chain, is discrete. In effect we have two degenerate ground states, with spins aligned in one of the two directions $\pm z$. The group consists of the identity and the inversion. The absence of a continuous symmetry group implies the possibility of a finite excitation energy from the ground state to the first excited one.

12.3.1   Dipole-exchange cross-over

From the dispersion relation (12.37) we conclude that at small q (*i.e.* at large distances) the dipolar terms dominate, while the opposite is true when q is a sizeable fraction of π/a. The relevant parameter is the ratio of the typical dipolar and exchange energies which we define as the *dipolar anisotropy E_d*:

$$E_d = \frac{\gamma^2}{Ja^3}$$

We can rewrite E_q for q small as

$$\frac{E_q}{SJ} = q^2 + E_d \;\; (6\zeta(3) - Aq^2 \log q^2 + Bq^2 \;) + \mathcal{O}(q^3)$$

(12.50)

where A, B = numerical coefficients of order 1. The first term in q^2 on the r. h. s. of Eq. 12.50 is the exchange contribution, while terms proportional to E_d are either dipolar-exchange or pure dipolar ones. The dipolar terms dominate due to the logarithmic factor in the limit $q \to 0$. When $E_d \mid \log q_c^2 \mid \approx 1$ the dipolar and exchange terms become comparable, and there is a cross-over to the exchange dominated regime. Typical values of E_d are 10^{-2} to 10^{-4}. The cross-over q_c is exponentially small, so for most values of q inside the BZ exchange dominates the dynamics of the one-dimensional magnons. However, the ground state properties, in particular the existence of a gap and the polarization of spins, are determined by the dipolar interactions, since they are controlled by the $q \rightarrow 0$ limit.

12.4 One dimensional instabilities

In spite of these results one must be cautious before asserting the existence of LRO in the FM chain. There is a well known argument due to R. E. Peierls [12] in this respect. The basic idea is that one possible source of disorder at low (but not zero) T is the spontaneous breaking of the chain into FM domains separated by kinks. For spins polarized basically along the chain at very low T , all spins before a kink located at the site n_i would point in, say, the $+z$ direction, while for $n > n_i$ they would point in the $-z$ direction. These defects will be randomly distributed along the chain on a set $\{n_i\}$ of points. Each such defect increases the interaction energy, but their contribution to the entropy decreases the Helmholtz free energy

$$F = U - TS \tag{12.51}$$

and we must determine the minimum of F in the presence of a distribution of kinks. The crucial hypothesis in Peierls' argument is the assumption of short range of $J(n)$. In such a case let us consider the change in the internal energy ΔU upon introducing $N_0 \ll N$ kinks. Since they are sparsely distributed along the chain, they are separated by an average distance much greater than the range of $J(n)$ and therefore they do not interfere with one another. Then ΔU is proportional to N_0:

$$\frac{\Delta U}{N} = C \frac{N_0}{N} \equiv C\rho \ , \tag{12.52}$$

$\rho = N_0/N$ being the average concentration of kinks and $C > 0$. If the probability of a kink being localized at any given point, very far from other kinks, is some constant P, then the total probability of finding N_0 kinks is proportional to WP, where $W =$ number of ways of selecting N_0 out of N sites is the combinatorial number

$$W = \left(\begin{array}{c} N_0 \\ N \end{array} \right) \tag{12.53}$$

We approximate W, for $(N_0, N) \rightarrow \infty$, with Stirling's formula, obtaining for the entropy the estimate

$$S = k_B \log W \approx k_B N_0 \log \rho \tag{12.54}$$

and for $\Delta F = F(T) - F(0)$

$$\frac{\Delta F}{N} = \frac{\Delta U}{N} - TS = C\rho - k_B T \rho \log \rho \qquad (12.55)$$

neglecting terms $\mathcal{O}(1/N)$. We now minimize ΔF with respect to ρ and find

$$\rho(T) = e^{-C/k_B T} \qquad (12.56)$$

So under the present assumptions, there will be a finite concentration of kinks, and consequently no LRO, at any finite T in the FM chain. The argument applies both to the Ising and the Heisenberg model. A similar argument led Peierls [12] to prove that in $2d$ the Ising FM *does have* a phase transition at a finite T_c.

However, no *long range* forces have been taken into account in the above argument. The problem of the existence of a phase transition in the presence of long range forces has been considered within the FM Ising model, defined by the Hamiltonian

$$H = -\sum_{i>j} J(i-j)\mu_i\mu_j \ , \ \ \mu_i = \pm 1 \ , \ \forall i \qquad (12.57)$$

with

$$J(n) > 0 \ , \ \ \forall n \ . \qquad (12.58)$$

For this case it can be proven [13] that the infinite system is a well defined limit of the finite one, with consistent definitions of the thermodynamic averages, *provided the zero order moment of the coupling satisfies*

$$M_0 \equiv \sum_{n=1}^{\infty} J(n) < \infty \qquad (12.59)$$

Exercise 12.5
Prove that when $M_0 = \infty$ in the Ising model there is an infinite energy gap separating the Weiss ground state from the excited states. Therefore at any finite T the system will remain ordered: there is no phase transition to a disordered state.

If M_0 satisfies (12.59) the system will be disordered at any finite T, as we have just seen [14]. A theorem by Ruelle [15] proves that an Ising chain which satisfies Eqs. (12.57) and (12.58) will be disordered at any finite T provided

$$M_1 \equiv \sum_{n=1}^{\infty} n J(n) < \infty \qquad (12.60)$$

Exercise 12.6
Prove that $2M_1$ is the energy for creating one kink in an Ising FM chain.

Kac and Thompson [16] made the

Conjecture 1 *A 1d system obeying (12.57) with*

$$M_0 < \infty, \quad M_1 = \infty$$

has a phase transition at some finite T.

From this follows

Corollary 1 *A 1d system obeying (12.57) with*

$$J(n) = n^{-\alpha} \tag{12.61}$$

has a phase transition at a finite T if and only if

$$1 < \alpha \leq 2 \tag{12.62}$$

Based on the preceding results Dyson [8] proved that an Ising system satisfying Eq. (12.61) has a phase transition if $1 < \alpha < 2$. Therefore there is no phase transition at finite T in an Ising system with dipolar interactions.

Dyson conjectures that the same is true for the Heisenberg case.

In conclusion, we probably should discard the existence of LRO in a FM chain at finite T. However, as we indicated at the beginning of this chapter, one finds excitations similar to one dimensional magnons in many experiments on quasi-one dimensional ferromagnets, indicating the existence of FM order over distances larger or at least comparable with the mean free path of these excitations, even at fairly high T.

12.5 Antiferromagnetic chain

We consider now a chain of spins with AFM short range exchange together with dipole-dipole magnetic interactions. We know that the one-dimensional Heisenberg AFM does not have LRO even at zero temperature. Just as for a FM, we shall ask ourselves whether dipolar interactions could lead to LRO [17]. We write again

$$H = -\sum_{a \neq b} J_{ab} \mathbf{S}_a \cdot \mathbf{S}_b + \sum_{a \neq b} S_a^\alpha S_b^\beta \mathcal{D}_{ab}^{\alpha\beta} \tag{12.63}$$

with $J_{ab} < 0$. At low T we take the Néel state as a good approximation to the ground state when exchange prevails. If dipolar interactions dominate, the minimum energy configuration is the FM ordering along the chain. To calculate the classical energy consider all spins substituted by c-vectors. We write the spin at point $z_n = na$ along the chain as:

$$\mathbf{S}_n = \mathbf{S}_0 e^{ik_0 na} \quad , \quad \mathbf{S}_0^2 = S(S+1) \tag{12.64}$$

where $q_0 \equiv k_0 a$ satisfies the condition

$$q_0 = (2m+1)\pi \quad , \quad m = \text{integer} \tag{12.65}$$

in the Néel state. The total classical energy with the assumption (12.64) is

$$\frac{U_{class}(q_0)}{NS(S+1)} = \frac{\gamma^2}{2} \sum_{n \neq 0} e^{iq_0 n} \mathcal{D}^{\alpha\alpha}(n) - \sum_{n \neq 0} e^{iq_0 n} J(n) \tag{12.66}$$

where α denotes the spin polarization axis. We consider now the longitudinal and transverse polarizations separately.

$\mathbf{S_0} \parallel$ chain.

For this case, in which the quantization axis z is along the chain,

$$\frac{U^{\parallel}_{class}(q_0)}{NS(S+1)} = \gamma^2 2 \mathcal{D}^{zz}(q_0) - J_{q_0} \tag{12.67}$$

With the condition (12.65) on q_0 we find:

$$\mathcal{D}^{zz}(q_0) = \frac{3\zeta(3)}{a^3} \tag{12.68}$$

and

$$\frac{U^{\parallel}_{class}(q_0)}{NS(S+1)} = \frac{3\zeta(3)}{a^3} - J_{q_0} \tag{12.69}$$

For $q_0 = 0$ (*FM* order) we have

$$\mathcal{D}^{zz}(0) = -\frac{4\zeta(3)}{a^3} \tag{12.70}$$

and

$$\frac{U^{\parallel}_{class}(0)}{NS(S+1)} = -\frac{2\gamma^2\zeta(3)}{a^3} - J_0 \tag{12.71}$$

which is the dipolar-dominated configuration as we already know (see Eq. (12.33)).

$\mathbf{S_0} \perp$ chain.

Now we take as in the FM case the x axis along the chain, and z along the quantization axis. Then

$$\mathcal{D}^{zz}(q_0) = -\frac{3\zeta(3)}{2a^3} \tag{12.72}$$

and

$$\frac{U^{\perp}_{class}(q_0)}{NS(S+1)} = -\frac{3\gamma^2\zeta(3)}{4a^3} - J_{q_0} \tag{12.73}$$

We can compare the dipolar dominated configuration of Eq. (12.71) with the exchange dominated one from Eq. (12.73). If we limit ourselves to only first n.n. exchange interactions, the critical dipolar anisotropy E_d^c for the cross-over from exchange- to dipolar-dominated regime is

$$E_d^c \equiv \frac{\gamma^2}{Ja^3} = \frac{16}{5\zeta(3)} \approx 2.66 \tag{12.74}$$

For $E_d > (<)E_d^c$ the FM (AFM) configuration is more stable.

12.6 RPA for the AFM chain

12.6.1 Exchange dominated regime

We assume that the ground state has the AFM configuration

$$\langle S_n^z \rangle = \sigma e^{iq_0 n} \tag{12.75}$$

with z perpendicular to the chain, which we take along the x axis. Following the same procedure as in Chap. 6, we obtain the algebraic equation satisfied by the retarded Green's function $G^{+-}(q, \omega)$ in the RPA [17]:

$$\frac{\omega}{\sigma} G^{+-}(q, \omega) = \frac{1}{\pi} - 2[\, J_{q+q_0} - J_{q_0}\,] G^{+-}(q, \omega)$$

$$-\gamma^2 [\, \mathcal{D}^{zz}(q_0) + \frac{1}{2}\mathcal{D}^{zz}(q + q_0)\,] G^{+-}(q + q_0, \omega) -$$

$$-\frac{3\gamma^2}{2} B^*(q + q_0) G^{--}(q + q_0, \omega) \tag{12.76}$$

where

$$B(q) = \sum_n{}' \frac{(x_n - iy_n)^2}{(x_n^2 + y_n^2 + z_n^2)^{5/2}} e^{iqn} \tag{12.77}$$

and we have $x_n = na, y_n = z_n = 0$, $\forall n$. Then $B(q) = B^*(q) = \mathcal{D}^{zz}(q)$. Let us choose the spin at b as $\uparrow$. A look at Eq. (12.63) tells us that a spin reversal at a given site couples with spin reversals in the other sublattice through exchange and also to spin ascending transitions due to dipolar forces, so that now we must deal with four Green's functions.

Let us choose the origin of x coordinates along the chain at an $\uparrow$ site. The alternative choice of origin at the $\downarrow$ sublattice only changes the signs of all the Fourier transforms, which does not alter the dispersion relations. We write now the equations for $G^{--}(q, \omega)$:

$$\frac{\omega}{\sigma} G^{--}(q, \omega) = 2[\, J_{q+q_0} - J_{q_0}\,] G^{--}(q, \omega) +$$

$$\gamma^2 [\, \mathcal{D}^{zz}(q_0) + \frac{1}{2}\mathcal{D}^{zz}(q + q_0)\,] G^{--}(q + q_0, \omega) +$$

$$\frac{3\gamma^2}{2} B(q + q_0) G^{+-}(q + q_0, \omega) \tag{12.78}$$

To simplify the notation we define

$$c_k = 2(J_{k+q_0} - J_{q_0}) + \gamma^2 \left[\mathcal{D}^{zz}(q_0) + \frac{1}{2}\mathcal{D}^{zz}(k + q_0) \right]$$

$$d_k = \mathcal{D}^{zz}(k)$$

$$\nu = \frac{\omega}{\sigma} \tag{12.79}$$

If we eliminate $G^{+-}(q + q_0, \omega)$ and $G^{--}(q + q_0, \omega)$ the resulting equations for fixed ν are

$$\frac{\nu - c_q}{\pi} = \left(\nu^2 - c_q c_{q+q_0} + \frac{9\gamma^2}{4} d_q d_{q+q_0} \right) G_q^{+-}$$

$$- \frac{3\gamma^2}{2}(d_q c_q - d_{q+q_0} c_{q+q_0}) G_q^{--} \tag{12.80}$$

$$\frac{3\gamma^2}{2\pi} d_{q+q_0} = \left(\nu^2 - c_q c_{q+q_0} + \frac{9\gamma^2}{4} d_q d_{q+q_0} \right) G_q^{--} -$$

$$\frac{3\gamma^2}{2}(d_q c_q - d_{q+q_0} c_{q+q_0}) G_q^{+-} \tag{12.81}$$

From the condition that the secular determinant of the system of Eqs. (12.80), (12.81) vanish we get two branches for ν^2, which we denote as $E_s(q)^2$, with $s = \pm 1$:

$$\nu^2 = E_s^2(q) = \left(c_q + \frac{3\gamma^2 s}{2} d_{q+q_o} \right)\left(c_{q+q_0} - \frac{3\gamma^2 s}{2} d_q \right) \tag{12.82}$$

The r. h. s. can be written as the product:

$$E_s(q)^2 = 4 \left[J_{q_0} - J_{q+q_0} - \frac{\gamma^2}{2} d_{q_0} - \gamma^2 \left(\frac{1 + 3s}{2} d_{q+q_o} \right) \right] \times$$

$$\left[J_{q_0} - J_q - \frac{\gamma^2}{2} d_{q_0} - \gamma^2 \left(\frac{1 - 3s}{2} d_q \right) \right] \tag{12.83}$$

and also as the difference of two squares:

$$E_s(q)^2 = (A_q^s)^2 - (B_q^s)^2 \tag{12.84}$$

where

$$A_q^s = 2J_{q_0} - J_{q+q_0} - J_q - \gamma^2 d_{q_0} - \frac{\gamma^2}{2} \left(\frac{1 - 3s}{2} \right) d_q$$

$$- \frac{\gamma^2}{2} \left(\frac{1 + 3s}{2} \right) d_{q+q_0}$$

$$B_q^s = J_q - J_{q+q_0} - \frac{\gamma^2}{2} \left(\frac{1 + 3s}{2} \right) d_{q+q_0}$$

$$+ \frac{\gamma^2}{2} \left(\frac{1 - 3s}{2} \right) d_q \tag{12.85}$$

Since we must have $\nu^2 \geq 0$ there is a stability condition:

$$(A_q^s)^2 \geq (B_q^s)^2 \tag{12.86}$$

The dispersion relations have the following properties:

$$E_s(q + q_0) = E_{-s}(q) \tag{12.87}$$

One verifies that

$$\lim_{q \to 0} E_-(q) = 0 \quad ; \quad \lim_{q \to 0} E_+(q) > 0$$

$$\lim_{q \to q_0} E_+(q) = 0$$

$$E_+(-q_0/2) = E_+(q_0/2) = E_-(q_0/2) \tag{12.88}$$

the last equation showing that both branches $E_\pm$ intersect at $\pm q_0/2$. This degeneracy is the consequence of the invariance of the AF ground state under the combination of time reversal and a primitive translation of the atomic lattice, and the result is that one can consider only one branch, with a continuous dispersion relation in the extended (atomic) BZ $[-\pi/a, \pi/a \}$. [17]

Exercise 12.7
Verify that if $q_0 = 0$ we get the same dispersion relation as in Chap. 6 for the FM chain. Notice that this case corresponds to the FM polarization perpendicular to the chain which is unstable for the dipolar forces.

For small q, the coefficient d_q has a logarithmic divergence, as we already found upon studying the FM chain:

$$d_q = a^3 \mathcal{D}^{zz}(q) = d_0 - c_2 q^2 + q^2 \log \frac{q^2}{4\pi} + \mathcal{O}(q^4) \tag{12.89}$$

where $d_0 = 2\zeta(3)$; $c_2 = 0.666$ [17]. This entails a logarithmic term in the dispersion relation of the acoustic mode. In summary, in the exchange-dominated regime, with magnetization perpendicular to the chain, we find an acoustic (Goldstone) mode with a logarithmic singularity in the second derivative, and and optical mode, which is obtained by folding the accoustic branch at $q = \pm q_0/2$. Both are degenerate at the magnetic zone border.

Exercise 12.8
Explain why we find a Goldstone mode in this case.

Let us simplify the notation: for any function $f(q)$ we call $f(q) = f$, and $f(q + q_0) = f'$. We complete now the calculation of the Green's function:

$$G^{+-}(q, \nu) = \frac{R(\nu)/\pi}{(\nu^2 - E_+^2)(\nu^2 - E_-^2)} \tag{12.90}$$

where

$$R(\nu) = (\nu - c)(\nu^2 - cc' + \frac{9\gamma^2}{4}dd') - \frac{9\gamma^2}{4}(d'c' - dc)d' \qquad (12.91)$$

Notice that

$$E_+^2 - E_-^2 = 3\gamma^2(dc - d'c') \qquad (12.92)$$

We can now obtain the sublattice magnetization. Let us limit ourselves for simplicity to the case $S = 1/2$. As we have done before, we calculate

$$\Phi = -\frac{1}{N\sigma}\sum_k \int d\omega \mathcal{I}m\ G^{+-}(q, \omega + i\epsilon)N(\omega) \qquad (12.93)$$

where $N(\omega)$ is the Bose-Einstein distribution function, and for $S = 1/2$

$$\sigma = \frac{1/2}{1 + 2\Phi} \qquad (12.94)$$

The sum over k in (12.93) is to be performed inside the magnetic BZ, $-q_0/2 < q \le q_0/2$, which is half the atomic one. The k summation can be simplified by exploiting relations (12.88), and we can integrate over only one of the branches by extending the integral to the whole atomic BZ, obtainig as final result the expression

$$\Phi = \frac{1}{2N}\sum_q{}'\{\frac{A_q^-}{E_-(q)} - 1\} + \frac{1}{N}\sum_q{}'\frac{A_q^-}{E_-(q)}N(\sigma E_-(q)) \qquad (12.95)$$

and the prime in the sum above is there to remind us that we integrate over the atomic BZ.

Exercise 12.9
Obtain Eq. (12.95).

The first term on the r. h. s. of Eq. (12.95) is the contribution to the magnetization deviation from the quantum zero-point fluctuations. The second term is the thermal contribution.
It turns out that the zero-point fluctuations diverge:

Exercise 12.10
Prove that the zero point fluctuations diverge due to the singular behaviour of E_q^- at $q = 0$.

The conclusion is that there is no LRO in the AFM chain, even at $T = 0$, in the exchange dominated regime.

12.6.2   Dipolar dominated regime

Now all spins are parallel to the chain. Then we are back in the FM case, with only a change of sign of J, and in our present notation the dispersion relation,

Eq. (12.37) reads

$$E_q = -4 \mid J \mid (1 - \cos q) - \gamma^2 (d_0 + \frac{d_q}{2}) \qquad (12.96)$$

We have $d_0 = -4\zeta(3)/a^3$ and $d_{q_0} = 3\zeta(3)/a^3$. We obtain the values

$$
\begin{aligned}
E_0 &= 6\gamma^2 \zeta(3)/a^3 \\
E_{q_0} &= -8 \mid J \mid + \frac{5}{2a^3}\gamma^2 \zeta(3) \qquad (12.97)
\end{aligned}
$$

Due to the competition between exchange and dipolar forces, the magnon near the zone border becomes soft. Eq. (12.97) shows that $E_{q_0} = 0$ for $E_d = E_d^c$. This instability leads to the perpendicular (dipole dominated) phase for $E_d > E_d^c$ that we have found before. For $E_d > E_d^c$ there is a finite gap in the spectrum, which is in tune with the fact that the choice of the Weiss ground state breaks a discrete symmetry in this case. There are no zero point fluctuations, as we have seen, in the FM configuration.

Exercise 12.11
Prove that in the dipolar dominated regime Φ converges as long as $\sigma \neq 0$.

Therefore in the dipolar regime we have LRO at $T = 0$, which, if Dyson's conjecture were right would disappear at any finite T (see Sec. 9.5).

12.7   Dipolar interaction in layers

Let us return to Eqs. (6.16) and (6.17) and suppose we consider a system with $2d$ translational symmetry. We can think of the system as consisting of a family, finite or infinite, of parallel crystal planes, like a multilayer system with two surfaces, a semi-infinite one with one surface, or a superlattice, consisting of an infinite periodic system with a repetition "super-cell" in the direction perpendicular to the family of planes. Then we can Fourier transform Eq. (6.16) with respect to time and to the position vectors of points on each plane, keeping an index which denotes the plane, since there is no translational symmetry in the direction perpendicular to the planes, except in the super-lattice case, which we shall exclude. The usual retarded Green's functions will be written now as

$$\langle\langle S^+_{l \, \mathbf{1}_\parallel}; S^-_{n \, \mathbf{n}_\parallel} \rangle\rangle \equiv G^{+-} \left(l \, \mathbf{1}_\parallel, n \, \mathbf{n}_\parallel; \ (t - t_0) \right) \qquad (12.98)$$

where we explicitly separate the plane index from the two-component position vectors of sites on the corresponding plane, and the index "r" for "retarded" is implicit. We define the $2d$ Fourier transform of G:

$$G^{+-}_{ln} (\mathbf{k}_\parallel, \omega + i\epsilon) = \sum_{\mathbf{R}_\parallel} e^{i\mathbf{k}_\parallel \cdot \mathbf{R}_\parallel} G^{+-}_{ln} (\mathbf{R}_\parallel, \omega + i\epsilon) \qquad (12.99)$$

where the $2d$ translational symmetry of the system has been taken into account. We also Fourier transform J:

$$J_{ln}(\mathbf{k}_{\parallel}) = \sum_{\mathbf{R}_{\parallel}} e^{i\mathbf{k}_{\parallel}\cdot\mathbf{R}_{\parallel}} J_{ln}(\mathbf{R}_{\parallel}) \tag{12.100}$$

Next step is to apply the RPA to the layered system. At this point we shall make the simplifying assumption that on each plane n the average $\langle S_{n,\mathbf{R}_{\parallel}}^z \rangle$ is translationally invariant on the plane, so that it becomes independent on $\mathbf{R}_{\parallel}$. Then the Green's functions for fixed $\mathbf{k}_{\parallel}, \omega$ are matrices with indices denoting the planes, which satisfy a system of linear equations. Non-linearity enters through the self-consistent calculation of the statistical averages $\langle S_n^z \rangle$. This leads to a generalization of the RPA for layered systems which we might call the *layer RPA* (LRPA) [18, 19].

We apply now this program to the exchange-dipolar Hamiltonian of Chap. 6. We consider the quantization axis z to be along the plane, for simplicity, in order to avoid depolarization effects. Then the axes x, z are parallel to the planes, and y is perpendicular to them. We shall also partially Fourier transform the dipolar tensor elements. We define the coefficients

$$A_{lm}(\mathbf{k}_{\parallel}) = \sum_{\mathbf{R}_{\parallel}} e^{i\mathbf{k}_{\parallel}\cdot\mathbf{R}_{\parallel}} \frac{1}{(R_{\parallel}^2 + y_{lm}^2)^{3/2}} \{ 1 - \frac{3z_{\parallel}^2}{R_{\parallel}^2 + y_{lm}^2} \} \tag{12.101}$$

and

$$B_{lm}^{\pm}(\mathbf{k}_{\parallel}) = \sum_{\mathbf{R}_{\parallel}} e^{i\mathbf{k}_{\parallel}\cdot\mathbf{R}_{\parallel}} \frac{(x_{lm} \pm iy_{lm})^2}{(R_{\parallel}^2 + y_{lm}^2)^{5/2}} \tag{12.102}$$

Let us simplify the notation and call for any F: $F_{ln}(\mathbf{k}_{\parallel} = 0) \equiv F_{ln}(0)$. We obtain the following equations for the Green's functions:

$$\left[\omega - \gamma B - 2 \sum_j \left(J_{lj}(0) - \gamma^2 A_{lj}(0) \right) \langle S_j^z \rangle \right] G_{lm}^{+-}(\mathbf{k}_{\parallel}, \omega + i\epsilon)$$

$$+ \langle S_l^z \rangle \sum_j \{ 2J_{lj}(\mathbf{k}_{\parallel}) + \frac{\gamma^2}{2} A_{lj}(\mathbf{k}_{\parallel}) \} G_{jm}^{+-}(\mathbf{k}_{\parallel}, \omega + i\epsilon)$$

$$+ \frac{3\gamma^2}{2} \langle S_l^z \rangle \sum_j B_{lj}^*(\mathbf{k}_{\parallel}) G_{jm}^{--}(\mathbf{k}_{\parallel}, \omega + i\epsilon)$$

$$= \frac{\langle S_l^z \rangle}{\pi} \tag{12.103}$$

and

$$\left[\omega + \gamma B + 2 \sum_j \left(J_{lj}(0) - \gamma^2 A_{lj}(0) \right) \langle S_j^z \rangle \right] G_{lm}^{--}(\mathbf{k}_{\parallel}, \omega + i\epsilon)$$

$$-\langle S_l^z\rangle \sum_j \{\ 2J_{lj}(\mathbf{k}_\parallel) + \frac{\gamma^2}{2}A_{lj}(\mathbf{k}_\parallel)\ \}G_{jm}^{--}(\mathbf{k}_\parallel,\omega+i\epsilon)$$

$$-\frac{3\gamma^2}{2}\langle S_l^z\rangle \sum_j B_{lj}(\mathbf{k}_\parallel)G_{jm}^{+-}(\mathbf{k}_\parallel,\omega+i\epsilon)=0 \qquad (12.104)$$

Exercise 12.12
Obtain Eqs. (12.103) and (12.104).

Let us now specialize to the case in which each plane is a square $2d$ lattice, and assume all planes register with each other, so that the resulting $3d$ lattice has tetragonal symmetry. We consider first n.n. interactions only so that

$$J_{ll}(\mathbf{k}_\parallel) = J\sum_{\delta_\parallel} e^{i\delta_\parallel \cdot \mathbf{k}_\parallel}$$

and

$$J_{l,l\pm1}(\mathbf{k}_\parallel) = J = J_{l,l\pm1}(\mathbf{k}_\parallel = 0)$$

12.7.1 Monolayer

It was shown [20, 21] that this model exhibits LRO at finite T and accordingly it has a phase transition. The current interest in magnetic multilayer systems [22, 23, 24] in view of technical applications arises from the present day capability of synthezising multilayers with widths varying from several hundred layers down to a monolayer. If a system is confined to a finite width in the direction perpendicular to the layers, one can still apply Mermin-Wagner theorem unless the forces have long range and/or are anisotropic, which is of course the case with dipolar forces.

For the monolayer there is only one order parameter $\langle S^z\rangle = \sigma$. We rewrite Eqs. (12.103) and (12.104), eliminating repeated indices, as

$$\Bigg[\ \omega - \gamma B$$

$$-\sigma\ 2(\ J(0)-J(k)\ )-\gamma^2\left(\ A(0)+\frac{1}{2}A(k)\ \right)\Bigg]G^{+-}(k,\omega)$$

$$+\frac{3\gamma^2}{2}\sigma B^*(k)G^{--}(k,\omega)=\frac{\sigma}{\pi} \qquad (12.105)$$

$$\Bigg[\ \omega + \gamma B$$

$$+\sigma\ 2(\ J(0)-J(k)\ )-\gamma^2\left(\ A(0)+\frac{1}{2}A(k)\ \right)\Bigg]G^{--}(k,\omega)$$

$$-\frac{3\gamma^2}{2}\sigma B(k)G^{+-}(k,\omega)=0 \qquad (12.106)$$

We define

$$b(k) = a^3 B(k) \; ; \quad d(k) = a^3 \mathcal{D}^{zz}(k) = a^3 A(k) \; ; \quad \nu = \omega/J$$

and

$$\lambda(k) = \sigma\{\ 2z\,(1 - \gamma(k)) - E_d\,(d(0) + d(k)/2)\ \}$$

where z = number of first n.n. and $\gamma(k)$ is the usual structure factor. The vanishing of the secular determinant of Eqs. (12.105) and (12.106) leads to the eigenvalues

$$\nu^2 = \lambda(k)^2 - \frac{9}{4}\mid b(k)\mid^2 E_d^2 \sigma^2 \tag{12.107}$$

We must require that the r. h. s. of (12.107) be ≥ 0 for stability.

For a system magnetized along the z axis on the x, z plane with y perpendicular to the plane, the coefficient $B \neq 0$, and there is coupling of G^{+-} and G^{--}. The calculation of $A(k)$ and $B(k)$ [20, 21] can be performed numerically very efficiently with Ewald's method specialized to $2d$. We quote now the main results.

Let us call ϕ the angle that $\mathbf{q}$ makes with the magnetization (*i.e* with the z axis). In the square lattice the dispersion relation for small q is:

$$\left(\frac{\nu}{\sigma E_d}\right)^2 = (608.0\sin^2\phi + 73.0\cos^2\phi)\mid \mathbf{q}\mid \tag{12.108}$$

The approximate calculation in the continuum limit yields only a term in $\sin^2$ [21].

For a layered finite or semi-infinite system the calculation of the dipolar coefficients can also be done using Ewald's method. We shall discuss the results for $q = 0$ near a surface in Chap. 13. Let us briefly mention the results obtained for the simple case of a bilayer, which corresponds to two magnetic layers separated by a few layers of a non-magnetic material.

12.7.2   Bilayer

We consider a system of spins localized on the sites of two identical infinite square lattices of constant a, parallel to the (z, x) plane and separated by a distance y. Both lattices will be assumed exactly registered on one another (atop geometry), corresponding to a 3d tetragonal structure, and in particular a (010) simple cubic stacking for $y = a$. We assume first n .n. ferromagnetic exchange interactions, both in and between the planes,

$$J_{ij} = \begin{cases} J_\parallel & \text{in plane} \\ J_\perp & \text{interplane} \end{cases} \tag{12.109}$$

We choose as before the z axis along the magnetization, in the (z, x) plane, and assume ferromagnetic order, both intra- and inter-planar. If all magnetic sites of the system are equivalent (symmetric bilayer) then $\langle S_l^z \rangle \equiv \sigma$ is independent on the plane index $l = (1, 2)$, We find the inhomogeneous linear system of equations for the Green's functions [25]:

$$\Omega G = \Sigma \tag{12.110}$$

Here, we have:

$$\mathbf{G} = \begin{pmatrix} G_{11}^{+-}(\mathbf{k}_{\parallel}, \omega) & G_{12}^{+-}(\mathbf{k}_{\parallel}, \omega) \\ G_{21}^{+-}(\mathbf{k}_{\parallel}, \omega) & G_{22}^{+-}(\mathbf{k}_{\parallel}, \omega) \\ G_{11}^{--}(\mathbf{k}_{\parallel}, \omega) & G_{12}^{--}(\mathbf{k}_{\parallel}, \omega) \\ G_{21}^{--}(\mathbf{k}_{\parallel}, \omega) & G_{22}^{--}(\mathbf{k}_{\parallel}, \omega) \end{pmatrix} \tag{12.111}$$

$$\Sigma = \frac{1}{\pi} \begin{pmatrix} \sigma & 0 \\ 0 & \sigma \\ 0 & 0 \\ 0 & 0 \end{pmatrix} \tag{12.112}$$

The subindices (1,2) in (12.111) refer to the different planes. The matrix Ω is:

$$\Omega = \begin{pmatrix} \omega + \lambda\sigma & \sigma\eta & \sigma b_1^* & \sigma b_2^* \\ \sigma\eta & \omega + \lambda\sigma & \sigma b_2^* & \sigma b_1^* \\ -\sigma b_1 & -\sigma b_2 & \omega - \lambda\sigma & -\sigma\eta \\ -\sigma b_2 & -\sigma b_1 & \sigma\eta & \omega - \lambda\sigma \end{pmatrix} \tag{12.113}$$

where

$$\lambda = -2\left[J_{11}(0) + J_{12}(0) - J_{11}(\mathbf{k}_{\parallel})\right] + \frac{\gamma^2}{2}\left[2(A_{11}(0) + A_{12}(0)) + A_{11}(\mathbf{k}_{\parallel})\right] \tag{12.114}$$

$$\eta = 2J_{12}(\mathbf{k}_{\parallel}) + \frac{\gamma^2}{2}A_{12}(\mathbf{k}_{\parallel})$$

$$b_1^* = \frac{3\gamma^2}{2}B_{11}^*(\mathbf{k}_{\parallel})$$

$$b_1 = \frac{3\gamma^2}{2}B_{11}(\mathbf{k}_{\parallel})$$

$$b_2^* = \frac{3\gamma^2}{2}B_{12}^*(\mathbf{k}_{\parallel})$$

$$b_2 = \frac{3\gamma^2}{2}B_{12}(\mathbf{k}_{\parallel}) \tag{12.115}$$

The Fourier transforms in (12.114) and (12.115) are:

$$A_{lj}(\mathbf{k}_{\parallel}) = \sum_{\mathbf{R}_{\parallel}} \frac{e^{i\mathbf{k}_{\parallel} \cdot \mathbf{R}_{\parallel}}}{(R_{\parallel}^2 + Y_{lj}^2)^{5/2}}\left[(R_{\parallel}^2 + Y_{lj}^2) - 3Z_{lj}^2\right] \tag{12.116}$$

$$B_{lj}(\mathbf{k}_{\parallel}) = \sum_{\mathbf{R}_{\parallel}} \frac{(R_{\parallel}^+)^2 e^{i\mathbf{k}_{\parallel} \cdot \mathbf{R}_{\parallel}}}{(R_{\parallel}^2 + Y_{lj}^2)^{5/2}} \tag{12.117}$$

$$J_{lj}(\mathbf{k}_\parallel) = \sum_\delta e^{i\mathbf{k}_\parallel \cdot \delta} \tag{12.118}$$

where indices (l, j) identify the planes, and δ is the star of first nearest neighbours of a given site.

The average $\langle S_l^z \rangle = \sigma$ is related as usual to the spectral distribution function of the local propagator, an for $S = 1/2$,

$$\sigma = \frac{1/2}{1 + 2\Phi(T)} \tag{12.119}$$

where:

$$\Phi(T) = -\frac{1}{\pi} \lim_{\epsilon \to 0^+} \int d\omega \left(\frac{a}{2\pi}\right)^2 \int d^2 k_\parallel \frac{\mathcal{I}m \left[\mathbf{\Omega}^{-1}(\mathbf{k}_\parallel, \omega + i\epsilon)_{11}\right]}{e^{\beta\omega} - 1} \tag{12.120}$$

and of course we can change index 1 by 2, since both planes are equivalent. The poles of $\mathbf{G}$, or correspondingly, the zeroes of $det\,\mathbf{\Omega}(\mathbf{k}_\parallel, \omega)$, give the dispersion relations of the acoustic and optical branches of the magnon spectrum. We find:

$$D(\mathbf{k}_\parallel, \omega) = det\,\mathbf{\Omega} = \omega^4 - 2\omega^2(\tilde{a}_0^2 + \tilde{a}_1^2 - \tilde{b}_0^2 - \tilde{b}_1^2) + \Delta \tag{12.121}$$

with

$$\Delta = (\tilde{a}_0^2 + \tilde{a}_1^2 - \tilde{b}_0^2 - \tilde{b}_1^2)^2 - 4(\tilde{b}_0\tilde{b}_1 - \tilde{a}_0\tilde{a}_1)^2 \tag{12.122}$$

The coefficients in (200) and (201) are defined as:

$$\begin{aligned}
\tilde{a}_0 &= \sigma\lambda\ , & \tilde{a}_1 &= \sigma\eta \\
\tilde{a}_1 &= \sigma\eta\ , & \tilde{b}_0 &= \sigma b_1 \\
\tilde{b}_0 &= \sigma b_1\ , & \tilde{b}_1 &= \sigma b_2
\end{aligned} \tag{12.123}$$

The residues of $\mathbf{G}$ at its poles are determined by the cofactor of $(\mathbf{\Omega}(\mathbf{k}_\parallel, \omega))_{11}$. In the neighbourhood of $T = 0$, we calculate the zero-point spin deviation, defined as:

$$\delta(0) = \frac{1}{2} - \sigma(0) = \frac{\Phi(0)}{1 + 2\Phi(0)} \tag{12.124}$$

where:

$$\Phi(0) = \lim_{T \to 0} \Phi = -\frac{a}{2\pi}^2 \int d^2\mathbf{k}_\parallel \sum_{\alpha=1}^2 \frac{C_{11}(-\mid \omega_\alpha \mid)}{D'(\omega_\alpha)} \tag{12.125}$$

$C_{11}(\omega_\alpha)$ is the cofactor of $(\mathbf{\Omega}(\mathbf{k}_\parallel, \omega_\alpha))_{11}$ at the pole ω_α for a fixed $\mathbf{k}_\parallel$, and $D'(\omega_\alpha)$ is the derivative of $det\,\mathbf{\Omega}$ at the pole. The index α denotes the magnon branch $(\alpha = +, -)$. Both the numerator and the denominator inside the sum in (12.125) are polynomials in ω_α of the same order, so it is convenient to introduce the variable

$$\nu_\alpha = \omega_\alpha/\sigma \tag{12.126}$$

to eliminate any explicit dependence on σ. We call

$$G(\nu_\alpha) \equiv \frac{C_{11}(-\omega_\alpha)}{D'(\omega_\alpha)} \tag{12.127}$$

and get for the inverse critical temperature the equation

$$\frac{J_\parallel}{k_B T_c} = \frac{2}{N} \sum_{\mathbf{k}_\parallel} \sum_{\alpha=1}^{2} \frac{G(\nu_\alpha)}{\nu_\alpha} \tag{12.128}$$

The numerical results can be fitted within 5% with the formula [25]

$$\frac{J_\parallel}{k_B T_c} = a + b|\ln E_d| \tag{12.129}$$

The coefficients a and b depend on the particular parameters and geometry. The logarithmic divergence as $E_d \to 0$ in Eq. (12.129) arises from the integral over $2d$ $\mathbf{k}_\parallel$ space in Eq. (12.128) which determines T_c. The vanishing of T_c in the absence of dipolar interactions is consistent with Mermin-Wagner theorem.

The spin deviation at $T = 0$, which measures the effect of the zero point spin fluctuations can be numerically calculated as a function of E_d for $0 < E_d \leq 2 \times 10^{-2}$ and it turns out to be linear in E_d [25] for small E_d.

From the point of view of the existence of LRO the interesting branch is the acoustic one. The important result is:

Exercise 12.13
Prove that for $\mathbf{k}_\parallel = 0$, $A_{12}(0) = -B_{12}(0)$ and that this implies, that

$$\lim_{\mathbf{k}_\parallel \to 0} \nu_-(\mathbf{k}_\parallel) = 0 \tag{12.130}$$

That is, the magnon spectrum has no gap. The dipolar forces stabilize the LRO exclusively because they change the functional form of the dispersion relation for small q. If the interplanar terms B_{12}, A_{12} and J_{12} are neglected, corresponding to the limit of infinite distance between the planes, then $\nu_-(\mathbf{k}_\parallel)$ is proportional to $\sqrt{|\mathbf{k}_\parallel|}$ for very small $|\mathbf{k}_\parallel|$. This is the dispersion relation of a monolayer we found before.

For the coupled planes, instead, $\nu_-^2(\mathbf{k}_\parallel)$ is quadratic in $|\mathbf{k}_\parallel|$. If we study $\nu_-^2(k_z)$, as a function of E_d, we find that the second derivative at $k_z = 0$ increases with E_d (it vanishes for $E_d = 0$), up to a maximum and then decreases, becoming eventually negative for $E_d > E_d^c$ which depends on the parameters. This implies that ν_- is pure imaginary for $E_d > E_d^c$. This instability is a consequence of the static (small k) dipolar field, which favours AFM alignment of the parallel planes, for the chosen geometry. As E_d grows, the stability of the FM order in $3d$ depends on the competition between interplane exchange and dipolar interactions. Although E_d is usually very small, at long distances, or small $|\mathbf{k}_\parallel|$, the dipolar interaction dominates the dispersion relations, and this is why the $3d$ FM inter-planar alignment is not stable in the s.c. structure [26]. Each plane will still be FM, but the relative orientation of the planes will not.

In conclusion, as E_d grows from 0 one first finds the stabilization of LRO, but when E_d reaches a critical value which depends on the system parameters, a crossover is observed from $2d$ to $3d$ behaviour, which for this geometry destabilizes the FM pairing of the planes.

Since the exchange interaction is clearly dominant for large $k \approx \pi/a$, we know that ν^2 must become positive as k grows. That means that if we are in the instability regime $(E_d \geq E_d^c)$ there must be some $k = k_c$ such that $\nu^2 \geq 0$ for $k \geq k_c$. This defines a length scale k_c^{-1} which could be the typical size of domains. Inside one of these the bilayer might have a FM alignment of both planes. This is one possible configuration that may compete to minimize the free energy of the system.

References

1. Mermin, N. D. and Wagner, H. (1966) *Phys. Rev. Lett.* **17**, 1133.

2. Hohenberg, P. C.(1967), *Phys. Rev.* **158**, 383.

3. Majlis, N., Selzer, S. and Strinati, G. C. (1992) *Phys. Rev.* **45**, 7872; (1993) *Phys. Rev.* **48**, 957.

4. Birgenau, R. J., Guggenheim, H. J. and Shirane, G. (1969) *Phys. Rev. Lett.* **22**, 720.

5. Tinus, A. M. C., de Jonge, W. J. M. and Kopinga, K. (1985) *Phys. Rev.* **B32**, 3154.

6. Dingle, R., Lines, M .E. and Holt, S. L. (1969), *Phys. Rev.* **187**, 643.

7. Bogoliubov, N.N. (1962) *Physik. Abhandl. Sowietunion* **6**, 1,113,229.

8. Dyson, Freeman J. (1969) *Commun. Math. Phys.* **12**, 91.

9. Ewald, P. P. (1921) *Ann. Phys. (Leipz.)* **64**, 253.

10. Born, Max and Huang, Kun (1956) *Dynamical Theory of Crystal Lattices*, Oxford at the Clarendon Press, London.

11. Abramowitz, Milton and Stegun, Irene A. (eds.) (1965) *Handbook of Mathematical Functions* Dover Publications, Inc., New York.

12. Peierls, R. E. (1936) *Proc. Cambridge Phil. Soc.* **32**, 477.

13. Gallavoti, G. and Miracle-Sole, S. (1967) *Commun. Math. Phys.* **5**, 317.

14. Rushbrooke, G. and Ursell, H. (1948) *Proc. Cambridge Phil. Soc.* **44**, 263.

15. Ruelle, D. (1968) *Commun. Math. Phys.* **9**, 267.

16. Kac, M. and Thompson, C. J. (1968) *Critical behaviour of several lattice models with long range interaction*, Preprint, Rockefeller University.

17. Humel, M., Pich, C. and Schwabl, F. (1997) Preprint, cond-mat/9703218, 25th March.

18. Selzer, S. and Majlis, N. (1980) *J. Mag. and Mag. Mat.* **15-18**, 1095.

19. The-Hung, Diep, Levy, J. C. S. and Nagai, O. (1979) *Phys. Stat. Sol. (b)* **93**, 351.

20. Maleev, S. V. (1976) *Sov. Phys. JETP* **43**, 1240.

21. Yafet, Y., Kwo J. and Gyorgy, E. M. (1988) *Phys.Rev.* **B33**, 6519.

22. Parkin, S. S. P. (1991) *Phys. Rev. Lett.* **67**, 3598.

23. Celotta, R. J. and Pierce, D. T. (1986) *Science* **234**, 249.

24. Baibich M. N. *et al.* (1988) *Phys. Rev. Lett.* **69**, 969.

25. de Arruda, A. S., Majlis, N., Selzer, S. and Figueiredo, W. (1955) *Phys. Rev.* **B51**, 3933.

26. Cohen, M. H. and Keffer, F. (1955) *Phys. Rev.* **99**, 1128;1135.

Chapter 13

Surface Magnetism

13.1   Introduction

The study of magnetic phenomena at surfaces has been greatly stimulated by the availability of several techniques which are capable to probe the first layers of a magnetic sample. Among them, the progress in methods for ultra high vacuum (UHV) deposition of epitaxial films , in particular molecular beam epitaxy (MBE) [1]. Other methods, as reflection high energy electron diffraction (RHEED), low energy electron diffraction (LEED) and Auger electron spectroscopy (AES) [2], allow us to determine the detailed atomic structure of the synthesized films, one of the critical problems in this field. Finally, the study of the surface magnetization is facilitated by the application of several techniques like spin polarized low energy electron diffraction (SPLEED), electron capture spectroscopy (ECS) [3], spin-analyzed photoemission, electron scanning spectroscopy with spin polarization analysis (SEMPA) [2] and the atomic force microscope (AFM) [4].

For insulators, an ordered surface at temperatures at which the bulk is disordered can only be expected as the result of anisotropy or long range of the interactions. By definition, an ordered surface phase extends only through a finite length into the bulk. Therefore we require LRO in a *two dimensional*, finite width, surface layer. According to the Mermin-Wagner theorem LRO in a $2d$ system cannot exist unless the interactions are anisotropic and/or have long range parallel to the surface. The theorem was in fact proven only for a strictly $2d$ system, but it is not difficult to realize that it can be extended to a finite number of layers. To see this, one can imagine that a perpendicular inter-plane coupling is turned on between a series of parallel two dimensional layers. Independently of the nature of the interactions, as long as they are uniform in the directions parallel to the surface, they cannot destroy the LRO on each plane, although they can lead to an infinite variety of relative orientations of the magnetization on the different planes. Therefore we conjecture that the same conditions for the establishment of LRO as in a strictly $2d$ system will apply.

We remark that in order to have surface magnetism at temperatures greater than the bulk transition temperature, the surface layer must not only be ordered, but its own critical temperature must be greater than the bulk one [5]. Néel [6] analyzed the possibility of surface anisotropy in a FM due to local changes in magneto-elastic and magneto-crystalline interactions at the surface. Similar arguments based on the difference in local environments between surface and bulk spins suggest that one can also expect variations of the exchange anisotropy in going from bulk to surface.

Surface magnetism is found in the rare earths Gd [7] and Tb [8]. These materials are metallic, but their magnetic behaviour is very well described by the localized spins of incomplete f ionic shells, which interact through indirect exchange of the RKKY type.

13.2 MFA treatment of surfaces

In the presence of a surface, the translation symmetry group is two dimensional. We can describe the semi-infinite lattice as a family of crystal planes parallel to the surface. We choose a pair of primitive translations on each plane, which for simplicity we assume identical on all planes, although this depends on the particular crystal structure. Let us call n_1, n_2 the pair of integer coordinates of a site on a particular plane, and denote by n_3 the coordinate of the plane. We choose $n_3 = 0$ at the surface and $n_3 > 0$ for the bulk. In Fig. 13.1 we show a semi-infinite s.c. FM lattice so described.

Let us assume that the translation symmetry along the planes applies as well to $\langle S^z_{\mathbf{n},n_3} \rangle$, so that this quantity only depends on n_3:

$$\langle S^z_{\mathbf{n},n_3} \rangle = \sigma_{n_3} \tag{13.1}$$

It is now straightforward to obtain the molecular field equations. This case is in principle similar to helimagnetism, in that we have many different planar sublattices, each of which has a given σ_{n_3}. The molecular field equations are

$$
\begin{aligned}
\sigma_{n_3} &= B_S(\gamma \beta S B^{mol}_{n_3}) \ , \quad \forall\, n_3 \geq 0 \\
B^{mol}_{n_3} &= \sum_{\mu=0,\pm 1} \sigma_{n_3+\mu} z_\mu J_{n_3,n_3+\mu}
\end{aligned}
\tag{13.2}
$$

where z_μ is the corresponding number of equidistant neighbours. Let us look at some simple examples to clarify the notation used in Eq. (13.2).

Simple cubic (s.c.) (100) surface

We take an exchange interaction of first n. n. range, with value J in the bulk, $J_\parallel$ for bonds on the surface plane, and $J_\perp$ for bonds between planes $n_3 = 0$ (surface) and $n_3 = 1$, as depicted in Fig. 13.1. In the absence of an external field and of anisotropy the orientation of the magnetization is immaterial. However,

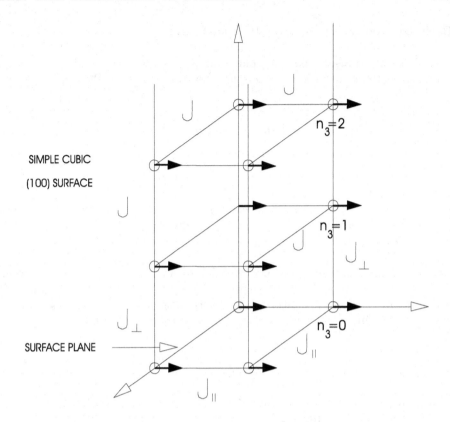

Figure 13.1: *Schematic description of the surface of a semi-infinite s.c. ferromagnet. Horizontal axes n_1, n_2. The vertical coordinate n_3 increases into the bulk. Different values of J_{ij} in the surface region are indicated. The bulk exchange constant is denoted J.*

in real systems, the presence of the dipolar interactions, in the absence of other anisotropies, orients the magnetization parallel to the surface. We assume this is the case, which results in a vanishing depolarizing field in the bulk. We sall neglect, for simplicity, the possibility of of a surface depolarizing field.

The molecular field on the different planes is:

$$
\begin{aligned}
B_0^{mol} &= 4J_\parallel \sigma_0 + J_\perp \sigma_1 \\
B_1^{mol} &= 4J\sigma_1 + J_\perp \sigma_0 + J\sigma_2 \\
B_{n_3}^{mol} &= J(\sigma_{n_3+1} + 4\sigma_{n_3} + \sigma_{n_3-1}) \ , \ \forall \, n_3 > 1.
\end{aligned} \tag{13.3}
$$

Body centred cubic (b.c.c.) (100) surface

In this case spins on the (100) planes are second nearest neighbours of each other, so that the corresponding coupling constant vanishes. We have:

$$B_0^{mol} = 4J_\perp \sigma_1$$
$$B_1^{mol} = 4J\sigma_2 + 4J_\perp \sigma_0$$
$$B_{n_3}^{mol} = 4J(\sigma_{n_3+1} + \sigma_{n_3-1}) \ , \ n_3 > 1 \tag{13.4}$$

Exercise 13.1
Obtain the expressions of the molecular fields for the (111) face of the f.c.c. lattice under the previous approximations. In this case we assume the magnetization is along $(1\bar{1}0)$.

At $T = 0$, if all the interactions are FM, we expect to find the system in the Weiss ground state, irrespective of the values of the parameters. At finite T we must solve the set of equations (13.2) self-consistently. One can pose oneself the questions (in the following s stands for surface, b for bulk):

1. Does σ_s vanish at $T < T_c^b$?

2. Is it possible that $\sigma_s > \sigma_b$? And, if so,

3. Is it possible that the surface stays magnetized at $T > T_c^b$ with the bulk in the PM phase ($\sigma_b = 0$)?

Within MFA the answer is *yes* to 2) and 3), *no* to 1) [9]. However, calculations which take into account thermal fluctuations of the transverse spin components, even within the FSWA or RPA yield different answers.

Let us start with 1). For n_3 large enough, the molecular field equations are identical for all successive planes and then we must get the same σ_b as without a surface and the same critical temperature T_c^b. At $T = 0$ in a FM we know that $\sigma_{n_3} = S$ on all planes, so that the profile is completely flat. At $T > 0$, in general, $\sigma_{n_3} \neq \sigma_b$ near the surface. If $J_\parallel > J$ and/or $J_\perp > J$ we could have $\sigma_s > \sigma_b$. In this case MFA can yield a non-zero surface magnetization at $T > T_c^b$. The magnetization decreases as n_3 grows and it vanishes deep inside the crystal. As already mentioned, this result is a consequence of the MFA, and we shall see that the picture can be different in the RPA. At any rate, when the surface remains ordered at $T > T_c^b$, the MFA predicts a second transition, called the *surface transition*, at some T_c^s at which $\sigma_s \to 0$. In the case that the transition to the PM phase at the surface occurs at $T_c^s < T_c^b$ we speak of the *ordinary* surface transition, and in this case the surface is controlled by the bulk. When both temperatures coincide the transition is multicritical and is called *special*. When there is a surface transition, so that $T_c^s > T_c^b$, the MFA predicts a discontinuity in the first derivative of the magnetization with respect to T, and this is called the *extraordinary* transition.

The critical behaviour of magnetic surfaces has been analyzed in a series of papers [9, 10]. We shall proceed now to describe the results obtained with the self consistent LRPA (Layer RPA) for ferromagnets, which was already introduced in the previous chapter [12].

13.3 Surface excitations

We extend now the LRPA by considering as many sublattices as necessary to describe reasonably the non uniformity of the magnetization near to a surface. We assume that the local average $\langle\,\sigma^z(\mathbf{n}_\parallel, n)\,\rangle\,=\,\sigma_n^z$ is uniform across each plane n, where n stands for n_3, so that $n = 0$ is the surface plane and $n \geq 0$ for the semi-infinite lattice. The argument that follows can be applied at low T to obtain a spin-wave theory of surface magnons. If however the magnetization is not saturated and we take it as a T dependent parameter, we shall need to extend the treatment and solve the self-consistency problem by application of the LRPA.

The translations perpendicular to the surface are not symmetry operations of the Hamiltonian. As already mentioned, those parallel to the surface can still be considered as symmetry operations if the system extends to infinity in both directions parallel to the surface plane, which we will assume throughout. If Bloch waves are used as a basis to expand the magnon wave functions, the $k_\perp$ component of the wave vector need not be real as in an infinite system. A Bloch wave with a real $k_\perp > 0$ arriving at the surface from the bulk will be reflected, with a perpendicular wave-vector $-k_\perp$, and in general with some possible change of phase and amplitude, so that as a result we expect that bulk solutions are standing waves, with a continuum spectrum which resembles very closely that of the infinite bulk.

On the other hand, since the system is bounded by the surface, we can consider stationary wave functions with $\mathcal{I}m(k_\perp) \neq 0$, which must vanish far into the bulk, since they must be normalizable over the whole semi-infinite space occupied by the system. These waves, whose amplitude decreases exponentially towards the bulk, are the basis for obtaining surface-localized magnons. Therefore we are again facing the problem of the analytic continuation of the bulk spectrum as in the previous discussion of indirect exchange (see Chap. 10). By imposing the correct boundary conditions at the surface and the reality of the energy of the magnon energy we can in fact obtain in some cases the surface magnon energy and wave function [13].

These general considerations are valid whatever the Hamiltonian we adopt as model for the system. In most of this chapter we shall leave aside, for simplicity, the dipolar interactions. In a macroscopic approach, the exchange-dipolar surface magnons were studied by Damon and Eshbach [11]. A microscopic calculation of dipolar-exchange spin excitations in a system with a surface, at finite T, is cumbersome but it does not present new complications, although we are not aware of any example in the literature.

13.4   LRPA method

Let us consider again the isotropic Heisenberg hamiltonian

$$H = -\sum_{l \neq m} J_{lm} \mathbf{S_l} \cdot \mathbf{S_m} \tag{13.5}$$

and the retarded Green's functions

$$\langle\langle S_{l\,\mathbf{1}_\parallel}^+ ; S_{n\,\mathbf{n}_\parallel}^- \rangle\rangle^{(r)} \equiv G^{(r)} \left(l\,\mathbf{1}_\parallel, n\,\mathbf{n}_\parallel; \ (t - t_0) \right) \tag{13.6}$$

with the Fourier transform

$$G_{ln}^{(r)}(\mathbf{k}_\parallel, \omega + i\epsilon) = \sum_{\mathbf{R}_\parallel} e^{i\mathbf{k}_\parallel \cdot \mathbf{R}_\parallel} G_{ln}^{(r)}(\mathbf{R}_\parallel, \omega + i\epsilon) \tag{13.7}$$

with $\mathbf{R}_\parallel = \mathbf{1}_\parallel - \mathbf{n}_\parallel$, using the translation invariance. We can also Fourier transform the exchange coupling:

$$J_{ln}(\mathbf{k}_\parallel) = \sum_{\mathbf{R}_\parallel} e^{i\mathbf{k}_\parallel \cdot \mathbf{R}_\parallel} J_{ln}(\mathbf{R}_\parallel) \tag{13.8}$$

We write the equation of motion for the retarded Green's function of Eq. (13.6) and make the usual factorization of the extra factor S^z in the Green's function with three operators. As in the MFA, we assume that in the present non-uniform case the average $\langle\, S_n^z\,\rangle$ is a function of the plane index. The resulting equations for the Green's functions are conveniently written in dimensionless form, by defining:

$$\nu = \omega/2J \ \ ; \ \ g_{ln} = JG_{ln} \ \ ; \epsilon_{ln} = J_{ln}/J \tag{13.9}$$

and we find that the matrix $\mathbf{g}$ (with plane indices l, n) satisfies the equation:

$$(\nu\mathbf{1} - \mathbf{h^{eff}})\mathbf{g} = \mathbf{\Sigma}/\pi \tag{13.10}$$

where

$$
\begin{aligned}
(\mathbf{\Sigma})_{lj} &= \delta_{lj}\sigma_l \ \ , \ \ \sigma_l = \langle\, S_l^z\,\rangle \\
(\mathbf{h}^{eff})_{lj} &= \delta_{lj} \sum_i \epsilon_{ij}(\mathbf{k}_\parallel = 0)\sigma_i - \sigma_l\epsilon_{jl}(\mathbf{k}_\parallel)
\end{aligned} \tag{13.11}
$$

The matrix equation (13.10) can be written

$$\mathbf{\Omega g} = \frac{1}{\pi}\mathbf{\Sigma} \tag{13.12}$$

For planes well inside the bulk, where there are no alterations of the exchange constants or of the local coordination, the diagonal matrix element of $\mathbf{h}^{eff}$ is

$$h_{mm}^{eff} = h_b^{eff} \equiv z_\perp \sigma_b + z_\parallel \sigma_b(1 - \gamma_0(\mathbf{k}_\parallel)) \tag{13.13}$$

Here $\gamma_0(\mathbf{k}_\parallel)$ and $z_\parallel$ are respectively the structure factor and the coordination number of the two-dimensional planar lattice, $z_\perp$ is the number of first n.n. of a given site which belong to adjacent layers and σ_b is the bulk magnetization. All diagonal matrix elements (13.13) are equal for large enough m. Therefore we shall write the diagonal term for all m as the bulk one (Eq. (13.13)) plus a local deviation and define

$$\nu - (\mathbf{h}^{eff})_{mm} \equiv 2t + \alpha_{mm} \tag{13.14}$$

where

$$2t \equiv \nu - \mathbf{h}_b^{eff} \quad , \quad \alpha_{mm} \equiv \mathbf{h}_b^{eff} - \mathbf{h}_{mm}^{eff} \tag{13.15}$$

We rewrite accordingly the elements of $\mathbf{\Omega}$ as

$$\mathbf{\Omega g} = \frac{1}{\pi}\mathbf{\Sigma} \tag{13.16}$$

where

$$\begin{aligned}
\Omega_{ll} &= 2t + \alpha_{ll} \\
\Omega_{lj} &\equiv \beta_{lj} = -\sigma_l \epsilon_{jl}(\mathbf{k}_\parallel) \quad , \quad j \neq l \ .
\end{aligned} \tag{13.17}$$

From Eq. (13.10) we get:

$$\mathbf{g} = \frac{1}{\pi}\mathbf{\Omega}^{-1}\mathbf{\Sigma} \tag{13.18}$$

so that the main mathematical problem is now the inversion of $\mathbf{\Omega}$.

The extension to this case of the fluctuation-dissipation theorem and of Callen's formulas for the self-consistent evaluation of σ_n (see Chap. 6) is immediate [12]. In Eq. (5.90) we express the local magnetization in terms of one function Ψ in a uniform FM, while in the non-uniform case we need one such function for each plane, so that we have, for plane n:

$$\Psi_n = -\frac{1}{N_s}\sum_{\mathbf{k}_\parallel}\int \frac{d\omega}{\pi}\mathcal{I}m(\mathbf{\Omega}^{-1})_{nn}(\mathbf{k}_\parallel, \omega + i\epsilon)N(\omega) \tag{13.19}$$

For a FM with spin S the layer magetization is given by Callen's formula for the layer:

$$\langle S_n^z \rangle = \frac{(S - \Psi_n)(1 + \Psi_n)^{2S+1} + (S + 1 + \Psi_n)\Psi_n^{2S+1}}{(1 + \Psi_n)^{2S+1} - \Psi_n^{2S+1}} \tag{13.20}$$

Let us now look at the (100) surface of a s.c. latice as a specific example. In principle there is an indeterminate number of different values of σ_n if the non-uniform profile extends to infinity. However, it is reasonable to assume that the influence of the surface only affects a finite number of planes. Although this may seem dubious for the dipolar interactions, one can verify that their contribution to the effective hamiltonian $\mathbf{h}^{eff}$ decays very rapidly, in fact exponentially fast, towards the interior of the system [14]. Therefore we make the following simplifying assumptions:

1. The coupling constants are different from the bulk only for intra-surface bonds and bonds between surface and the next plane;

2. The magnetization profile saturates to the bulk value at the third plane of the semi-infinite system, *i.e.* $\sigma_n = \sigma_b$, $n \geq 2$.

This situation is indicated in Fig. 13.1. For this case one can easily work out the matrix elements of $\mathbf{\Omega}$:

$$\alpha_{00x} = 2\sigma_b - \sigma_1\epsilon_\perp + (\sigma_b - \sigma_0)\Lambda_k$$
$$\alpha_{11} = \sigma_b - \sigma_0\epsilon_\perp + (\sigma_b - \sigma_1)\Lambda_k$$
$$\alpha_{22} = (\sigma_b - \sigma_1)$$
$$\beta_{01} = \sigma_0\epsilon_\perp$$
$$\beta_{10} = \sigma_0\epsilon_\perp$$
$$\beta_{12} = \sigma_1$$
$$\beta_{21} = 1 \tag{13.21}$$

where we defined
$$\Lambda_k = 4\left(1 - \gamma(\mathbf{k}_\|)\right) \tag{13.22}$$

All β's following the ones above are $= 1$.
The matrices $\mathbf{\Omega}$ and $\mathbf{\Sigma}$ are:

$$\mathbf{\Omega} = \begin{pmatrix} 2t + \alpha_{00} & \beta_{01} & 0 & 0 & 0 & \cdots \\ \beta_{10} & 2t + \alpha_{11} & \beta_{12} & 0 & 0 & \cdots \\ 0 & \beta_{21} & 2t + \alpha_{22} & 1 & 0 & \cdots \\ 0 & 0 & 1 & 2t & 1 & \cdots \\ \vdots & \vdots & \vdots & \ddots & \ddots & \ddots \end{pmatrix} \tag{13.23}$$

$$\mathbf{\Sigma} = \frac{1}{\pi} \begin{pmatrix} \sigma_0 & 0 & 0 & 0 & \cdots \\ 0 & \sigma_1 & 0 & 0 & \cdots \\ 0 & 0 & \sigma_b & 0 & \cdots \\ 0 & 0 & 0 & \sigma_b & 0 \\ \vdots & \vdots & \vdots & \vdots & \ddots \end{pmatrix} \tag{13.24}$$

The three-diagonal matrix (13.23) can be directly inverted:

$$(\mathbf{\Omega}^{-1})_{nm} = \frac{Adj(\Omega_{nm})}{det\mathbf{\Omega}} \tag{13.25}$$

Let us for the moment admit that we have a finite number of planes in the system, say N. Then $\mathbf{\Omega}$ is $N \times N$. Let us call D_{N-p} , $0 \leq p \leq N - 1$, the determinants of the lower minor of $\mathbf{\Omega}$ of order successively decreasing with increasing p, so that $D_N = det\mathbf{\Omega}$. In particular, if we want to calculate the surface magnetization we need

$$(\mathbf{\Omega}^{-1})_{00} = \frac{D_{N-1}}{D_N} \tag{13.26}$$

The direct expansion of D_N by the first row gives

$$D_N = (2t + \alpha_{00})D_{N-1} - \beta_{10}\beta_{01}D_{N-2} \tag{13.27}$$

For $p \geq 3$, in our special example, the recursive relations between these determinants have coefficients independent on p :

$$D_{N-p} = 2tD_{N-p-1} - D_{N-p-2} \tag{13.28}$$

Suppose that in the interval of t we shall be interested in,
$D_{N-p} \neq 0$, $\forall\, p \geq p_0$ for some integer p_0. Then for $p > p_0 + 1$ let us divide
Eq. (13.28) by D_{N-p-1}:

$$\frac{D_{N-p}}{D_{N-p-1}} = 2t - \frac{D_{N-p-2}}{D_{N-p-1}} \tag{13.29}$$

We take the thermodynamic limit and define

$$\lim_{N \to \infty} \frac{D_{N-p-1}}{D_{N-p}} = \xi \tag{13.30}$$

assuming it exists, in which case it must be obviously p -independent. We remark that the ratio in Eq. (13.29) can be expressed as a continued fraction:

$$\frac{D_{N-p-1}}{D_{N-p}} = \frac{1}{2t - \frac{D_{N-p-2}}{D_{N-p-1}}} \tag{13.31}$$

which as $N \to \infty$ can be written as

$$\xi = \frac{1}{2t - \xi} \tag{13.32}$$

or

$$\xi + \xi^{-1} = 2t \tag{13.33}$$

In the more general case in which the quantities in Eq. (13.28) are matrices, ξ is called the *transfer matrix*, and we shall use this name for the case in hand in which it is a scalar.

The solutions of Eq. (13.33) are

$$\xi = t \pm \sqrt{t^2 - 1} \tag{13.34}$$

The continued fraction which results from the repeated iteration of (13.32),

$$\xi = \frac{1}{2t - \frac{1}{2t - \cdots}} \tag{13.35}$$

converges for $|t| > 1$ to the solution of (13.33) which has the minimum absolute value, which implies $|\xi| < 1$ [15]. The other solution is spurious. This is crucial in searching for the poles of the Green's function, as we shall see. In

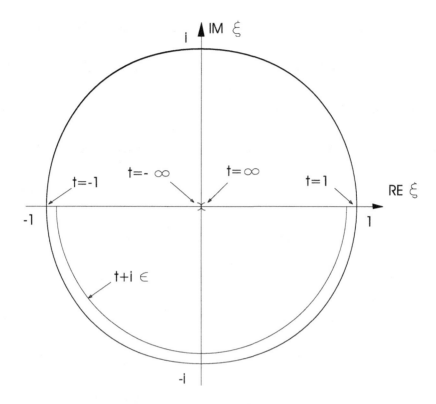

Figure 13.2: *Convergence region of the continued fraccion expansion for the transfer matrix .*

Fig. 13.2 we show the convergence circle for the continued fraction (13.32). In order to understand the physical meaning of the transfer matrix let us return to the Green's function equations (13.10), and write in particular an off-diagonal element, in which case the r. h. s. vanishes, and we write

$$\Omega_{n,n-1}\, g_{n-1,m} + \Omega_{nn}\, g_{n,m} + \Omega_{n,n+1}\, g_{n+1,m} = 0 \qquad (13.36)$$

For n, m large enough, we would have

$$g_{n-1,m} + 2tg_{n,m} + g_{n+1,m} = 0 \qquad (13.37)$$

Suppose now that

$$g_{n+1,m} = \rho g_{n,m} \qquad (13.38)$$

If we use this relation in (13.37) we find that ρ satisfies the quadratic equation

$$\rho^2 + 2t\rho + 1 = 0 \qquad (13.39)$$

so that $-\rho = \xi$. We get further insight into the role played by the transfer matrix if we expand $\mathbf{h}^{eff}$ in terms of its eigenvectors and eigenvalues, as

$$\mathbf{h}^{eff} = \sum_{\mu} E_{\mu} \mid \mu \rangle \langle \mu \mid \qquad (13.40)$$

Then the resolvent operator $\mathbf{\Omega}^{-1}$ is

$$(\mathbf{\Omega}^{-1})(\nu + i\epsilon) = \sum_{\mu} \frac{\mid \mu \rangle \langle \mu \mid}{\nu + i\epsilon - E_{\mu}} \qquad (13.41)$$

and the generic matrix element of the retarded Green's function is

$$g_{n,m} = \sum_{\mu} \frac{\langle n \mid \mu \rangle \langle \mu \mid m \rangle}{\nu + i\epsilon - E_{\mu}} \frac{\sigma_m}{\pi} \qquad (13.42)$$

where $\langle n \mid \mu \rangle \equiv \psi_n^{(\mu)}$ = amplitude of the wave function of the magnetic excitation with energy E_{μ} at site n . If we assume now that for fixed m (13.38) is valid, we must have as well

$$\psi_{n+1}^{(\mu)} = \rho^{(\mu)} \psi_n^{(\mu)} \qquad (13.43)$$

With this result in hand, let us suppose that $\mid \mu \rangle$ is a surface magnon. Then we must require $\mid \rho^{(\mu)} \mid < 1$, since we know that the amplitude of a surface wave must decrease towards the bulk. If on the contrary $\mid \mu \rangle$ where a bulk magnon, we require $\mid \rho^{(\mu)} \mid = 1$, because a bulk excitation must propagate throughout the whole system. We can then write for a bulk magnon

$$\rho = e^{ik_{\perp}a} \qquad (13.44)$$

with $k_{\perp}$ = real. As a matter of fact, we can also express ρ for surface states as in (13.44), but then $\mathcal{I}mk_{\perp} > 0$.
Substituting ρ from Eq. (13.44) into Eq. (13.39) we have

$$t = -\cos k_{\perp}a \qquad (13.45)$$

Suppose we choose for the s.c. lattice the axes x, z along the surface, and $k_{\perp} = k_y$. With this choice, we can go back to Eqs. (13.13) and (13.15) and we find, for bulk states, that the poles of the Green's function are

$$\nu = 6 - 2(\cos k_x a + \cos k_y a + \cos k_z a) \qquad (13.46)$$

which is exactly the dispersion relation for bulk magnons in a s.c. ferromagnet (see Chap. 4). We verify that the bulk states have a uni-modular transfer matrix and that their dispersion relation is the same for the semi-infinite as for the infinite system.

13.5   Wave functions for bulk and surface

The matrix (13.10) has the same form as the tight-binding Hamiltonian for a semi-infinite $1d$ metal with a first n.n. range hopping term, or, with the change $\nu \to \omega^2$, that of phonons in a chain with short range elastic coupling between neighbouring atoms, etc. We must however keep in mind that the matrix elements in (13.10) depend on $\mathbf{k}_\parallel$, which restores the three-dimensional behaviour of the system. We can say that the representation chosen has transformed a semi-infinite $3d$ system into a family of semi-infinite chains , each one dependent parametrically on the corresponding $\mathbf{k}_\parallel$ within the $2d$ Brillouin zone of the planar lattice. A given eigenstate of the semi-infinite chain can be labelled by $\mathbf{k}_\parallel$ and some other quantum numbers that we can call generically μ, so that the corresponding ket can be denoted as $\mid \mu \mathbf{k}_\parallel \rangle$. Let us simplify the notation and call this state $\mid \psi \rangle$ and the eigenvalue $E(\mu \mathbf{k}_\parallel) = \nu$. The layer index in this representation can be interpreted as the site index in the one dimensional mapping of the problem. Let us then write the Schrödinger equation

$$\mathbf{h}^{eff} \mid \psi \rangle = \nu \mid \psi \rangle \tag{13.47}$$

and expand the eigenket in terms of site eigenkets $\{ \mid n \rangle \}$:

$$\mid \psi \rangle = \sum_n \psi_n \mid n \rangle \tag{13.48}$$

Let us now consider the simple case of the s.c. lattice without change of surface parameters.

Exercise 13.2
Show that if there is no change of parameters at the surface, and if T is very low, so that the magnetization profile is flat, one finds

$$\alpha_{00} = 1 \ , \quad \alpha_{nn} = 0 \ , \quad \forall \, n > 0 \tag{13.49}$$

$$\beta_{i,i\pm1} = 1 \ , \quad \beta_{i,j} = 0 \ \text{ otherwise.} \tag{13.50}$$

Notice that this Hamiltonian is equivalent to that of a semi-infinite chain *with the diagonal end parameter altered*, since in the case without surface perturbation one should have $\alpha_{00} = 0$ aswell. Therefore the set of linear equations satisfied by the amplitudes $\{\psi_n\}$ in (13.47) are:

$$
\begin{aligned}
(2t + 1)\psi_0 + \psi_1 &= 0 \\
\psi_0 + 2t\psi_1 + \psi_2 &= 0 \\
&\cdots
\end{aligned}
$$

$$\psi_{n-1} + 2t\psi_n + \psi_{n+1} = 0 \tag{13.51}$$

We already know that the Ansatz $\psi_n = \rho \psi_{n-1}$ solves these equations. Let us first consider the bulk states, so that

$$\rho = e^{i\theta} \tag{13.52}$$

From Eq. (13.39) we also know that $\rho^{-1} = e^{-i\theta}$ is on the same footing as ρ, so that we can construct a general solution of the last of (13.51) by combining these two solutions. Since we can iterate the application of the transfer matrix we finally arrive at the general solution

$$\psi_n = A \left(e^{in\theta} + \eta e^{i\phi} e^{-in\theta} \right) \tag{13.53}$$

with the coefficient η and the phase ϕ to be determined by the boundary condition provided by the first equation of the system (13.51), which is obviously not automatically satisfied by (13.53). The constant A will be fixed by the condition that the wave function be normalized to one. From Eq. (13.39) we also get $t = -\cos\theta$.

Exercise 13.3
Show that the solution of the system (13.51) with $A = 1$ is

$$\psi_n = 2e^{-i\theta/2} \cos{(n + 1/2)\theta} \tag{13.54}$$

13.6 Surface density of magnon states

The density of states of a Hamiltonian H is defined as

$$\rho(\nu) = Tr\delta(\nu - H) \tag{13.55}$$

In the case of our effective one dimensional Hamiltonian, we can calculate the trace for a particular $\mathbf{k}_\parallel$, which yields a partial density of states, so that we just need to sum over all $\mathbf{k}_\parallel$ within the first Brillouin zone (BZ) of the planar lattice to get the total density of states. We can calculate the trace (which we know is invariant under unitary transformations) in the basis of the layer states, defined as

$$| n \rangle = S^-_{\mathbf{k}_\parallel, n} | 0 \rangle \tag{13.56}$$

where we assume that the Weiss ground state $| 0 \rangle$ is polarized $\uparrow$ and we simplified the notation avoiding the explicit reference to $\mathbf{k}_\parallel$. Then

$$\rho(\nu) = \sum_n \sum_\mu \langle n | \mu \rangle \delta(\nu - E_\mu) \langle \mu | n \rangle \tag{13.57}$$

where we have inserted twice the unit operator

$$\sum_\mu | \mu \rangle \langle \mu | = 1 \tag{13.58}$$

We can also write (13.57) as

$$\rho(\nu) = \lim_{\epsilon \to 0} \mp \frac{1}{\pi} \mathcal{I}m \left\{ \frac{1}{\nu - H \pm i\epsilon} \right\} \tag{13.59}$$

or

$$\rho(\nu) = \lim_{\epsilon \to 0} \mp \frac{1}{\pi} \sum_n \mathcal{I}m \left\{ \langle n \mid \frac{1}{\nu - H \pm i\epsilon} \mid n \rangle \right\} \qquad (13.60)$$

The expression between curly brackets above is precisely

$$(\boldsymbol{\Omega}^{-1})_{nn} \quad .$$

From Eq. (13.60) we infer that this quantity is the local projection of the total density of states on that particular layer, for that particular $\mathbf{k}_{\parallel}$. Let us consider the previous example of a s.c. (001) surface at very low temperatures, so that $\sigma_n = \sigma_b$, $\forall$ $n=$. Let us for a moment consider also the case in which the parameters at the surface coincide with the bulk ones. This does not mean however that the system is uniform, since near the surface there are less bonds, and this is equivalent to a perturbation potential, as we have just seen. In this particular example, all the diagonal perturbation terms $\alpha_{nn} = 0$ with the exception of $\alpha_{00} = 1$.

Exercise 13.4
Show that in this case we obtain

$$(\boldsymbol{\Omega}^{-1})_{00}(\nu + i\epsilon) = \frac{1}{2t + 1 - \xi} \qquad (13.61)$$

We just found that for bulk magnons $\xi = e^{-i\phi}$, where we made a convenient choice of sign in the exponent. We also have $t = \cos \phi$ from our previous results, so that we find

$$\rho_0(\nu) = -\frac{1}{\pi} \mathcal{I}m \ (\boldsymbol{\Omega}^{-1})_{00}(\nu + i\epsilon) = \frac{1}{2\pi} \sqrt{\frac{1 - t}{1 + t}} \qquad (13.62)$$

The bulk states contribute to the imaginary part of the resolvent operator in the interval of energies where the continued fraction for ξ has a cut along the interval $-1 \le t \le 1$ (see Fig. 13.2). As a consequence, in this continuum range of bulk energies the Green's function also has a cut on the real axis. On the other hand, for $\mid \xi \mid < 1$ we can find real poles of g_{00}. Each localized eigenstate (surface magnon) adds a delta function contribution to the density of states. We see from Eq. (13.61), which corresponds to the special case in which the surface parameters are the same as in the bulk, that the resolvent(or equivalently g_{00}) has only one real pole, at $\xi = -1$, which according to the discussion in the previous section does not describe a surface state, for which we need $\mid \xi \mid < 1$.

In the general case in which we have different parameters in the surface region, we may find one or more real poles of the Green's functions, besides the continuum of bulk sates. A real pole with $\mid \xi \mid < 1$ determines the energy and the penetration length $\lambda = -(\log \mid \xi \mid)^{-1}$ of the corresponding surface magnon.

This example indicates how one can construct an algorithm to obtain the surface states spectrum as a function of $\mathbf{k}_{\parallel}$, and to calculate the resolvent operator both in the bulk continuum and in the discrete sector of the spectrum contributed by the surface magnons. With this we can in turn calculate the function Ψ_n and therefrom the layer magnetization. This program for the simple cubic (010) surface leads to results shown in Fig. 13.3, where we plot $\omega/(2J\sigma_b)$ for the different branches of surface magnons, together with the limits of the continuum spectrum, vs. $\Lambda = 1 - \gamma(\mathbf{k}_\parallel)$, where γ is the 2d structure factor defined before. In Fig. 13.4 we show a profile obtained for the case in which

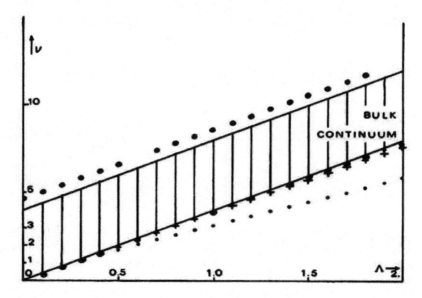

Figure 13.3: *Dispersion relation of surface magnons in a s.c. FM. $\nu = \omega/(2J\sigma_b(T))$ vs. $\Lambda = 1 - \gamma(\mathbf{k}_\parallel)$. $S = 1/2$, $\epsilon_\parallel = 1$. For $T/T_c = 0.17$, crosses correspond to $\epsilon_\perp = 1$, open circles to $\epsilon_\perp = 2$. Full circles correspond to $T/T_c = 0.3$ and $\epsilon_\perp = 1$. From [16]*

the magnetization of the third plane has been fixed at the bulk value. The temperature is below the bulk critical temperature. Note that even in the case $\epsilon_\perp = \epsilon_\parallel = 1$ the surface magnetization is smaller than the bulk one at finite T, as expected.

13.7 Surface phase-transitions

LRPA calculations for several lattice structures show that as the coupling $J_\perp$ of the surface to the bulk decreases the surface magnetization also decreases. This tendency can be clearly verified in Fig. 13.5. The possibility that the surface be magnetized for $T > T_c^b$ and accordingly that there be a phase transition localized at the surface at $T_c^s > T_c^b$ has been studied with a Heisenberg Hamiltonian with exchange anisotropy [5]. One may conveniently write such a Hamiltonian with

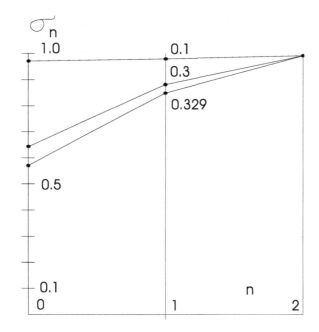

Figure 13.4: *Magnetization profile near the surface of a s.c.* $S = 1/2$ *FM for* $\epsilon_\perp = \epsilon_\parallel = 1$. *We plot* $\sigma_n = \langle S_n^z(T)\rangle/\langle S_b^z(T)\rangle$ *vs.* $n = $ *plane index. The surface plane is* $n = 0$. *The values of* $\tau = k_B T/6J$ *are indicated on the curves, and the critical value is* $\tau_c = 0.33$. [16]

the restriction to first n.n. interactions as

$$H = -(J^I/\eta) \sum_{\langle ij \rangle} \left(\mathbf{S_i} \cdot \mathbf{S_j} + (\eta - 1)S_i^z S_j^z \right) \tag{13.63}$$

Then when $\eta \to 1$ we obtain the isotropic Heisenberg Hamiltonian, while if $\eta \to \infty$ we obtain the Ising Hamiltonian. For any finite $\eta > 1$, calling $J = J^I/\eta$, we get the bulk spin-wave dispersion relation

$$\frac{\omega(k)}{2Jz\sigma} = \Delta + (1 - \gamma(k)) \tag{13.64}$$

with $\Delta = (\eta - 1)=$ gap at $k = 0$. This gap will occur in any dimension of the lattice, so that in this case Mermin-Wagner theorem does not discard LRO at finite temperatures for $d < 3$. The LRO however might still be absent in 1d at finite T due to kinks which increase the entropy, as discussed in Chap. 12. In 3d, when $\eta > 1$ we find for $J > 0$ a FM ground state with spins aligned along z (easy axis). If $\eta < 1$ instead, spins tend to align on the plane perpendicular to z (easy plane). In 2d, if we take $z \perp$ to the lattice plane, we obtain the "XY" model. In this case, at low T $\langle S_i^z \rangle \approx 0$ and one can treat the spins as two-dimensional vector operators. This model has an interesting phase transition as

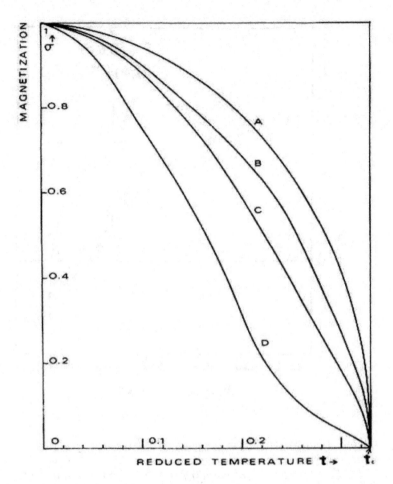

Figure 13.5: *Bulk $\sigma_b(T)/S$ (A) and surface $\sigma_s(T)/S$ for a s.c. FM vs. $t = k_B T/6J$ for $S = 1/2$ and $\epsilon_\parallel = 1$. The values of $\epsilon_\perp$ for the different curves are: $1(B); 0.5(C)$ and $0.1(D)$ [16].*

T decreases, from the PM disordered phase to a phase which has no long range order but in which the magnetization on the plane has long-range correlations. This phase, described by Kosterlitz and Thouless [17] has topological defects in the form of vortices of the magnetization field.

In summary, systems with $\eta > 1$ belong to the Ising universality class, those with $\eta < 1$ to the XY universality class, and those with $eta = 1$ to the Heisenberg one.

In the anisotropic case with $\eta > 1$, or at any rate with $\eta_s > 1$ on the surface plane, we can expect a surface phase transition. If η_s is large enough there is the possibility of a surface critical temperature $T_c^s > T_c^b$. This behaviour is

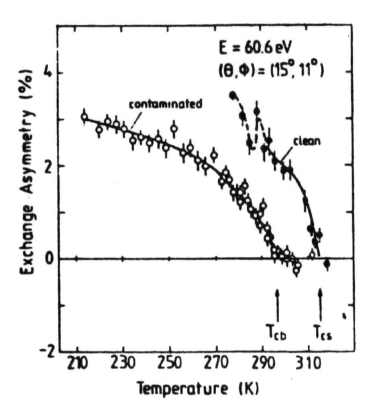

Figure 13.6: *Temperature dependence of the SPLEED asymmetry A, proportional to the surface magnetization, in a Gd (0001) film of 140 Å thick. $T_s = 315$ K, $T_b = 293$ K. No asymmetry is detected on a contaminated surface (open circles).* [7]

observed in Gd [7] (see Fig. 13.6) and Tb [8] (see Fig. 13.7). The cause of the sudden increase of the surface magnetization in these metals at T near the surface transition to the PM phase is not clear [18].

An ordinary surface phase transition was observed in the EuS (111) surface [19]. These insulator is a model bulk Heisenberg FM, but the measured critical exponent of the surface, different from the bulk, could signal the presence of surface anisotropy.

Calculations of the magnetization profile within the LRPA yield a surface phase transition for a sufficiently anisotropic surface, with $\eta_s > \eta_c$ [5]. The minimum surface anisotropy η_c depends on the geometry and on the other surface parameters.

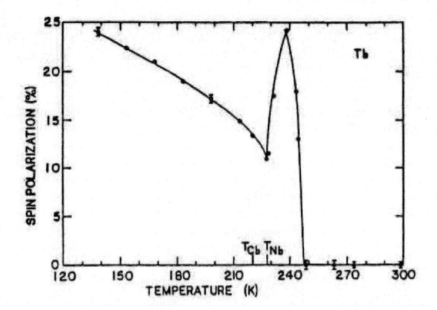

Figure 13.7: *Electron spin polarization of the topmost surface layer of 1mm thick Tb samples, as function of T. T_{cb} is the bulk Curie temperature and T_{Nb} the bulk Néel temperature [3].[8]*

13.8 Dipolar surface effects

Let us consider now the effect of dipolar interactions at the surface. Until now we have explicitly assumed (see Chapter 7) that all points at which the dipolar fields were calculated were very far from the surface. We shall now consider points on or near the surface of a semi-infinite cubic lattice of localized spins. In this case the dipolar sums extend over a set of point-dipole sources which are not symmetrically distributed around the field point. We shall quote the results, for the uniform case $\mathbf{k}_{\parallel} = 0$, of the calculation of the magnetic field components B^{α} at a given lattice point chosen as the origin $\mathbf{r}_{\parallel} = 0$, on the layer l of the semi-infinite lattice. If the $m - th$ layer magnetization is $\mathbf{M}_m$, one has

$$B^{\alpha}(l, \mathbf{0}) = - \sum_{m, \mathbf{r}_{\parallel}}{}' \mathcal{D}^{\alpha\beta}(m, \mathbf{r}_{\parallel}) M_m^{\beta} \qquad (13.65)$$

where the prime on the summation sign means that self-interaction is excluded. This sum was calculated by Kar and Bagchi [14] for the three cubic lattices. The results are conveniently displayed in terms of inter-layer coupling coefficients ξ_m:

$$B_l^z = - \sum_{m} \xi_m M_{l+m} \qquad (13.66)$$

Here M_j is the magnetic moment per basic cell volume on layer j, so that the coefficients are dimensionless. Let us consider for example $\alpha = \beta = (1\bar{1}0)$ and the (111) surface of an f.c.c. cubic lattice. The coefficients are displayed in Table 13.1.

Table 13.1: *Coefficients for the layer by layer calculation of the dipolar field in an f.c.c. lattice with a (111) surface. The magnetization is parallel to the surface.*

ξ_n	n
3.9013	0
0.1434	1
0.0004	2
-2.063×10^{-4}	3
3.0×10^{-9}	4

The very fast decrease with m of the field amplitude contributed by parallel planes m layers away must be attributed to the interference between the infinite point dipole fields. The same fast convergence is found for the Fourier components of the field.

These results can be related to the macroscopic theory. For the slab geometry, the macroscopic local Lorentz field with M parallel to the slab is $B = 4\pi M/3$, so that the sum

$$S = \xi_0 + 2\sum_{n=1}^{\infty} \xi_n = \frac{4\pi}{3} \tag{13.67}$$

which can be verified from Table 13.1.

13.9   Surface magnetism in metals

One can understand the main processes leading to the determination of the magnetic properties of a transition metal surface by appealing again to the Hubbard Hamiltonian. We incorporate the changes brought about by the symmetry breaking due to the surface in a similar way to what we did for insulators. The coordination at the surface is lower, and the boundary conditions on the electron wave functions of the surface atoms are different from the bulk ones. This implies that the degeneracy and the energy levels of the atomic orbitals which determine the Hubbard Hamiltonian parameters change at the surface. If we restrict the model Hamiltonian to a non degenerate basis, we expect to be able to model these changes phenomenologically simply by allowing for different local parameters at the surface. In the usual tight-binding basis, the choice of perturbed parameters is often reduced to those involving specifically surface atoms. For the one-orbital atomic basis, we assume that the atomic energy level

E_l and the intra-atomic Coulomb repulsion parameter U_l of the $l - th$ atom at the surface, as well as the hopping integrals t_{ln} which connect surface atoms with each other or with interior atoms, are different in principle from those of the bulk. The d-band electrons of a transition metal will be described in what follows, within the HF approximation, by the effective one-electron Hamiltonian

$$H = \sum_{l\sigma} \epsilon_l a_{l\sigma}^\dagger a_{l\sigma} + \sum_{l\sigma} U_l \hat{n}_{l\sigma} \langle \hat{n}_{1-\sigma} \rangle +$$

$$\sum_{l,n\sigma} t_{l,n} a_{l\sigma}^\dagger a_{n\sigma} - \sum_l U_l \langle \hat{n}_{1\uparrow} \rangle \langle \hat{n}_{1\downarrow} \rangle \tag{13.68}$$

We can now calculate the matrix elements of the resolvent operator, which for a one-electron Hamiltonian, coincides with the Green's function:

$$\mathbf{G}(\omega + i\epsilon) = (\omega + i\epsilon - H)^{-1} \tag{13.69}$$

We already know that we can use the translational symmetry parallel to the surface, so we go over to the 2d-Fourier-transformed, mixed Bloch-Wannier, basis as we did in the case of insulators and assume also that the hopping has only nearest neighbor range. We define the dimensionless matrix $g \equiv tG$ and divide all energies by t. We call E the ground state energy of a bulk atom and $E + \Delta\epsilon$ that corresponding to a surface atom . We also call $U_1/t = u_1$. The intra-surface hopping integral will be called $t_\parallel$ and the hopping from the surface to the next plane t_{01}. The structure factor of the two-dimensional lattice of each plane is $\Lambda(\mathbf{k}_\parallel)$, which for a s.c. lattice and a (010) surface is

$$\Lambda(\mathbf{k}_\parallel) = 2(\cos k_x a + \cos k_z a) \tag{13.70}$$

(the y axis is perpendicular to the planes). Finally we define $\alpha_\parallel(\mathbf{k}_\parallel) = (t_\parallel/t)\Lambda(\mathbf{k}_\parallel)$, $\alpha = t_{01}/t$, $\nu = (\omega - E)/t$ and $\Delta E = \Delta\epsilon/t$. With this notation we find:

$$g_{00}(\nu, \mathbf{k}_\parallel) = \frac{1}{F_0^\sigma(\nu, \mathbf{k}_\parallel) - \xi^\sigma(\mathbf{k}_\parallel)\alpha^2} \tag{13.71}$$

where

$$F_0^\sigma(\nu, \mathbf{k}_\parallel) = \nu - n_{0-\sigma} u_s - \alpha_\parallel(\mathbf{k}_\parallel) - \Delta E \tag{13.72}$$

The transfer matrix ξ^σ is the solution of

$$\xi^\sigma = 1/(F^\sigma - \xi^\sigma) \tag{13.73}$$

Here $F^\sigma = \nu - un_{-\sigma} - \Lambda(\mathbf{k}_\parallel)$ only depends on bulk parameters and the bulk average $\langle \hat{n}_\sigma^b \rangle = n_\sigma^b$. As we already know, for states in the bulk continuum spectrum $| \xi |= 1$ and it is complex, while for surface states $| \xi |< 1$. The effective one-dimensional perturbation potential at the surface will contain the parameters $\Delta^\sigma(\mathbf{k}_\parallel) = F_0^\sigma - F^\sigma$, $\alpha_\parallel(\mathbf{k}_\parallel)$ and α.

The layer electron charge and spin,

$$
\begin{aligned}
n_l &= n_{l\uparrow} + n_{l\downarrow} \\
m_l &= n_{l\uparrow} - n_{l\downarrow}
\end{aligned}
\tag{13.74}
$$

must be determined self-consistently. At $T = 0$ the self-consistency conditions are

$$n_l = -\frac{1}{\pi}\frac{A}{4\pi^2}\sum_\sigma \int_{BZ} \mathrm{d}^2\mathbf{k}_\parallel \int_{-\infty}^{\nu_F} \mathcal{I}m\ g_{ll}^\sigma(\nu + i\epsilon, \mathbf{k}_\parallel)$$

$$m_l = -\frac{1}{\pi}\frac{A}{4\pi^2}\sum_\sigma \int_{BZ} \mathrm{d}^2\mathbf{k}_\parallel \int_{-\infty}^{\nu_F} \sigma \mathcal{I}m\ g_{ll}^\sigma(\nu + i\epsilon, \mathbf{k}_\parallel) \qquad (13.75)$$

where A is the area of the two dimensional lattice cell. For the bulk we know the electron concentration and the saturation magnetization, from which we can determine the set of atomic parameters which inserted in the respective equations (13.75) must yield the known values of n and m. The Fermi energy ν_F and the parameters t and u must be consistent with the bulk quantities. The d-band electron concentration can be estimated from experiments, as well as the bulk magnetization contribution from the d-bands. One must allow for the charge and the magnetic polarization of the s, p bands. The parameters u, t and ϵ were obtained in this way by Hasegawa within this model [20]. The parameters for the surface must fulfill the requirement of local, or quasi-local, neutrality. This condition reduces the number of free surface parameters. In Ref. [21] the values of ΔE and u_s were fixed by the self-consistency HF conditions and the requirement of overall charge neutrality in a region of a few planes near the surface. The results are not sensitive to the width of the neutrality layer, and one obtains essentially the same parameters upon varying this width from 1 to 20 planes. The magnetization m_s of the surface plane is the main output of the calculation.

The effect of the surface on the layer density of states is shown in the curves obtained by Kalkstein and Soven with the model described here, in a non-self-consistent calculation [22]. Figure 13.8 shows that the local reduction of coordination narrows the band at the surface plane, as should be expected. Note that, in agreement with our previous remarks, the influence of the surface has practically disappeared already in the third plane. The layer density of states in the bulk is independent on the layer and has the same form as the bulk density of states. Note also that the surface density of states is lower than the bulk one at the band edges, while it peaks at the band center. Finally we remark that the density of states at the third layer shows some small oscillations around the bulk curve, which reflect the oscillations of the wave functions according to Eq. (13.54). These are the equivalent, for the surface, of the Friedel oscillations near a point charge in a metal. The main features described so far are maintained in more elaborate calculations within the LCAO (Linear Combination of Atomic Orbitals) with degenerate bands. The concentration of states near the band center increases the Stoner parameter $U_s N(E_F)$ and favours ferromagnetism at the surface in metals like Fe, Cr and V with E_F near the band center, while the narrowing of the band has the opposite effect in Ni, Pd or Pt where E_F is near the band edge.

A theory of surface spin waves in metals [23] can be developed with Green's functions methods similar to the ones we presented above for insulators.

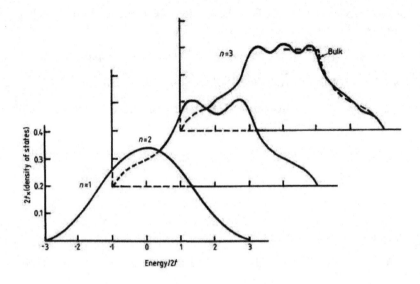

Figure 13.8: *Layer density of states in tight-binding model of a s.c. transition metal. The figure shows the density of states of the (100) surface plane (n=1) and the next two planes.*[22]

References

1. Yang, K. Y. , Homa, H. and Schuller, I. K.,(1980) *J. Appl. Phys.* **63** 4066.

2. Celotta, R. J. and Pierce, D. T., (1986) *Science* **2324** 249.

3. Rau, C. and Ecke, G.,(1980) *Nuclear Physics Methods in Materials Research*, edited by Bethge, K. *et al.*, Vieweg, Braunshwewig,

4. Sáenz *et al.*, (1987) *J. Appl. Phys.* **62**4293.

5. Selzer, S. and Majlis, N. (1983) *Phys. Rev.* **B27**, 544.

6. Néel, L.,(1953) *Comptes Rendus Acad. Sc. (Paris)* **237**, 1468.

7. Weller, D., Alvarado, S. F., Gudat, W., Schröder, K. and Campagna, M. (1985) *Phys. Rev. Lett.* **63**, 155.

8. Rau, C. and Jin, C. (1988) *J. Appl. Phys.* **54**, 3667.

9. Binder, K. (1983) *"Phase Transitions and CriticalPhenomena"* **8**, Edited by Domb, C. and Lebowitz, J., Academic Press, London.

10. Diehl, H. W. (1982) *J. Appl. Phys.* **53**, 7914.

11. Damon, R. W. and Eshbach, J. R. (1981) *J. Phys. Chem. Solids* **19**, 308.

12. Selzer, Silvia (1982) Ph.D. Thesis, unpublished, Centro Brasileiro de Pesquisas Físicas, Rio de Janeiro, Brazil.

13. Harriague, S. and Majlis, N. (1971) *J. Phys. Soc. (Japan)* **31**, 1350.

14. Kar, N. and Bagchi, A. (1980) *Sol. State. Comm.* **33**, 645.

15. Wall, H. S. (1967) *"Analytic Theory of Continued Fractions"*, Chelsea Publishing Co., New York.

16. Selzer, S. and Majlis, N. (1980) *J. Mag. and Mag. Mat.* **15-18**, 1095.

17. Kosterlitz, J. M. and Thouless, D. J. (1973) *J. Phys. C* **6**, 1181.

18. Majlis, N. (1990) *"Surface Science"*, *Proceedings of the 6th. Latin American Symposium on Surface Physics*, *Springer Proceedings in Physics* **62**, Springer–Verlag, p.443.

19. Dauth, B., Dürr W. and Alvarado, S. F. (1987) *Surface Science* **189/190**, 729.

20. Hasegawa, H (1980) *J. Phys. Soc. Japan* **49**, 963.

21. Rodrigues, A. M., Majlis, N. and Ure, J. E. (1986) *Il Nuovo Cimento* **7D**, 528.

22. Kalkstein, D. and Soven, P. (1981) *Surf. Sci.* **26**, 85.

23. Mathon, J. (1988) *Reps. Prog. Phys.* **51**, 1.

Chapter 14

Two-Magnon Eigenstates

14.1   Introduction

Wortis [1] made a detailed study of the scattering and bound states of two magnons in a FM, which we shall now survey. As we have mentioned before, this problem had been first solved in $1d$ by Bethe [2]. The general feature in any dimension is that two spin waves interact through an effective *attractive* potential. The question is then whether this potential can sustain bound states. Wortis' conclusion is that in $1d$ there are bound states for any value of the total wave vector of the pair of excitations, in agreement with the results obtained by Bethe. For two and three dimensions the bound state eigenvalue equations must be solved numerically, which makes the complete solution of the problem practically impossible, since there are too many parameters involved. However, it is possible to obtain definite results for some restricted regions of the parameter space, and then make some reasonable conjectures on the general behaviour of the solutions. At the end of this chapter we shall review the conclusions of this work.

Let us consider the states of the Heisenberg Hamiltonian for a ferromagnetic insulator with two spin-flips located at sites (i, j):

$$|\psi_{ij}\rangle = S_i^+ S_j^+ |0\rangle \tag{14.1}$$

where $|0\rangle$ is the Weiss ground state, which we shall choose with all spins down. We know already that the one-magnon states with a given wave vector k,

$$|k\rangle = \frac{1}{\sqrt{N}} \sum_i e^{i\mathbf{k} \cdot \mathbf{R_i}} S_i^+ |0\rangle \tag{14.2}$$

are exact orthogonal eigenstates with total z component of spin $S_{tot}^z = -NS+1$ and energy

$$\omega(k) = 2SJ(\gamma(0) - \gamma(k)) + \gamma B + E_0$$

in an external field, where E_0 is the ground state energy and $\gamma(k)$ is the structure factor. Consider now the family of two plane-wave states

$$|kk'\rangle = A \sum_{i,j} e^{i(\mathbf{k} \cdot \mathbf{R_i} + \mathbf{k'} \cdot \mathbf{R_j})} |ij\rangle \qquad (14.3)$$

For simplicity we shall not use boldface notation for vectors in this chapter when not necessary for clarity. One can verify that $|k0\rangle$ and $|0k'\rangle$ are exact eigenstates of H.

Exercise 14.1
Prove that $|k0\rangle$ and $|0k'\rangle$ are exact eigenstates of H with eigenvalues

$$E(0, k) = E(k, 0) = E_0 + \omega(0) + \omega(k) \qquad (14.4)$$

This implies that in this particular case we can superpose the two excitations as if they were mutually independent. However if both $k, k' \neq 0$ we have a more interesting situation.

In the first place the wave functions (14.3) are not orthogonal. To see this, consider first the scalar product of the localized two-spin-flip states:

Exercise 14.2
Prove that

$$\langle 12|34\rangle = 4S^2[\delta_{13}\delta_{24} + \delta_{14}\delta_{23}](1 - \frac{\delta_{12}}{2S}) \qquad (14.5)$$

In particular we can obtain the normalization constant:

$$\langle 12|12\rangle = 4S^2(1 + \delta_{12})(1 - \frac{\delta_{12}}{2S}) \qquad (14.6)$$

and then the normalized state can be written as

$$|\tilde{12}\rangle = \frac{S^+(1)S^+(2)|0\rangle}{(2S)[(1 + \delta_{12})h_2(12)]^{1/2}} \qquad (14.7)$$

where

$$h_2(12) \equiv (1 - \delta_{12}/2S) \qquad (14.8)$$

The expresssion (14.7) is to be compared with the one-flip state

$$|\tilde{1}\rangle = \frac{1}{\sqrt{2S}} S^+(1)|0\rangle \qquad (14.9)$$

With the help of Eq. (14.6) we can now obtain the scalar product we are looking for:

Exercise 14.3
Prove that

$$\langle k\ k'|k''\ k'''\rangle = (\delta_{kk''}\delta_{k'k'''} + \delta_{kk'''}\delta_{k'k''}) - \frac{1}{NS}(\delta_{k+k',k''+k'''}) \qquad (14.10)$$

For the case $k = k' = k'' = k'''$ we have

$$\langle k\ k|k\ k\rangle = 8S^2 - \frac{4S}{N} \qquad (14.11)$$

We see from Eq. 14.10 that different states with the same total wave vector have an overlap $\approx \frac{1}{N}$. This is what Dyson [3] called *kinematical interaction*. If we impose, as usual, periodic boundary conditions, the number of linearly independent states $\{k\ k'\}$ is the same as that of the states i, j. If $i \neq j$ we have $N(N-1)/2$ different states. If $i = j$ and $S > 1/2$ we add N more, so the total is $N(N+1)/2$ for $S > 1/2$ or $N(N-1)/2$ for $S = 1/2$. Observe that since we can choose $N(N+1)/2$ pairs for (k, k') because there is no special restriction for $k = k'$, then this set is overcomplete in the space of two-spin-flips, at least for $S = 1/2$, so that in this case the set cannot be orthogonalized: N states must be linear combinations of the $N(N-3)/2$ remaining ones.
Let us see what is the effect of H on these states. If neither k nor k' is zero, then $|k\ k'\rangle$ is not an eigenstate. The Hamiltonian will connect states $|k_1\ k_2\rangle$ with $|k_3\ k_4\rangle$ with the only restriction that $k_1 + k_2 = k_3 + k_4$, which is due to the assumed translation invariance of the infinite lattice. These are the *dynamical interactions* according to Dyson. Their effect is the scattering of the spin-flip excitations off one another. These processes eventually can lead to bound states of two spin flips. Let us now study this problem within the Green's function formalism [1].

14.2 Green's function formalism

Since we want to study the eigenstates of the Hamiltonian in the susbpace of two-spin flips, we shall calculate expectation values in the ground state (Weiss state for a FM). We follow Wortis and define the one-spin flip retarded Green's function in a slightly different way from the one we were using so far, as

$$G_1(1; 2; t) = -i\langle 0|S^-(1; t)S^+(2; 0)|0\rangle\theta(t) \qquad (14.12)$$

whose Fourier transform satisfies the equation

$$[\omega - \gamma B + 2S(J(k) - J(0))]G_1(k, \omega) = 2S \qquad (14.13)$$

In this chapter, angular brackets indicate the expectation value in the ground state.

In a hypercubic lattice of d dimensions with constant a, and with only first n.n. exchange we have

$$J(k) = 2\sum_{i=1}^{d} \cos k_i a \qquad (14.14)$$

and the magnon frequency is

$$E(k) = 4S \sum_{i=1}^{d} (1 - \cos k_i a) \qquad (14.15)$$

The two-site Green's function, which will be called G_2, is defined as

$$G_2(12; 1'2'; t) = (-i)^2 \theta(t) \times$$
$$\langle 0 \mid S^-(1;t) S^-(2;t) S^+(1';0) S^+(2';0) \mid 0 \rangle \qquad (14.16)$$

When use is made of the scalar product (14.6) and the definition of $h_2(12)$ in Eq. (14.8), we obtain the equation of motion

$$\left[i\frac{\partial}{\partial t} - 2(\gamma B + 2SdJ) \right] G_2(12; 1'2'; t)$$
$$+ \sum_3 2SJ(13) G_2(32; 1'2') + 2S \sum_3 J(23) G_2(13; 1'2')$$
$$+ J(12) G_2(12; 1'2') - \delta_{12} \sum_3 J(13) G_2(32; 1'2')$$
$$= (-i)(2S)^2 [\delta_{11'} \delta_{22'} + \delta_{12'} \delta_{21'}] h_2(1'2') \qquad (14.17)$$

The last two terms on the l. h. s. of Eq. (14.17) involve explicitly the coordinates $1, 2$ and are therefore responsible for the interaction of these two sites. The first terms are proportional to the probability amplitude for the independent propagation of flips which were present at $t = 0$ on points $1', 2'$ to the points $1, 2$, at a later time t. Then it is natural to introduce the symmetric non-interacting propagator

$$\Gamma_2(12; 1'2') = G_1(1, 1') G_1(2, 2') + G_1(1, 2') G_1(2, 1') \qquad (14.18)$$

Exercise 14.4
Verify that Γ_2 satisfies Eq. (14.17) without the last two terms of the l. h. s., and with $h(12)$ substituted by 1 on the r. h. s.

Let us write Eq. (14.17) more concisely as

$$(\boldsymbol{\Omega}_0 + \delta \boldsymbol{\Omega}) \mathbf{G}_2 = -4iS^2 \boldsymbol{\Delta} \mathbf{h}_2 \qquad (14.19)$$

where

$$\delta \Omega(12t; 34t') = 2[\, J(12)\delta_{13}\delta_{24} - J(23)\delta_{12}\delta_{24} \,] \delta(t - t')$$
$$\Delta(12, t; 34) = [\, \delta_{13}\delta_{24} + \delta_{14}\delta_{23} \,] \delta(t - t') \qquad (14.20)$$

and Ω_0 contains the rest of the integro-differential operator acting on G_2 on the l. h. s. of Eq. (14.17):

$$(\boldsymbol{\Omega}_0)(12t; 34t') =$$
$$\delta(t - t')[\, i\frac{\partial}{\partial t} + \delta_{13}\delta_{24}(-2\gamma B + 2SdJ)$$
$$+ \quad 2SJ(13)\delta_{24} + 2SJ(24)\delta_{13}] \qquad (14.21)$$

Then one finds

$$\boldsymbol{\Omega_0}\boldsymbol{\Gamma_2} = -4S^2 i\boldsymbol{\Delta} \tag{14.22}$$

Notice that the matrix products above involve two lattice sums and one integration over a time variable, so for two arbitrary matrices $\mathbf{A}, \mathbf{B}$:

$$(\mathbf{AB})(12t; 1'2', t') =$$
$$\int ds \sum_{34} \mathbf{A}(12, t; 34, s)\mathbf{B}(34, s; 1'2', t') \tag{14.23}$$

Exercise 14.5
Prove that

$$\boldsymbol{\Delta}^2 = 2\boldsymbol{\Delta} \tag{14.24}$$

From Eq. (14.22) we obtain

$$\boldsymbol{\Omega_0} = -4S^2 i\boldsymbol{\Delta}\boldsymbol{\Gamma_2}^{-1} \tag{14.25}$$

so that

$$-4S^2 i\boldsymbol{\Delta}\boldsymbol{\Gamma_2}^{-1}\mathbf{G_2} = -4S^2 i\boldsymbol{\Delta}\mathbf{h_2} - \delta\boldsymbol{\Omega}\mathbf{G_2} \tag{14.26}$$

Now we multiply both sides of the previous equation by $\boldsymbol{\Delta}\boldsymbol{\Gamma_2}$ from the left and use Eq. (14.24), obtaining

$$\mathbf{G_2} = \boldsymbol{\Gamma_2}\mathbf{h_2} - \frac{i}{8S^2}\boldsymbol{\Gamma_2}\delta\boldsymbol{\Omega}\mathbf{G_2} \tag{14.27}$$

Let us now introduce an interaction kernel by defining:

$$\boldsymbol{\Gamma_2}\delta\boldsymbol{\Omega} \equiv \mathbf{K_2}\mathbf{J} \tag{14.28}$$

where

$$K_2(12t; 34s) =$$
$$\Gamma_2(12t; 34s) - \frac{1}{2}[\Gamma_2(12t; 33s) + \Gamma_2(12t; 44s)] \tag{14.29}$$

Finally we get

$$\mathbf{G_2} = \boldsymbol{\Gamma_2}\mathbf{h_2} - \frac{i}{8S^2}\mathbf{K_2}\mathbf{J}\mathbf{G_2} \tag{14.30}$$

In order to arrive at (14.30) we have used the symmetry

$$G_2(abt; \cdot) = G_2(bat; \cdot) \quad .$$

We shall now Fourier analyze all the functions involved. Let us start with the free-magnon-pair propagator Γ_2 defined in Eq. (14.18):

$$\Gamma_2(120; 1'2't) = \frac{1}{N^2}\sum_k \int \frac{d\omega}{2\pi} G_1(k, \omega)e^{-i\omega t}e^{ik(1-1')} \times$$
$$\sum_{k'} \int \frac{d\omega'}{2\pi} G_1(k', \omega')e^{-i\omega' t} \times$$
$$\{e^{ik'(2-2')}e^{ik(1-1')} + e^{ik(1-2')}e^{ik(2-1')}\} \tag{14.31}$$

The frequency transform of Γ_2 is

$$\Gamma_2(12; 1'2'; \Omega) = \int \Gamma_2(120; 1'2't)dt\ e^{i\Omega t} \qquad (14.32)$$

If we substitute both factors G_1 in (14.31) by their expression given in Eq. (14.13), we can perform the time integration, obtaining the factor $2\pi\delta(\Omega - \omega - \omega')$, so that we are left with only one integration over ω in (14.31). The ω integral is

$$\int d\omega \frac{1}{(\omega + i\epsilon - E_k)(\Omega - \omega + i\epsilon - E_{k'})} \qquad (14.33)$$

The integrand is analytic in the upper half plane, where we can close the contour and use Cauchy formula to obtain

$$\Gamma_2(12; 1'2'; \Omega) = \frac{8S^2}{N^2} \sum_k \sum_{k'} \frac{1}{\Omega - (E_k + E_{k'}) + i\epsilon} \times$$
$$\{e^{ik'(2-2')}e^{ik(1-1')} + e^{ik(1-2')}e^{ik(2-1')}\} \qquad (14.34)$$

To simplify we introduce the *center of mass* (c.o.m.) and *relative* coordinates of any pair of points 1 and 2, and correspondingly, the c.o.m. and relative wave vectors, defined respectively as:

$$\begin{aligned} \mathbf{R} &= (\mathbf{1} + \mathbf{2})/2\ , \quad \mathbf{r} = \mathbf{1} - \mathbf{2} \\ \mathbf{K} &= \mathbf{k_1} + \mathbf{k_2}\ , \quad \mathbf{k} = (\mathbf{k_1} - \mathbf{k_2})/2 \end{aligned} \qquad (14.35)$$

Due to the symmetry under permutations $1 \leftrightarrow 2$, $1' \leftrightarrow 2'$, G_2 is even in both arguments r, r'. Since K is the sum of two vectors inside the first BZ, it belongs to the extended zone, and in the hypercube this implies $-2\pi/a \le K_i < 2\pi/a$. Call the extended zone $\overline{BZ}$. We can now write Γ_2 as a function of the relative and c.o.m. coordinates:

$$\Gamma_2(r_1 r_2 t; r_1' r_2' t') =$$
$$\sum_{K \in \overline{BZ}} e^{iK(R-R')} \int \frac{d\Omega}{2\pi} e^{-i\Omega(t-t')} i\Gamma_2(r, r', K, \Omega) \qquad (14.36)$$

We analyze K_2 in a similar way:

$$K_2(r_1 r_2 t; r_1' r_2' t') =$$
$$\sum_{K \in \overline{BZ}} e^{iK(R-R')} \int \frac{d\Omega}{2\pi} e^{-i\Omega(t-t')} i K_2(r, r', K, \Omega) \qquad (14.37)$$

A phase factor i was introduced into Eqs. 14.36 and 14.37 for convenience.

Exercise 14.6
Show that

$$\Gamma_2(r, r', K, \omega) =$$
$$- 8S^2 N \sum_k \frac{\cos(\mathbf{k} \cdot \mathbf{r}) \cos(\mathbf{k} \cdot \mathbf{r}')}{\omega - 2\gamma B - S(k, K) + i\epsilon} \qquad (14.38)$$

and

$$K_2(r, r', K, \omega) = -8S^2 N \times$$
$$\sum_k \frac{\cos{(\mathbf{k} \cdot \mathbf{r})}\left[\ \cos{(\mathbf{k} \cdot \mathbf{r}')} - \cos{(1/2\mathbf{K} \cdot \mathbf{r}')}\right]}{\omega - 2\gamma B - S(k, K) + i\epsilon} \tag{14.39}$$

where for a hypercubic d-dimensional lattice

$$S(k, K) = 8SJ \sum_{i=1}^{d} (\ 1 - \cos{(K_i a/2)}\cos{(k_i a)}\ ) \tag{14.40}$$

and k is the relative wave vector defined before.

Substituting (14.38) and (14.39) into the integral equation (14.27) one gets:

$$G_2(r, r', K, \omega) = \Gamma_2(r, r', K, \omega)$$
$$+ \frac{J}{2S^2} \sum_\delta K_2(r, \delta; K, \omega) G_2(\delta, , r', K, \omega) \tag{14.41}$$

where $\{\delta\}$ = star of the first n.n. of a site in the lattice.

The real poles of G_2 are the excitation energies of the ferromagnet, in the two-spin-flip subspace. We can discard poles of the free propagator Γ_2 or those of the kernel K_2. The last ones, according to Eq. (14.27), are also those of Γ_2. In turn, poles of Γ_2 are those of the single spin wave propagator G_1. The reason to discard these one-magnon poles is that one can prove that the residues of G_2 at the poles of G_1 vanish, so that the only remaining poles of G_2 are the excitation energies of a bound pair of spin waves arising mathematically from the zeroes of the secular determinant of the linear system (14.41) as a function of ω, for fixed total wave vector K. Since by assumption $J(\delta)$ has short range it suffices to consider one star of nearest neighbour sites. Since $G_2(r, \cdot), G_2(\cdot, r')$ are even, we only consider those vectors in $\{\delta\}$ with $\delta_i > 0$. We obtain the linear system $(l, n, m = 1, \cdots, d)$:

$$\sum_{m=1}^{d} \left[\ \delta_{l,m} - \frac{2J}{S^2} K_2(l, m)\ i\right]\ G_2(m, n) = \Gamma_2(l, n) h_2(n) \tag{14.42}$$

If we substitute K_2 from Eq. (14.39) we find

$$det\ [\ 2S\delta_{l,m} - B_2(l, m)\ ]\ = 0 \tag{14.43}$$

with

$$B_2(l, m) = \frac{1}{\pi^d} \int_0^\pi dk_1 \cdots \int_0^\pi dk_d \frac{\cos k_i \cos k_j}{\Delta} \tag{14.44}$$

where

$$\Delta\ =\ t - i\epsilon - \sum_{l=1}^{d} \alpha_l \cos k_l$$

$$t \;=\; d - \frac{\omega - 2\gamma B}{8SJ}$$

$$\alpha_l \;=\; \cos\left(K_l/2\right) \tag{14.45}$$

We have simplified the notation: $ka \to k$, $Ka \to K$. We first remark that $\mathbf{B_2}$ has a cut on the real axis in the interval

$$S(0, K) \leq \omega - 2\gamma B \leq S(\pi, K) \tag{14.46}$$

This interval contains all the zeroes of Δ, that is all excitation energies

$$\omega(k, K) = 2\gamma B + S(k, K) \tag{14.47}$$

of scattering eigenstates which result from the superposition of two magnons. In fact the exact energies of the scattering eigenstates differ from the sum of the energies of the individual magnons by an amount of order $\approx \frac{1}{N}$, where $N =$ number of spins in the system, so that both excitation spectra coincide in the thermodynamic limit. G_2 has the same cut as B_2, and both are complex on the corresponding interval of t. For instance,

$$\mathcal{I}m(B_2(i, j)) =$$
$$\frac{1}{\pi^{d-1}} \int_0^{\pi} \mathrm{d}k_1 \cdots \int_0^{\pi} \mathrm{d}k_d \cos k_i \cos k_j \; \delta(\Delta) \tag{14.48}$$

This cut is the support of the density of two-magnon scattering states. Any bound state must lie therefore in one of the intervals

$$-\infty < \quad t \quad < -\sum_{l=1}^{d} \alpha_l$$

$$\sum_{l=1}^{d} \alpha_l < \quad t \quad < \infty \tag{14.49}$$

One can define a set of integrals in terms of which $B_2(i, j)$ can be expressed. They are

$$D_0(t) = \frac{1}{\pi^d} \int_0^{\pi} \cdots \int_0^{\pi} \frac{\mathrm{d}^{(d)}k}{\Delta}$$

$$D_i(t) = \frac{1}{\pi^d} \int_0^{\pi} \cdots \int_0^{\pi} \frac{\mathrm{d}^{(d)}k \cos k_i}{\Delta}$$

$$D_{i,j}(t) = D_{ji}(t) =$$

$$\frac{1}{\pi^d} \int_0^{\pi} \cdots \int_0^{\pi} \frac{\mathrm{d}^{(d)}k \cos k_i \cos k_j}{\Delta} \tag{14.50}$$

We find

$$B_2(i, j) = D_{i,j}(t) - D_i(t)\alpha_j \tag{14.51}$$

The D integrals satisfy the recurrence relations

$$tD_0(t) - \sum_{l=1}^{d} \alpha_l D_l(t) = 1$$

$$D_i(t) - \sum_{l=1}^{d} \alpha_l D_{i,l}(t) = 0 \tag{14.52}$$

When t is outside the continuum interval, we can expand B_2 in inverse powers of t, obtaining:

$$B_2(i,i) = \frac{1}{2t} - \frac{\alpha_i^2}{2t^2} + \frac{1}{t^3}\left(\frac{\alpha_i^2}{8} + \frac{1}{4}\sum_{l=1}^{d} \alpha_l^2 \right) + \mathcal{O}\left(\frac{\alpha}{t}\right)^4$$

$$B_2(i,j) = \alpha_i \alpha_j \left(\frac{1}{2t^3} - \frac{1}{2t^2} \right) + \mathcal{O}\left(\frac{\alpha}{t}\right)^4 \tag{14.53}$$

For small α_i there is a d-times degenerate solution of the secular equation 14.40:

$$t_B = \frac{1}{4S} \tag{14.54}$$

14.3 One dimension

In one dimension, the upper and lower limits of the continuum are

$$t_\pm = \pm\alpha \equiv \pm\cos\left(K/2\right)$$

The D integrals are easily evaluated. One obtains:

$$D_0 = \frac{\text{sign}(t-\alpha)}{(t^2 - \alpha^2)^{1/2}} \tag{14.55}$$

With the recurrent relations (14.52) one obtains the bound state eigenvalue equation:

$$2S = \frac{t - \alpha^2}{\alpha^2}\left[\frac{\pm t}{(t^2 - \alpha^2)^{1/2}} - 1 \right] \tag{14.56}$$

where

$$0 \le \alpha \le 1 \ , \quad |t| \ge \alpha \ .$$

Exercise 14.7
Verify Eqs. (14.55) and (14.56).

To $t < -\alpha$ corresponds the lower sign in Eq. (14.56), so that in this case there is no solution. In conclusion, there is only one solution for each α for $t > \alpha > 0$.

Exercise 14.8
Show that for $S = 1/2$ the solution is the singlet

$$t_B = \frac{1}{2}(1 + \alpha^2) \tag{14.57}$$

Then the corresponding excitation energy, referred to the Weiss ground state energy and in units of $8SJ$, is

$$\omega_B/(8SJ) = 1/2(1 - \alpha) \ .$$

The upper (u) and lower (l) limits of the scattering continuum for two spin waves in $1d$ are

$$\frac{\omega_c^{u,l}}{8SJ} = 1 \pm \alpha \tag{14.58}$$

In Fig. 14.1 we show the region of the (ω, α) plane which contains the continuum

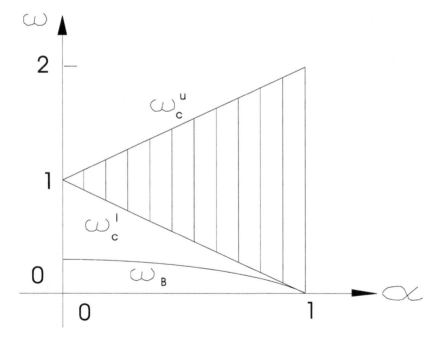

Figure 14.1: *Scattering and bound states in a 1d ferromagnet as a function of the total wave vector K of the pair of magnons ($\alpha = \cos K/2$) .*

and the dispersion relation of the bound state. The bound state energy lies below the continuum, so that clearly the superposition hypothesis, which is the basis of Bloch's [4] theory of spin waves, is not applicable to the $1d$ FM.

We define the binding energy of the bound pair as the difference for fixed K, which is equivalent to fixed α, between the bound state energy and the lower limit of the continuum, which is

$$\frac{\Delta_B}{8SJ} = -[\ \frac{1}{2}(1 + \alpha^2) - \alpha\ ] \leq 0\ , \quad 0 < \alpha < 1 \tag{14.59}$$

In one dimension the bound state energy depends only on one parameter, the c.o.m. wave vector. The wave function of the bound state for a given K has a form which depends on the relative wave vector k, which in the $1d$ case is completely determined for each eigenstate. Since the pair of spin waves is bound the wave function must be exponentially decreasing as a function of r for large r. This implies that the relative wave vector k must have an imaginary part. One elegant way to obtain the complex k is to consider the bound state energy as the analytic continuation of $S(k,K)$ to complex k. In $1d$ we have

$$S(k,\alpha) = 8SJ(1 - \alpha \cos k)\ \ .$$

We can determine for which k this expression has the same value as the bound state energy for the same α :

$$S(k,\alpha) = \omega_B(\alpha) = 4SJ(1 - \alpha^2) \tag{14.60}$$

and we find

$$\cos k = \frac{3 - \alpha^2}{2\alpha} \tag{14.61}$$

Since $|\alpha| \leq 1$, $\cos k \geq 1$. Then $k = iq$ for some real q. It is convenient to define an inverse length $q = a\lambda^{-1}$ so that

$$\cosh\left(\frac{a}{\lambda}\right) = \frac{3 - \alpha^2}{2\alpha} \tag{14.62}$$

and we interpret λ as the distance between both spin-flips for which the wave function $\Psi_B(r)$ of the bound pair is appreciable: for $r > \lambda$, $\Psi_B(r)$ is exponentially small. We see that for $\alpha \to 0$ $(K \to \pm\pi)$ $\lambda \to 0$: in this limit the support of $\Psi_B(r)$ reduces to a point. In fact, since the amplitude at $r = 0$ must vanish for $S = 1/2$, in this case $\Psi_B(r) \equiv 0$: there is no bound state for $K = \pm\pi$ if $S = 1/2$. However, if $\alpha > 0$, $\Psi_B(r)$ has a finite extension in space, which increases as $\alpha \to 1$, and diverges at $\alpha = 1$, which again correspods to zero amplitude of $\Psi_B(r)$ if the length of the chain $\to \infty$.

14.4   Two dimensions

For $d = 2$ the D integrals can be conveniently expressed in terms of the variable ξ defined as:

$$\xi^2 = \frac{4\alpha_1\alpha_2}{t^2 - (\alpha_1 - \alpha_2)^2}\ , \quad 0 \leq \xi \leq 1 \tag{14.63}$$

The results of the integrals are:

$$D_0(t) = \frac{\xi K(\xi)}{\pi(\alpha_1\alpha_2)^{1/2}} \tag{14.64}$$

$$D_i(t) = \frac{\xi}{\pi\alpha_i(\alpha_1\alpha_2)^{1/2}} \left[\, (t+\alpha_j)K(\xi) \right.$$
$$\left. -(t+\alpha_1+\alpha_2)\Pi(\beta_i^2,\xi) \, \right] \tag{14.65}$$

$$D_{12}(t) = \frac{1}{\xi\pi(\alpha_1\alpha_2)^{1/2}} \left[\, (2-\xi^2)K(\xi) - 2E(\xi) \, \right] \tag{14.66}$$

where

$$\beta_i^2 = \frac{-2\alpha_i}{t-\alpha_i+\alpha_j} \leq 0 \; , \quad \beta_1^2\beta_2^2 = \xi^2 \tag{14.67}$$

and $i,j = 1,2$, $i \neq j$. The functions K, E, Π are respectively complete elliptic integrals of the first, second and third kind [5].

In the special case $\alpha_1 = \alpha_2$, we have $\beta_i^2 = -\xi$. In this case the bound state equation yields a unique solution for $\alpha = 0$, with $t_B > 0$ which for $J > 0$ implies that ω_B is below the continuum. The analysis of the results for general $\alpha_1 \neq \alpha_2$ is complicated. Wortis discusses the special case $\alpha_1 = \alpha_2 = \alpha$. In Fig. 14.2 we show a sketch of the continuum region and the bound state dispersion relations in the $(\, t, \alpha)$ plane.

14.5   Summary of results

- In one dimension there is always one bound state. At $K = 0$ the binding energy vanishes but it increases monotonically with $|\, \mathbf{K} \,|$. As a consequence, the state with two free magnons is unstable against the creation of a bound state, and the superposition hypothesis of free magnons fails.

- In two dimensions, there is one bound state at $K = 0$ below the lower limit of the continuum, so that the conclusion is the same as in $1d$. For $\alpha_1 = \alpha_2$ there is a second bound state which requires a threshold of K.

- In three dimensions there are no bound states near $K = 0$ but there can be up to 3 bound states for $K \neq 0$, although numerical estimates so far indicate that this happens at relatively large values of K. One concludes that in $3d$ magnons behave as free excitations if the one-magnon-state wave vectors are not near the BZ boundary. Since this is the case at not too high temperatures, the superposition hypothesis of Bloch, and accordingly spin wave theory, are also valid at low T.

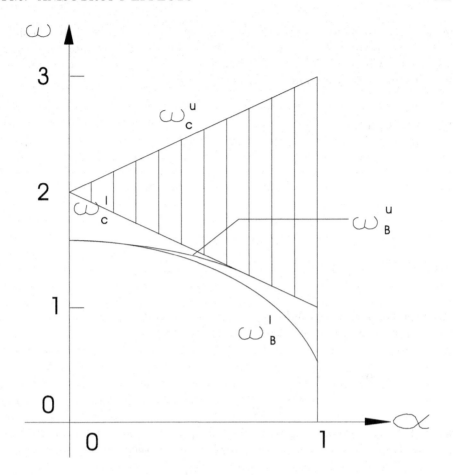

Figure 14.2: *Two magnon states in a 2d FM, for $\alpha_1 = \alpha_2 = \alpha$. The two bound states are degenerate at $\alpha = 0$. The lower branch merges with the continuum at $\alpha = \frac{1}{2S}[\frac{4}{\pi} - 1]$. For $t < 0$, which corresponds to states above the continuum, there are no bound states.*

14.6 Anisotropy effects

We consider now materials with highly anisotropic exchange interactions, described by

$$H = -\gamma B \sum_1 S^z(1)$$

$$- \sum_{12} J(12)[S^z(1)S^z(2) + \Delta S^+(1)S^-(2)] \tag{14.68}$$

We see that the limit $\Delta = 0$ is the Ising Hamiltonian while the Heisenberg isotropic case is obtained with $\Delta = 1$. For all intermediate values in that in-

terval the one-magnon spectrum has a gap, which would make LRO possible
at finite temperatures even in $1d$, were it not for the kink instability already
discussed in Chap. 12. For the Ising model it is easy to show that two separate
spin flips interact with an attractive effective force:

Exercise 14.9
*Show that for an Ising hypercubic lattice with first n.n. interactions the excita-
tion energy for two spin flips separated by more than a lattice constant is*

$$E_{cont} = 2\gamma B + 2\nu J S \qquad (14.69)$$

while if they lie on adjacent sites the energy is

$$E_{bound} = E_{cont} - 2JS \qquad (14.70)$$

which has degeneracy d.

Since the binding energy is d-independent, we expect these excitations to play
a role in particular in $3d$ sytems with high anisotropy.
For $0 < \Delta < 1$ one can basically repeat all the previous algebra. We can define

$$K_2(12t; 34s) =$$
$$\Gamma_2(12t; 34S) - \frac{\Delta}{2}[\ \Gamma_2(12t; 33s) + \Gamma_2(12t; 44s)\ ] \qquad (14.71)$$

where the one-magnon propagator factors in Γ_2 are those of the anisotropic
system. The only change in the previous formulae is the substitution

$$\alpha_i \rightarrow \Delta \cos\left(\frac{K_i}{2}\right)$$
$$0 \le \alpha_i < \Delta \le 1 \qquad (14.72)$$

Observe that in $1d$ the bound state is always below the lower limit of the continuum, which now has a finite gap proportional to $\eta = 1 - \Delta$. Then the population
of bound states is *exponentially higher* at low non-zero T than that of the continuum states.

Exercise 14.10
Show that in $1d$

$$E_{bound} - E_0 = 2\gamma B + 2SJ\left(1 - \Delta^2 \cos^2\frac{K}{2}\right) \qquad (14.73)$$

*which is the result obtained by Orbach [6] in his extension of the Bethe-Ansatz
theory to include anisotropic exchange.*

In $3d$ Wortis [1] finds that for

$$0 \leq \Delta < \frac{0.5163}{2S + 0.5163} \qquad (14.74)$$

there will be a bound state for $K = 0$ with energy lower than the lower continuum limit (with anisotropy) so that bound states dominate the thermodynamics at low T in a $3d$ highly anisotropic FM.

References

1. Wortis, M. (1963) *Phys. Rev.* **132**, 85.

2. Bethe, H. A. (1931) *Z. Physik* **71**, 205.

3. Dyson, F.J. (1956) *Phys. Rev.* **102**, 1217.

4. Bloch, F. (1931) *Z. Physik* **71**, 205.

5. Abramowitz M. and I. Stegun (1965) *Handbook of Mathematical Functions*, Dover Publications, Inc., New York, U.S.A.

6. Orbach, R. (1958) *Phys. Rev.* **112**, 309.

Chapter 15

Other Interactions

15.1   Introduction

In Chapter 4 we described the Holstein-Primakoff expansion of the spin operators in a series of products of boson operators, and we explicitly wrote that expansion up to fourth order terms. Clearly terms of order higher that the second give rise to magnon-magnon scattering and accordingly to a finite magnon lifetime. In the classic paper by Dyson [1] the low temperature scattering cross section is calculated to various orders. We remark that from the expansion of the factors $\tilde{f}$ (Eq. (4.17)) in the exchange Hamiltonian in Chap. 4 one can only get terms which contain the same number of creation and annihilation operators, so that each term conserves the total $\mathcal{S}^z$. The HP expansion of the *dipolar* Hamiltonian, instead, can generate terms with either an even or odd number of operators. Clearly the odd terms do not conserve $\mathcal{S}^z$.

If we classify the interactions which are responsible for the finite lifetime of magnons according to the number of magnon operators they involve, we have in the first place interactions of magnons with static impurities or structural defects in general. In particular surface pits are the basic source of the low order *two-magnon* processes in which the total pseudo-linear-momentum of the spin system is not conserved, the excess being transferred to the linear momentum of the lattice. An interaction Hamiltonian with impurities of the form

$$H_{imp} = \sum_{k,q} \{F_q \, b^{\dagger}_{k+q} b_k + c.c.\} \tag{15.1}$$

was studied by Sparks *et al.* [2]. We see that (15.1) conserves the z component of the total spin $\mathcal{S}^z$ of the system. In the case that one of the magnons involved is a uniform precession mode, this interaction does not conserve $\mathcal{S}^2$, instead.

The magnon-phonon interaction is also quadratic in the magnon operators. In dielectrics it can arise either from exchange or single-site anisotropy terms in the Hamiltonian, since the parameters of these interactions can be modulated by the ionic displacements. In the rigid-ion approximation, one assumes that the electrons follow the nuclear displacements but the ions stay in the ground state, although in fact one could consider as well the possibility that they make virtual transitions to excited states, a correction to the magnon-phonon matrix element which has been considered by Sinha and Upadhayaya [3].   In metals, the electron-phonon interaction gives rise to an effective magnon-phonon coupling which was obtained in the RPA [4, 5]

The next order is provided by the dipolar interaction. From the expansion of the terms in $H_{dip}^{(2)}$ in Eq. (6.32) we obtain cubic terms. These generate the processes of *confluence* and *splitting* of magnons. In a confluence, two magnons are destroyed and another one is created with conservation of the total wave vector. The time-reversed process is the splitting of one magnon into a pair. These processes do not conserve either $\mathcal{S}^z$ or, in general, $\mathcal{S}^2$.

In fourth order we have processes of scattering of two magnons .

All the above mentioned interactions are eventually responsible for the relaxation of the perturbed magnetization (*ferromagnetic relaxation*) to its equilibrium value, that is, they provide equilibration processes. One convenient way to model relaxation phenomena in the long wave limit is the use of a macroscopic differential equation for the magnetization based on Bloch's torque equation which was introduced in Chap. 4.:

$$\frac{d\mathbf{M}(\mathbf{r}, t)}{dt} = \gamma \mathbf{M}(\mathbf{r}, t) \wedge \mathbf{B}_{eff}(\mathbf{r}, t) \tag{15.2}$$

supplemented with phenomenological damping terms adequate to the various processes involved.

Several forms of the damping terms have been proposed. Landau and Lifshitz [6] in 1935 chose the form

$$-\frac{1}{2T}\frac{1}{\mid \mathbf{M} \mid^2}\mathbf{M}(\mathbf{r}, t) \wedge [\ \mathbf{M}(\mathbf{r}, t) \wedge \mathbf{B}_{eff}(\mathbf{r}, t)\ ] \tag{15.3}$$

The factor 2 in the time constant is included because in general T is reserved for the time constant of the energy, which is proportional to the square of the magnetization. This rate of change is perpendicular to $\mathbf{M}$, so that $d\mathbf{M}^2 = 0$. Therefore, the tip of $\mathbf{M}$ spirals on the surface $\mid \mathbf{M} \mid=$ constant. Let us write

$$\mathbf{M} = \mathbf{M}_{\parallel} + \mathbf{m} \tag{15.4}$$

separating the longitudinal component, parallel to the external field B, and the - assumed small- transverse part and rewrite (15.3) as

$$-\frac{1}{2T}\frac{M_{\parallel}B_{eff}}{\mid \mathbf{M} \mid^2}\mathbf{m} + \frac{m^2}{2TM^2}\mathbf{B}_{\mathbf{eff}} \tag{15.5}$$

Then to first order in m/M the motion of $\mathbf{M}$ is a spiralling approach to the direction of $\mathbf{B}_{eff}$ during which M is constant, while the transverse component decreases and the longitudinal one increases.

Another form initially proposed by Felix Bloch to describe nuclear magnetic relaxation was later adapted by N. Bloembergen to describe ferromagnetic relaxation [7]:

$$\frac{dM_z}{dt} = -\gamma(\mathbf{M} \wedge \mathbf{B_{eff}})_\mathbf{z} - \frac{M_z - M}{2T_1}$$

$$\frac{dM_{x,y}}{dt} = -\gamma(\mathbf{M} \wedge \mathbf{B_{eff}})_{x,y} - \frac{M_{x,y}}{2T_2} \qquad (15.6)$$

Let us consider a few examples of cases where one or the other of these forms is adequate to describe the relaxation processes.

In a two-magnon process in which a uniform precession magnon is destroyed and another with a finite k is created, $\mathbf{M}$ and $\mathbf{m}$ change while M_z is kept constant. This contributes to the Bloch-Bloembergen (BB) time T_2.

Consider now a 3-magnon process in which a $k = 0$ and a $k \neq 0$ magnons are destroyed, while another is created with $k \neq 0$. Then the total number of uniform magnons n_0 decreases and the total number of magnons changes, so that M_z changes (see Chap. 7). At the same time, the total number of magnons with $k \neq 0$ does not change, so that $|\mathbf{M}|$ remains constant. This is correctly described by the Landau-Lifshitz (LL) term.

Let us consider now a 3 magnon confluence process, in which two magnons with respective wave vectors k_1, k_2 are annihilated and one with wave vector $k_1 + k_2$ is created. If none of these wave vectors is zero, n_0 does not change, and accordingly (see Chap. 6) m^2 remains the same, while M^2 and M_z change, so that we can describe this process with Bloembergen's T_2 but not with the LL form.

Exercise 15.1

Analyze the following examples and establish which of the above forms of damping term applies:

1. *In a 3 magnon process, one magnon with* $\mathbf{k} = 0$ *and one with* $\mathbf{k} \neq 0$ *are annihilated, while another one with* $\mathbf{k} \neq 0$ *is created.*

2. *In a two-magnon-one-phonon process one uniform precession magnon is annihilated, one* $\mathbf{k} \neq 0$ *magnon is created and one phonon is either created or annihilated.*

We shall now consider the different interactions separately.

15.2 Two-magnon interaction

In this section we want to describe briefly the kind of processes which are mainly responsible for the relaxation of the uniform mode. An interaction Hamiltonian based on the assumption that the main cause is the scattering with imperfections at the surface was given in Eq. (15.1). The matrix elements F_q were obtained by Sparks *et al.* [8].

Let us consider the possible channels for relaxation of the uniform mode. In the experiments a ferromagnetic sample is placed in an external magnetic field B large enough to saturate the magnetization along the z axis. The total magnetic moment will precess around the field at the Larmoor angular frequency $\omega_c = \gamma B$. If B is typically a few thousand oersteds $\omega_c \approx 10^{11}$ sec^{-1}, or a frequency $\nu = \omega_c/(2\pi) \approx 10$ kMc, which lies within the X band. A transverse r. f. field **h** of the same frequency can transfer energy to the magnet, increasing its potential energy $W = -V\mathbf{M} \cdot \mathbf{B}$ in the static field, thereby increasing the inclination of the magnetization in relation to the z axis. If at some instant the r. f. field is turned off, the system is left in a state out of equilibrium. The interactions we are considering are responsible for the subsequent relaxation to equilibrium, which implies the recovery of the initial precessional motion. Clearly the interactions which decrease the transverse component **m** and/or increase the longitudinal component M_z of the magnetization will contribute to the equilibration process. Experimentally, transitions of the first kind, which conserve M_z, are found to be faster, with a characeristic time constant which it is customary to call T_2. In a subsequent, slower stage, M_z increases, with another time constant called T_1. In Fig. 15.1 we show schematically the spiralling motion of the vector **M** in the relaxation stage with $M_z = constant$ and the subsequent helical motion to recover the saturation value $M_z = M_s$ a process which is compatible with the BB damping term.

Fig. 15.2 shows a different path to equilibrium, in which $| \mathbf{M} |= $ constant, as in LL damping. We notice that the reverse path to that in Fig. 15.2 would be followed by **M** as the r. f. field is turned on, since a uniform field can only excite uniform precession magnons, and they conserve $\mathbf{M}^2$ (see Chap. 7). M_z

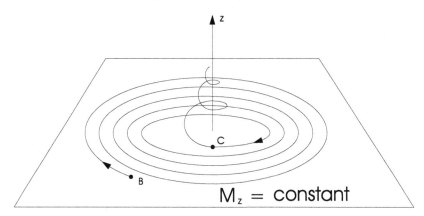

Figure 15.1: *Relaxation of* **M**. *From point B to C M_z is approximately conserved (T_2 BB stage). After instant C the basic process is the recovery of the saturated M_z, as in the T_1 BB stage.*

decreases because the spatially uniform r. f. field excites $\mathbf{k} = \mathbf{0}$ magnons. Each new magnon decreases VM_z by one unit of $\hbar$. Immediately after the r. f. field

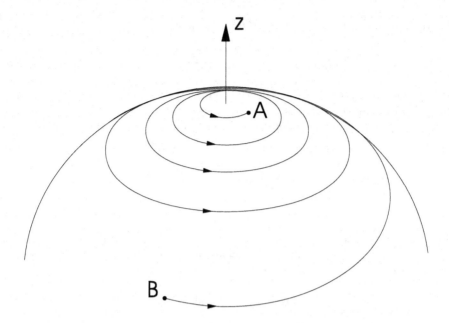

Figure 15.2: *Relaxation of* **M** *with conservation of* **M**^{2} *as in the LL damping. At B the r. f. field is switched off. At A the equilibrium saturated* M_s *has been recovered.*

is switched off the magnet begins to relax to the equilibrium uniform precession following in different scales of time one among several alternative competing channels. One such is the two-magnon process we described before. After a time comparable with T_2 other processes begin to dominate the relaxation. Experiments show that the fast relaxation is effectively dominated by the surface pit scattering. In effect, the resonance linewidth increases with the radius of the surface pits, which can be controlled by the polishing treatment of the sample surface, in agreement with the theory [2] based on this mechanism. A detailed discussion of the calculation of the linewidth based on this model can be found in Sparks' book [8].

15.3 Three-magnon processes

The dipolar interaction is the only source of terms with three magnon operators. From $H_d^{(1)}$ in Eq. (6.31) we get a cubic term

$$H_{cubic} = \frac{3\gamma^2\sqrt{2S}}{2} \sum_{l \neq m} F_{lm}[\, a_l^\dagger a_l a_l - a_l a_m^\dagger a_m \,] + \text{h. c.} \qquad (15.7)$$

We remind that the expresion of $F(R)$ is, from Eq. (6.14),

$$F(\mathbf{R}) = \frac{(X - iY)Z}{R^5} \tag{15.8}$$

We assume our system has the reflection symmetry $z \to -z$, so that the first sum in Eq. (15.7) vanishes, unless the point considered be very near to the surface (at a distance of only a few lattice planes from it). With this simplification we are left with only the second sum in (15.7), which in terms of spin wave operators reads:

$$H_{cubic} = -\frac{3\gamma^2\sqrt{2S}}{2} \sum_{q,k} \left[F_q b^\dagger_{k+q} b_q b_k + F^*_q b^\dagger_k b^\dagger_q b_{k+q} \right] \tag{15.9}$$

This describes processes of *confluence* of two magnons into one or *splitting* of one into two, where q is the pseudo-linear-momentum transfer. Let us simplify the notation, by calling the wave vectors, k_i, $i = 1, 2, 3$:

$$H_{cubic} = \sum_{1,2,3} \Delta(k_1 - k_2 - k_3) \left[C_{123} b^\dagger_1 b_2 b_3 + C^*_{123} b^\dagger_3 b^\dagger_2 b_1 \right] \tag{15.10}$$

where C is the matrix element of the perturbing Hamiltonian and $\Delta(k)$ is the Kronecker's $\delta_{k,0}$. We neglect the effects of the Bogoliubov transformation, so we calculate the matrix elements of H_{cubic} between states characterized by a given number of spin-waves of each wave-vector k_i. Let us consider for example the confluence event represented in Fig. 15.3(a):

$$\langle n_1 + 1, n_2 - 1, n_3 - 1 \mid H_{cubic} \mid n_1, n_2, n_3 \rangle$$
$$= C^*_{123} b^\dagger_1 b_2 b_3 + C^*_{132} b^\dagger_1 b_3 b_2 \tag{15.11}$$

The transition probability per unit time (TP) from a given initial state i to a final state f due to a perturbation potential V is given in the first Born approximation by the Golden Rule [9]

$$TP = \frac{2\pi}{\hbar} \mid \langle i \mid V \mid f \rangle \mid^2 \delta(\hbar\omega_f - \hbar\omega_i) \tag{15.12}$$

Let us consider now a given k_1 and an initial state where the number of magnons k_1 at time t is $n_1(t)$ and n_2, n_3 are the thermal equilibrium values given by Bose-Einstein distribution function. Then the total transition probability to any final state is the sum

$$TP = \frac{2\pi}{\hbar^2} \sum_f \mid \langle i \mid V \mid f \rangle \mid^2 \delta(\omega_f - \omega_i)(n)\Delta(K) \tag{15.13}$$

where we changed from energy to frequency in the argument of δ. The factor $\Delta(K)$ is the Kronecker Δ which takes account of wave-vector conservation, while the notation (n) contains all thermodynamic weight factors for the process

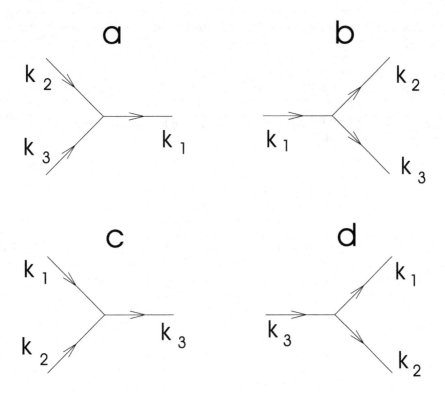

Figure 15.3: *Three magnon processes involving states k_1,k_2 and k_3.*

involved. Those contributions which increase (decrease) n_1 must be added to (subtracted from) the time rate of change of $n_1(t)$, so that the contributions from confluence and splitting events involving two given sates can be added together, with their corresponding sign. Therefore we arrive at

$$\frac{dn_1}{dt} = \frac{1}{2}\sum_{2,3} \left[(n_1 + 1)n_2 n_3 - n_1(n_2 + 1)(n_3 + 1) \right] \times$$

$$| \, C_{312} + C_{321} \, |^2 \, \Delta(k_1 - k_2 - k_3)\delta(\omega_1 - \omega_2 - \omega_3)$$

$$+ \sum_{2,3} \left[(n_1 + 1)(n_2 + 1)n_3 - n_1 n_2(n_3 + 1) \right] \times$$

$$| \, C_{312} + C_{321} \, |^2 \, \Delta(k_3 - k_1 - k_2)\delta(\omega_3 - \omega_1 - \omega_2) \qquad (15.14)$$

where the factor $1/2$ in the first sum corrects overcounting. The terms in square brackets are what we called (n) in Eq. (15.13). One verifies that if the three magnon populations are in equilibrium the expression (n) vanishes in both terms. If wew call $(\bar{n})$ the result of substituting in (n) the Bose-Einstein distributions, we have $(\bar{n}) \equiv 0$, so that we can always subtract $(\bar{n})$ from (n) in all expressions above. Let us consider processes described by Fig. 15.3 (a) and (b), which are a confluence of k_1 and k_2 into k_3 and the time reversed splitting process.

Exercise 15.2
Verify the vanishing of the statistical factor $(\overline{n})$ due to energy conservation.

Upon subtracting $(\overline{n})$ from the original statistical factor, and using again energy conservation, one finds that

$$(n)_{conf} = \overline{n}_{1+2}\overline{n}_2 e^{\beta\hbar\omega_2}\left(1 - e^{\beta\hbar\omega_1}\right)(\overline{n}_1 - n_1) \tag{15.15}$$

Notice that the expression above is independent of k_3. One verifies that the cases (c) and (d) of Fig. 15.3 can also be cast in a similar form, which yields

$$\frac{dn_1}{dt} = \left(\frac{1}{T_{conf}} + \frac{1}{T_{split}}\right)(n_1 - \overline{n}_1) \tag{15.16}$$

where (substituting the sum over k_2 by an integration over the BZ, as usual)

$$\frac{1}{T_{conf}} = V\left(\frac{1}{2\pi\hbar}\right)^2 \left(e^{\beta\hbar\omega_1} - 1\right)\int d\mathbf{k}_2 \times$$
$$|\, C_{k_1} + C_{k_2}\,|^2\, e^{\beta\hbar\omega_{k_2}}\delta(\omega_{k_1} + \omega_{k_2} - \omega_{k_1-k_2}) \tag{15.17}$$

and

$$\frac{1}{T_{split}} = \left(\frac{1}{2\pi\hbar}\right)^2 \left(e^{\beta\hbar\omega_1} - 1\right)\times$$
$$\int d\mathbf{k}_2 V\,|\, C_{k_2} + C_{k_1-k_2}\,|^2\,\delta(\omega_{k_1} - \omega_{k_2} - \omega_{k_1-k_2}) \tag{15.18}$$

Exercise 15.3
Verify Eqs. (15.17) and (15.18).

Each of these relaxation times is predominant for given intervals of values of k_1 which depend, in turn, on the temperature range. Detailed calculations were made by Sparks *et al.* [2] and by Schlömann *et al.* [10]. It turns out that phase space restrictions imposed by energy and momentum conservation lead to the vanishing of the transition probability for the confluence process as $k_1 \to 0$ [8]. We close this section with the calculation of the Fourier transform

$$F_k = \frac{1}{N}\sum_R{}' e^{i\mathbf{k}\cdot\mathbf{R}}\frac{(X - iY)Z}{R^5} \tag{15.19}$$

As we did in Chapter 7 for similar calculations, we expand the plane wave in spherical harmonics and change the sum into an integral, since we are interested in the long-wave limit $ka \ll 1$, a = lattice constant. We shall also assume $kR \gg 1$, where R is the sample diameter, and we shall neglect surface effects, so that every point is very far from the surface and we can assume translation symmetry. We recognize that

$$\frac{(X - iY)Z}{R^2} = AY_2^{-1}(\Omega)$$

where A is a coefficient, and substitute the upper limit in the r integral by ∞, obtaining

$$F_k \approx 4\pi AY_2^{-1}(\Omega_k) \int_a^\infty dr \frac{j_2(kr)}{r} \tag{15.20}$$

Finally, we get

$$C_k = -\frac{8\pi\mu_B}{\sqrt{N}}\sqrt{\frac{\mu_B M_s}{a^3}} \tag{15.21}$$

15.4 Magnon-phonon interaction

Both the effective exchange integral and the dipolar interaction tensor depend on the distance between interacting spins, which oscillates, even at $T = 0$, due to thermal fluctuations of the positions of ions. The distance R_{ab} between neighbouring sites a, b can be written as

$$\mathbf{R_{ab}} = \mathbf{R}_b - \mathbf{R}_a = \mathbf{R}_{ab}^{(0)} + \delta\mathbf{R}_b - \delta\mathbf{R}_a \tag{15.22}$$

where $\delta\mathbf{R}_{a,b}$ are the instantaneous displacements of the respective ions. We expand these displacements as linear combinations of phonon creation and annihilation operators. For simplicity consider that the crystal has only one ion per unit cell. Then the α component of the displacement of ion l is [12]

$$\begin{aligned} \delta R_{\alpha,l} &= \sqrt{\left(\frac{\hbar}{NM_{at}}\right)} \sum_{s,k} \frac{1}{\sqrt{2\,\omega_s(k)}} \left[e_\alpha^{(s)}(k)e^{i\mathbf{k}\cdot\mathbf{R}_l^0} a_s(k) \right. \\ &\quad \left. + e_\alpha^{(s)}(k)e^{-i\mathbf{k}\cdot\mathbf{R}_l^0} a_s^\dagger(k) \right] \end{aligned} \tag{15.23}$$

where $a_s^\dagger(k)$ ($a_s(k)$) creates (annihilates) a phonon of wave-vector k inside the BZ, with polarization s and corresponding dispersion relation $\omega_s(k)$. We have taken for simplicity a lattice with only one kind of atoms, so all ionic masses M_{at} are equal. The phases of the eigenvectors of the dynamical crystal matrix can be chosen so that

$$e_\alpha^{(s)*}(k) = e_\alpha^{(s)}(-k) \tag{15.24}$$

Then if we call

$$\mathbf{g}_s(k) = -i\mathbf{e}^{(s)}(k)\sqrt{\left(\frac{\hbar}{M_{at}}\right)} \tag{15.25}$$

we can simplify the expansion (15.23):

$$\delta\mathbf{R}_l = \frac{1}{\sqrt{N}} \sum_{s,k} \mathbf{g}_s(k)\left(a_s^\dagger(k) - a_s(-k)\right) e^{i\mathbf{k}\cdot\mathbf{R}_l^0} \tag{15.26}$$

We return to the Heisenberg Hamiltonian and expand the effective exchange integral $J(R_{lm})$ in a Taylor series around the equilibrium distance R_{lm}^0:

$$J(R_{lm}) = J(R_{lm}^0) + \nabla_R J(R_{lm})|_{R^0} \cdot (\delta \mathbf{R}_l - \delta \mathbf{R}_m) + \cdots \qquad (15.27)$$

We substitute (15.27) and the expansion (15.26) in the exchange Hamiltonian, to obtain the magnon-phonon interaction Hamiltonian:

$$H_{mp} = \sum_{k,q} \phi_{k,q}^s b_{k-q}^\dagger b_k \left(a_{qs}^\dagger - a_{-qs} \right) \qquad (15.28)$$

where

$$\phi_{k,q}^s = \frac{4S}{\sqrt{N}} \sum_\Delta \nabla J(\Delta) \cdot \mathbf{g}_s(k) \left(e^{i\mathbf{k}\cdot\boldsymbol{\Delta}} - 1 \right) \left(1 - e^{-i\mathbf{q}\cdot\boldsymbol{\Delta}} \right) \qquad (15.29)$$

and $\{\Delta\}$ is as usual the star of nearest neighbours. With this interaction we can calculate the lifetime and the energy renormalization of both phonons and magnons. This is particularly relevant in a non-equilibrium situation such that, due to some external perturbation, there is a deviation of the magnon population from the thermal equilibrium average, which can decay via the emission of phonons. Likewise, a deviation of the phonon population can decay through scattering with magnons. The relaxation of the energy excess accumulated in either the phonon or magnon system would be controlled by the spin-lattice (or magnon-phonon) τ_{mp} relaxation time if the equilibration times of the separate subsystems, τ_{mm} and τ_{pp} , were considerably longer than τ_{mp} [13, 14]. Perturbation theory calculations of the magnon-phonon lifetime were carried out by Sinha and Upadhyaya [3] and by Pytte [11]. In a two-sublattice ferrimagnet or in an AFM the magnon-phonon interaction involves also the optical phonon branches.[17, 18]

Since the coupling Hamiltonian of Eq. (15.28) is linear in the phonon operators the lowest order in which one finds a self energy correction for the magnon is the second. The problem is formally similar to the electron-phonon system [15]. One can use perturbation theory to second order and find the complex correction to the unperturbed energy of a magnon. It is also possible to extend the range of temperatures for this calculation if one maintains the original form of the Heisenberg Hamiltonian in terms of the spin operators and treats the problem with the Green function or equation of motion formalism, applying some decoupling scheme to the -by now familiar- hierarchy of equations (see Chap. 6) for the spin operators, and treats the phonon system in the usual perturbation theory. This program was developed by Pytte [11], who obtained the energy renormalization and the lifetime of magnons and phonons both at low temperatures and near the critical temperature of a FM. In both cases the unperturbed energies of both types of excitations are very different from each other. There is however some interval of values of temperature and k such that the dispersion relations of magnons and phonons intersect, and in this case one cannot apply the usual perturbation theory without first eliminating this degeneracy. This implies the consideration of the mixed magnon-phonon modes (or magneto-elastic modes in the long k limit). We address this problem in next section. We shall see that in this case a simpler Hamiltonian that (15.28) applies.

The local single-ion anisotropy energy generates a one-phonon-two-magnon interaction [16]. Consider the simplest case of uniaxial anisotropy

$$V_a = - \sum_i D(R_i) \, (S_i^z)^2 \tag{15.30}$$

Suppose we have a cubic crystal. Then if the $\{R_i\}$ are equilibrium coordinates, this term should be absent, so that

$$D(R_i^{(0)}) = 0, \quad \forall \, i$$

However if all sites are in motion, there may be distorsions at each site, and some anisotropy might appear as a result. If site μ is displaced by $\delta \mathbf{R}_\mu$, we can expand the anisotropy parameter in the relative displacements of the first nearest neighbours of each site as

$$D = \sum_{\mu, \nu} D_{\mu\nu} \Delta^{\mu\nu}, \quad \Delta^{\mu\nu} = \delta \mathbf{R}_\mu^\nu - \delta \mathbf{R}_0^\nu \tag{15.31}$$

Here index μ runs over the star of first n. n. and $\nu = x, y, z$. We may consider the case in which a uniaxial distorsion is generated, that is, when only $D_{zz} \neq 0$. There may be a small static uniaxial distortion, so that we write in general

$$\Delta^{zz} = \Delta^{zz}_{stat} + \Delta^{zz}_{phon} \tag{15.32}$$

Now we expand $(S_i^z)^2$ in magnon operators up to second order terms. For the strain component we can take the small k limit

$$\Delta^{zz} = \frac{\partial \delta \mathbf{R}_z}{\partial z} \tag{15.33}$$

and expand $\delta \mathbf{R}_z$ as in Eq. (15.23) to obtain a two-magnon-one-phonon interaction. This interaction was experimentally observed in the paramagnetic resonance of Fe^{3+} in Ytrium Gallium Garnet [16].

15.5 Bilinear magnon-phonon interaction

Let us return to the dipolar Hamiltonian. From Eq. (6.31) we get a linear term in the spin-wave operators:

$$-\frac{3\gamma^2 S \sqrt{2S}}{4} \sum_{n,m} {}' \{ F(R_{lm}) \, a_l + \text{h. c.} \} \tag{15.34}$$

The sum over m vanishes when the sites are in their equibibrium positions, which is why we disregarded the linear term until now. However, if we expand $F(R)$ in Taylor series around the equilibrium position

$$F(R_{lm}) = F(R_{lm}^0) + \nabla_R F(R_{lm})|_{R^0} \cdot (\delta \mathbf{R}_l - \delta \mathbf{R}_m) + \cdots \qquad (15.35)$$

and expand $\delta \mathbf{R}$ in phonon operators we obtain a Hamiltonian which is bilinear in the magnon and phonon operators:

$$H_{mp}^{(1)} = \sum_q b_q^\dagger \left(G_{-q}\, a_q^\dagger + G_q^*\, a_q \right) \qquad (15.36)$$

This interaction leads to mixing of both types of excitation involved. Insofar as the difference in their energies is larger than the coupling constant G one can treat its effects as a perturbation. This is not correct, however, in a region of q where both dispersion relations cross. The point in the ω, q plane where this occurs, if it were the case, is called the *nominal* crossing, since in reality the crossing does not occur, precisely because both modes mix and the dispersion relations resulting from the exact treatment of Hamiltonian (15.36) avoid each other in that region.

The acoustic-phonon dispersion relations are linear in the wave vector at small k, and along crystalline symmetry axes they can be written as

$$\omega_k^{(s)} \approx c^s k$$

In general one has three different phonon energies for an arbitrary $\mathbf{k}$, but along a symmetry axis two polarizations are degenerate, and for small k correspond to shear waves, while the remaining one is a longitudinal mode describing compression waves. A typical sound velocity is $c^s \approx 3 \cdot 10^5$ cm/sec. On the other hand, Eq. (6.58) gives for a spherical sample at long wavelengths, and for $\theta_k = \pi/2$ the expression

$$\omega_k^2 \approx \gamma^2 B^2 + \frac{\gamma D}{\hbar} k^2 \left(2B + \frac{4\pi M}{3} \right) \qquad (15.37)$$

where B is the external field and the exchange energy is

$$\omega_e(k) \approx D k^2.$$

If there is no demagnetization field, as in a disk with M on the plane of the disk, and if $\mathbf{k} \parallel \mathbf{M}$, we have a pure exchange dispersion relation if $B = 0$. One verifies in this case that both curves intersect if $k \approx 10^6 cm^{-1}$ at a frequency $\omega \approx 10^{12}$ 1/sec. Suppose B is not zero, there is no demagnetization field, and $\mathbf{k} \parallel \mathbf{B}$. Then the frequency for nominal crossing can increase. For $B = 100$ oersteds, the cross-over occurs at $\omega \approx 2 \cdot 10^9$ rad/cm , or a frequency of 300 Mc/sec, at $k \approx 10^6$ cm^{-1}, where the exchange contribution is negligible.

In general, the magnon frequency is finite at $k \to 0$ while the acoustic phonons have a dispersion relation linear in k, so that for very small k both excitations remain well defined, and magnons are the upper branch. For intervals of k as analyzed above there is a region where they turn into mixed magneto-elastic waves. For higher k they become again separate excitations but they have interchanged roles: the upper branch is now phonon-type, and the lower magnon-type.

One can easily verify this description. Call the non-interacting dispersion relation ω_k for magnons and Ω_k for phonons. Then we write the non-interacting Hamiltonian for both excitations as

$$H_0 = \sum_k \omega_k b_k^\dagger b_k + \sum_k \Omega_k a_k^\dagger a_k \qquad (15.38)$$

If we add the bilinear magnon-phonon Hamiltonian of Eq. (15.36) to Eq. (15.38) we can easily diagonalize the perturbed Hamiltonian.

Exercise 15.4
Add the bilinear magnon-phonon Hamiltonian of Eq. (15.36) to the harmonic magnon and phonon unperturbed Hamiltonian (15.38) and diagonalize the perturbed Hamiltonian. Analyze the resultant modes both near and far from the nominal crossing.

References

1. Dyson, F. J. (1956) *Phys. Rev.* **102**, 127.

2. Sparks, M., Loudon, R. and Kittel, C. (1961) *Phys. Rev.* **122**, 791.

3. Sinha, K. P. and Upadhyaya, U. N. (1962) *Phys. Rev.* **127**, 432.

4. Rajagopal, A. K. and Joshi, S. K. (1967) *Phys. Lett.* **24A**, 95.

5. Alascio, Blas and López, Arturo, (1970) *J. Phys. Chem. Solids* **31**, 1647.

6. Landau, L. and Lifshitz, E. (1935) *Phys. Sov. Union* **8**, 153.

7. Bloembergen, N. (1956) *Proc. IRE* **44**, 1259.

8. Sparks, M. (1964) *Ferromagnetic Relaxation Theory*, Mc Graw-Hill Book Co., New York.

9. Messiah, A. (1960) *Mécanique Quantique*, Dunod, Paris.

10. Schlömann, E. and Joseph, R. I. (1961) *J. Appl. Phys.* **32**, 165S and *idem*, 1006.

11. Pytte, E. (1965) *Ann. Phys. (N. Y.* **32**, 377.

12. Callaway, Joseph (1976) *"Quantum Theory of the Solid State"*, Academic Press, Inc., New York.

13. Sinha, K. P. and Kumar, N. (1980) *"Interactions in Magnetically Ordered Solids"*, Oxford University Press.

14. Akhiezer, A. I. (1946) *J. Phys. (USSR)* **10**, 217.

15. Abrikosov, A. A., Gorkov, L. P. and Dzyaloshinski, I. E. *"Methods of Quantum Field Theory in Statistical Physics"*, (1975), Dover Publications Inc., New York.

16. Geschwind, S. (1961) *Phys. Rev.* **12**,363.

17. Kasuya, T. and LeCraw, R. C., (1961) *Phys. Rev. Lett.* **6**, 223.

18. Shukla, G. C. and Sinha, K. P. (1967) *Can. J. Phys.* **45**, 2719.

Appendix A

Group Theory

A.1 Definition of group

Definition A.1 — Grupoid
A groupoid is a pair consisting of a non-empty set G called the carrier and a binary operator μ in G: if $g, h \in G$ then $(g, h)_\mu$, the result of the binary operator on that pair of elements of G, is also in G, and we write simply $gh = f \in G$. The order of G is the number of its elements. We say that the groupoid is associative if for any threee elements $a, b, c \in G$ we have

$$(ab)c = a(bc) \tag{A.1}$$

Definition A.2 — Semigroup
An associative groupoid is called a semigroup

An example is the so called *renormalization group* in statistical physics. If for any pair $(a, b) \in G$ we have

$$ab = ba \tag{A.2}$$

then G is commutative.
An element e is called an *identity* element if

$$\forall g \in G, \quad eg = ge = g \quad .$$

Definition A.3 — Monoid
A semigroup with an identity is called a monoid.

Let us call the identity simply 1. In a monoid, we call an *inverse* of g an h which satisfies

$$gh = 1 = hg \tag{A.3}$$

and this implies that g is reciprocally the inverse of h. An element cannot have more than one inverse.

Example A.1 — Example of monoid
Consider the set M_X of all the mappings of a set X on itself, and the binary operation consisting of the composition of mappings. Then if X is a non-empty set, M_X is a monoid.

Definition A.4 *A mapping $S \to T$ is onto if every $t \in T$ has at least one pre-image $s \in S$. If every t has one and only one pre-image we say it is onto and one-to-one, and the mapping is called matching or bijective.*

Definition A.5 — Group
A group is a monoid in which every element has an inverse.

Definition A.6 — Homomorphism
A homomorphism is a mapping θ of a grupoid (G, α) into a grupoid (H, β) such that

$$\big((g_1, g_2)\alpha\big)_\theta = (g_1\theta, g_2\theta)_\beta \ , \quad \forall g_1, g_2 \in G \tag{A.4}$$

where θ is the law of transformation from elements of G to elements of H, and $\alpha(\beta)$ are respectively the association operations between elements in $G(H)$.

Definition A.7 — Isomorphism
An isomorphism is a homomorphism which is both one to one and onto.

Clearly the definition applies also to groups, and in this case the identity of G corresponds to the identity of H and inverses of elements are transformed into inverses of the transformed elements. All mutually isomorphic groups can be thought of as copies or *realizations* of an abstract group, which is completely characterized by the multiplication table between its elements. In particular we shall deal with mappings of a set of points into itself as a realization of an abstract group.

A.2 Group representations

Let us consider a real vector space R_n of n dimensions and all the possible linear homogeneous transformations of this space into itself. The set $\Sigma(G)$ of linear transformations which is isomorphous to a given abstract group G is a special realization thereof which is called a *representation*. To each $s \in G$ we associate a linear transformation $U(s) : R_n \to R_n$ of the vector space such that

1. $U(s)U(t) = U(st)$

2. $U(e) = 1$

3. $U(s^{-1}) = (U(s))^{-1}$

Upon choosing a particular basis for the space R_n, all $n \times n$ matrices which represent the transformations must have determinant $\neq 0$ due to condition 3) above. A change of basis of the space R induces a cordinate transformation C whereby

$$U(s) \Rightarrow CU(s)C^{-1} \ .$$

A.2.1 Reducibility

Let us consider now such a family $\Sigma(G)$ of transformations $\Sigma :\ R \to R$. A subspace R' which is a subset of R is *invariant* under Σ if $\forall U \in \Sigma$, $U(R') \to R'$. If the representation is unitary, that is if $U^\dagger = U^{-1}$ $\forall U \in \Sigma$, the complement subspace $R'' = R - R'$ is also invariant under Σ (but in general this might not be the case). If both R' and R'' are invariant all matrices $U \in \Sigma$ contain two independent blocks which have respectively elements only in R' or R''. We say that Σ *reduces* to the sum of Σ' and Σ''. The original representation is called *reducible*. If one chooses bases for R' and R'' which are mutually orthogonal each matrix of Σ deblocks simultaneously into two blocks of dimensions n', n'', where $n = n' + n''$ and n' is the dimension of R', n'' that of R''. For complex vector spaces, these concepts can be applied to the case of unitary vector spaces.

Definition A.8 *A unitary vector space is one in which a scalar product of two vectors r, v exists, defined in general as*

$$(r, v) = r_i A_{ik} v_k \tag{A.5}$$

with $A^\dagger = A$, which ensures that the norm of a vector is real: (r, r)=real. If for any vector $r \neq 0$ the norm is positive, the eigenvalues of A are all positive. We shall consider this to be the case.

Theorem A.1 *If R is a unitary vector space invariant under a group of transformations then either R is irreducible under G or it can be completely reduced to a sum of mutually orthogonal subspaces which are also invariant under G. This property is invariant under unitary transformations of the basis of R.*

In a representation D of a group G each $s \in G$ is mapped into a matrix $D(s)$.

Definition A.9 — Characters of a representation
We call

$$\chi(s) = TrD(s) \tag{A.6}$$

the character of s in the representation D.

Since the trace is invariant under unitary transformations, so are the characters. If the representation is reducible such that

$$D = \sum_{i=1}^{r} D^{(i)} \tag{A.7}$$

then the characters satisfy

$$\chi_D(s) = \sum_{i=1}^{r} \chi_{D^{(i)}}(s) \tag{A.8}$$

Definition A.10 — Irreducible representations
If a representation D cannot be de-blocked into smaller dimensional matrices, it is called irreducible.

A.3   Orthogonality relations

Consider two different representations D^λ, D^μ of a group G of transformations of a unitary space R and let h be the order (assumed finite) of G. Then we take for every transformation $S \in G$ and for fixed pairs of indices $(kl), (ij)$ the product $D^\lambda(S^{-1})_{kl}D^\mu(S)_{ij}$ and sum S over the whole group to obtain the orthogonality relation

$$\sum_S D^\lambda(S^{-1})_{kl}D^\mu(S)_{ij} = \delta_{\lambda\mu}\delta_{kj}\delta_{il}\frac{h}{n_\lambda} \tag{A.9}$$

where (n_λ, n_μ) are the respective dimensions of both representations. If they are unitary, Eq. (A.9) reads

$$\sum_S D^\lambda(S)_{kl}^\dagger D^\mu(S)_{ij} = \delta_{\lambda\mu}\delta_{kj}\delta_{il}\frac{h}{n_\lambda} \tag{A.10}$$

which is called the *fundamental orthogonality relation of the first kind*. If we take $i = j$, $k = l$ and we sum over i and k we find an orthogonality relation for characters of irreducible representations:

$$\sum_S \chi_\lambda^*(S)\chi_\mu(S) = h\delta_{\lambda\mu} \tag{A.11}$$

If we consider an arbitrary representation D we can obtain its decomposition into irreducible components:

$$D = \sum_\lambda c_\lambda D^\lambda , \quad c_\lambda = \text{integers} \geq 0 \tag{A.12}$$

Upon applying relation (A.11) we have a formula for c_λ :

$$c_\lambda = \frac{1}{h}\sum_S \chi_\lambda^*(S)\chi_D(S) \tag{A.13}$$

c_λ is the number of times that the representation λ is contained in D. This decomposition is unique, except for unitary transformations: if two representations yield the same set c_λ they can only differ by a unitary transformation. Two representations related by a unitary transformation are called *equivalent*.

A.4   Projection operators

Given a unitary space R invariant under a group G we can project it onto all the subspaces it contains which are irreducible under G. We obtain a basis for the irreducible representation λ as the set of vectors $\{r_i, \quad i = 1, \cdots, n_\lambda\}$ defined as

$$r_i^\lambda = \sum_S D^\lambda(S)_{ij}^\dagger Sr \tag{A.14}$$

where n_λ = dimension of D^λ, j is fixed but arbitrary and r is any vector of the space R. Of course it may happen that the result of applying the *projection operator*

$$P_i^\lambda = \sum_S D^\lambda(S)_{ij}^\dagger S$$

on r be zero , in which case one has to choose another vector to operate upon. After acting with P_i^λ for $i = 1, \cdots, n_\lambda$ on one or more (if necessary) vectors r one obtains a basis for the representation λ.

A.5   Coordinate transformations

A change of coordinates, in general a composition of a translation and a rotation of the laboratory coordinate system (affine transformation), changes in principle the values of all the physical variables. If there is a set of such transformations that leave the Hamiltonian invariant (symmetry transformations), they constitute a group, since H is invariant under any composition of two or more symmetry transformations, the identity is obviously contained in the set and all rotations and translations have an inverse. The transformations can also include the inversion at a fixed point or reflections on symmetry planes. If we leave translations aside, the symmetry group is called a *point group*, since it contains rotations or reflections which leave one point fixed, which we choose as the origin. Consider now the stationary Schrödinger equation for an ion

$$H(x)\psi_\lambda^i(x) = E_\lambda\psi_\lambda^i(x) \tag{A.15}$$

The index $i = 1, \cdots,$ n_λ denotes one of the linearly independent wave functions belonging to the same eigenvalue E_λ with degeneracy n_λ. We apply now a transformation of the coordinates, of the form $x = Rx'$, which leaves H invariant, so that:

$$H(x)\psi_\lambda^i(Rx) = E_\lambda\psi_\lambda^i(Rx) \tag{A.16}$$

Changing the coordinates, however, does change the wave function, in general, and we can write

$$\psi_\lambda^i(Rx) = (P_R\psi_\lambda^i)(x) \tag{A.17}$$

which is a different function of x. This function, since it satisfies the same eigenvalue equation, must be a linear combination of the basis of the degenerate manifold of E_λ. Then we write

$$\psi_\lambda^i(Rx) = \sum_j D_\lambda^{ji}(R)\psi_\lambda^j(x) \tag{A.18}$$

It is clear that the matrices $\mathbf{D}(\mathbf{R})$ in Eq. (A.18) form a representation of the symmetry group G of the Hamiltonian. Before we conclude that this is an irreducible representation we must discard the possibility that the origin of the degeneracy be other than symmetry. In fact it is possible that there exist accidental degeneracies due to special properties of the potential, which would imply that the degenerate manifold contains more than one irreducible representation of G. If this situation is discarded, we have the

Theorem A.2 *The manifold of n_λ linearly independent degenerate eigenfunctions of a given eigenvalue E_λ constitutes a basis for an irreducible representation of the symmetry group of H.*

A.6   Wigner–Eckart theorem

In order to prove Wigner-Eckart theorem we shall make use of a very important lemma on linear transformations:

Lemma A.1 (Schur) *Given $\mathbf{D}^{(1)}(T), \mathbf{D}^{(2)}(T)$, $\forall T \in G$, two irreducible representations of dimensions n_1, n_2 and a matrix $\mathbf{P}$ such that*

$$\mathbf{P}\mathbf{D}^{(1)}(T) = \mathbf{D}^{(2)}(T)\mathbf{P} , \quad \forall T \in G \tag{A.19}$$

then:

1. *if $D^{(1)}$ is not equivalent to $D^{(2)}$, $\mathbf{P} = 0$;*

2. *if $D^{(1)}$ and $D^{(2)}$ are equivalent, either $\mathbf{P} = 0$ or $det\mathbf{P} \neq 0$;*

3. *if $\mathbf{D}^{(1)}(T) = \mathbf{D}^{(2)}(T)$, $\forall T \in G$, either $\mathbf{P} = 0$ or $\mathbf{P} = \lambda\mathbf{1}$.*

A given irreducible representation D^μ can appear several, or even an infinite number of times in the spectrum of a Hamiltonian H with symmetry group G. Let us call the different equivalent representations $D^{\mu r}$, where we shall assume that we choose always bases such that the matrices for the same μ are not only equivalent but in fact identical. We are looking for a general selection rule for matrix elements. Let us start by considering a set of functions $\{\phi_j^\lambda\}$, $j = 1, \cdots, n_\lambda$ which transform according to D^λ under G, and expanding them in terms of the eigenfunctions of H, which form a complete basis set of the Hilbert space:

$$\phi_j^{(\lambda)} = \sum_{\mu r} M_{ij}^{(\mu r)} \psi_i^{(\mu r)} \tag{A.20}$$

where

$$M_{ij}^{(\mu r)} = \int \psi_i^{(\mu r)*} \phi_j^\lambda \, \mathrm{d}^3\mathbf{r} \tag{A.21}$$

If $T \in G$, $T\phi_j^{(\lambda)}$ can be wrtten in two ways, according to whether we choose or not to expand ϕ in terms of ψ_i:

$$T\phi_j^{(\lambda)} = \sum_{\mu r} M_{ij}^{(\mu r)} D_{ki}^{(\lambda)}(T)\psi_k^{(\mu r)}$$

$$T\phi_j^{(\lambda)} = \sum_i D_{ij}^{(\lambda)}(T)\phi_i^{(\lambda)}$$

$$= \sum_{\mu r} \sum_{ik} D_{ij}^{(\lambda)}(T) M_{ki}^{(\mu r)} \psi_k^{(\mu r)} \tag{A.22}$$

Since the l. h. s. is the same in these equations, we have

$$\mathbf{D}^{(\mu)}\mathbf{M}^{(\mu r)} = \mathbf{M}^{(\mu r)}\mathbf{D}^{(\lambda)} \ , \ \forall T \in G. \tag{A.23}$$

We are in the conditions stated as the hypotheses for Schur's lemma with $\mathbf{M}$ in the role of the lemma's $\mathbf{P}$, so that the conclusions are:

Theorem A.3 (Wigner–Eckart)

1. *When $\lambda \neq \mu$, $\mathbf{M} = \mathbf{0}$: there cannot be contributions to the expansion (A.20) from functions which do not transform according to D^{λ};*

2. *When $\lambda = \mu$, $\mathbf{M} = a_{\lambda r}\mathbf{1}$: inside each particular set of basis functions $\{\psi_i^{\lambda r}\}$ the matrix $\mathbf{M}$ is a multiple of the unit matrix. The constant of proportionality depends on the representation and on the particular degenerate eigenvalue r.*

A.7 Space groups

A crystal lattice is invariant under a group S of symmetry operations $\{R \mid \mathbf{t_n}\}$ which are compositions of point group transformations R and lattice translations $t_{\mathbf{n}}$. S is called the *space group* of the lattice. The point group operations can be either pure rotations, called *"proper rotations"*, or *"improper rotations"*, the latter being the product of a proper rotation times the inversion at a lattice point.

The translations which leave a lattice of d dimensions invariant have the form

$$\mathbf{t_n} = \sum_{i=1}^{d} n_i \mathbf{a}_i \tag{A.24}$$

where $\{\mathbf{a}_i\}$ is a primitive vector basis of the lattice and $\{n_i\}$ are integers. In fact the infinite lattice itself can be defined as the set of points generated by substituting all the integers for the $\{n_i\}$ in Eq. (A.24). The primitive basis in Eq. (A.24) is not unique, although there is always a way of defining the simplest one.

The set of the translations by the vectors $\{\mathbf{t_n}\}$ is a group T, the translation group of the lattice, which is in all cases a subgroup of S.

Under a point group transformation one primitive translation is transformed into a new translation:

$$\mathbf{t}' = R(\mathbf{t_n}) \tag{A.25}$$

which must again have the form (A.24), so that $\{n_i\} \rightarrow \{n_i'\}$. This is a very strong restriction on the possible point operations. The main result in this connection is that the only proper or improper rotations compatible with a lattice in the above sense are integral multiples of $\pi/2$ or $\pi/3$.

In some cases, the space group operations are such that the point and translation parts do not exist separately as symmetry operations, but their composition does. In such cases the translation is not contained in the set (A.24). These are called *non-symmorphic space groups*. In a symmorphic space group the point symmetry operators constitute a group, which is a subgroup P of the space group S. A symmorphic space group is the direct product of the point and translation subgroups:

$$S = P \otimes T \ .$$

In non-symmorphic groups some elements of S have the form $\{R \mid \mathbf{v_R}\}$, with $\mathbf{v_R} \neq \mathbf{t_n}$ (we remark that $\mathbf{0} \in \{t_n\}$). Since $\{E \mid \mathbf{0}\} \in S$, the point operator R above $\neq E$. Then there is a cyclic subgroup of S of order $m = integer > 0$ and

$$\{R \mid \mathbf{v_R}\}^m = \{E \mid \mathbf{0}\} \ .$$

Therefore, $\mathbf{v_R}$ is a rational fraction $(q/m)\mathbf{t_n}$, $q/m < 1$, of a basic translation.

The possible primitive bases that fill the space in 3d with points of the form (A.24) are 14 primitive triads, which generate an equal number of so called *Bravais lattices*. All these restrictions limit the number of the possible space groups to 2 in 1d, 17 in 2d and 230 in 3d. There are 73 symmorphic space groups in 3d, the rest of the 230 being non-symmorphic.

Finally, there are only 32 point groups compatible with crystal lattices. To each of them corresponds a family of crystals called a *"crystalline class"*.

A.8   Bloch's theorem

We consider now specifically the translation subgroup T of a space group S. T is a commutative group, because the composition of translations is equivalent to an algebraic addition, so that T is isomorphous to the commutative group of addition of numbers.

Commutative groups are also called *Abelian*. They have only one dimensional (scalar) irreducible representations (reps). Call $\chi(t)$ the character of the translation t for a given rep. In a 1d representation, the matrix $D(t) = \chi(t)$.

We must have $\chi(-t) = \chi^{-1}(t)$ and $\chi(Nt) = \chi(t)^N$. It is clear that

$$\chi(t) = e^{i\mathbf{k}\cdot\mathbf{t}}$$

satisfies both conditions, with any $\mathbf{k}$.

Consider now a discrete Hamiltonian defined on the sites of a periodic infinite lattice. A solution of the stationary Schrödinger equation

$$H\psi_\alpha = E_\alpha \psi_\alpha \tag{A.26}$$

must be a basis for a rep of the translation group T, since H is invariant under that group. Since all reps of T are 1d, we have

$$HD(\mathbf{t_n})\psi_\alpha = H\chi_\alpha(\mathbf{t_n})\psi_\alpha = E_\alpha \chi_\alpha(\mathbf{t_n})\psi_\alpha \tag{A.27}$$

We already found $\chi_\alpha(\mathbf{t_n}) = e^{i\mathbf{k}\cdot\mathbf{t_n}}$ so that we can use $\mathbf{k}$ as a label of the eigen-solution, and write $\alpha = (\mathbf{k}, \mu)$, μ being the remaining set of quantum numbers necessary to characterize the eigenvalue. The wave function satisfies

$$D(\mathbf{t_n})\psi_\alpha = e^{i\mathbf{k}\cdot\mathbf{t_n}}\psi_\alpha \qquad (A.28)$$

Clearly the form

$$\psi_\alpha = \psi_{\mathbf{k}\mu} = A\sum_{\mathbf{n}} e^{i\mathbf{k}\cdot\mathbf{t_n}}\phi_\mu(\mathbf{R_n}) \qquad (A.29)$$

satisfies (A.28). Eq. (A.29) is Bloch's theorem for this case.

For an electron in a crystalline periodic potential the stationary Schrödinger equation reads

$$H(\mathbf{r})\psi_\alpha(\mathbf{r}) = E_\alpha\psi_\alpha(\mathbf{r}) \qquad (A.30)$$

Performing a lattice translation leaves H invariant:

$$H(\mathbf{r})\psi_\alpha(\mathbf{r} + \mathbf{t_n}) = E_\alpha\psi_\alpha(\mathbf{r} + \mathbf{t_n}) \qquad (A.31)$$

Repeating the argument above, we find that α must contain a given wave-vector $\mathbf{k}$ and other quantum numbers not related to translational symmetry, so that

$$\psi_\alpha(\mathbf{r}) = e^{i\mathbf{k}\cdot\mathbf{r}}u_{\mathbf{k}\mu}(\mathbf{r}) \qquad (A.32)$$

where $u_{\mathbf{k}\mu}(\mathbf{r} + \mathbf{t_n}) = u_{\mathbf{k}\mu}(\mathbf{r})$ is periodic. This form automatically satisfies

$$\psi_{\mathbf{k}\mu}(\mathbf{r} + \mathbf{t_n}) = e^{i\mathbf{k}\cdot\mathbf{t_n}}\psi_{\mathbf{k}\mu}(\mathbf{r}) \qquad (A.33)$$

Eq. (A.32) is Bloch's theorem for this case.

Appendix B

Time Reversal

B.1 Antilinear operators

We shall always consider linear or anti-linear operators acting on a Hilbert vector space. Let us first enunciate two useful theorems.

Theorem B.1 *Two linear operators A, B are equal iff*

$$\langle\, u \mid A \mid u \,\rangle = \langle\, u \mid B \mid u \,\rangle \ \ \forall \mid u \,\rangle \tag{B.1}$$

Theorem B.2 *Two linear operators A, B are equal up to a phase factor*

$$A = Be^{i\alpha} \tag{B.2}$$

iff

$$|\langle\, u \mid A \mid v \,\rangle| = |\langle\, u \mid B \mid v \,\rangle| \ \ \forall \mid u \,\rangle, \ \mid v \,\rangle \tag{B.3}$$

Definition B.1 *A is an anti-linear operator if for any pair of kets $\mid u \,\rangle$, $\mid v \,\rangle$, and any pair of complex numbers λ, μ*

$$A\,(\lambda \mid u \,\rangle + \mu \mid v \,\rangle) = \lambda^* A \mid u \,\rangle + \mu^* A \mid v \,\rangle \tag{B.4}$$

In particular we observe that an anti-linear operator does not commute with a c-number:

$$A\,(c \mid u \,\rangle) = c^* A \mid u \,\rangle \tag{B.5}$$

We recall the definition of the hermitian adjoint of a linear operator A. First consider the action of A on a bra. Let us take any bra $\langle \chi \mid$. The scalar product $\langle \chi \mid (A \mid u \rangle)$ is a linear function of $\mid u \rangle$ because $A = $ linear. We know that a linear function of the vectors (kets) belonging to a Hilbert space defines a vector (bra) of the dual vector space, so we call this bra $\langle \eta \mid$. There is a one-to-one linear correspondence between $\langle \chi \mid$ and $\langle \eta \mid$, because the scalar product is also a linear operation. We then define:

$$\langle \eta \mid \equiv \langle \chi \mid A \tag{B.6}$$

369

and we have the identity

$$(\langle \chi \mid A) \mid u \rangle = \langle \chi \mid (A \mid u \rangle) \tag{B.7}$$

Let us take now the ket $\mid v \rangle = (\langle u \mid A)^{\dagger}$. This is an anti-linear function of the bra $\langle u \mid$:

$$\left(\sum_i \lambda_i \langle u_i \mid A \right)^{\dagger} = \sum_i \lambda_i^* \mid v_i \rangle \tag{B.8}$$

But

$$\left\{ \left(\sum_i \lambda_i \mid u_i \rangle \right)^{\dagger} A \right\}^{\dagger} = \left\{ \left(\sum_i \lambda_i^* \langle u_i \mid \right) A \right\}^{\dagger} =$$
$$\left(\sum_i \lambda_i^* \langle v_i \mid \right)^{\dagger} = \sum_i \lambda_i \mid v_i \rangle \tag{B.9}$$

is a linear function of $\mid u \rangle$, which we shall define as the operator $A^{\dagger}$:

Definition B.2 $A^{\dagger}$ *is the hermitian adjoint (h. a.) of A if*

$$A^{\dagger} \mid u \ \rangle = (\langle \ u \mid A)^{\dagger} \ , \ \forall \mid u \ \rangle \tag{B.10}$$

Let us now consider the effect an anti-linear operator has on a bra. If A is anti-linear and $\langle \chi \mid$ is any bra, the scalar product

$$\langle \chi \mid (A \mid u \ \rangle)$$

defines an anti-linear function of the ket $\mid u \ \rangle$. Therefore, its complex conjugate is a *linear* function, which defines a bra η such that

$$\{ \langle \chi \mid (A \mid u \ \rangle) \}^* \equiv \langle \eta \mid u \ \rangle \equiv (\langle \chi \mid A) \mid u \ \rangle \tag{B.11}$$

We have obtained the general identity, for anti-linear operators:

$$(\langle v \mid A) \mid u \rangle \equiv \{ \langle v \mid (A \mid u \rangle) \}^* \tag{B.12}$$

This defines an anti-linear function of the bra $\langle \chi \mid$:

$$\left(\sum_i \lambda_i \langle \chi_i \mid \right) A = \sum_i \lambda_i^* \langle \chi_i \mid A \tag{B.13}$$

If we use definition (B.10) for the h. a. of any operator

$$A^{\dagger} \mid u \rangle \equiv (\langle u \mid A)^{\dagger} \tag{B.14}$$

we find that for an anti-linear A the h. a. satisfies the relation

$$\langle t \mid (A^\dagger \mid u) \rangle = \langle u \, (A \mid t) \rangle \tag{B.15}$$

to be compared with the corresponding relation for a linear operator:

$$\langle t \mid (A^\dagger \mid u) \rangle = \{\langle u \, (A \mid t) \rangle\}^* \tag{B.16}$$

Clearly if A is anti-linear $A^\dagger$ is also anti-linear.

B.2   Anti-unitary operators

Definition B.3 *An anti-linear operator A is anti-unitary if its h. a. equals its inverse:*

$$AA^\dagger = A^\dagger A = 1$$

An anti-unitary operator V induces a mapping

$$\mid \hat{u} \, \rangle = V \mid u \, \rangle \tag{B.17}$$

An operator B is transformed by V as

$$\hat{B} = VBV^\dagger \tag{B.18}$$

and a bra as

$$\langle \, \hat{v} \mid = \langle \, v \mid V^\dagger \tag{B.19}$$

Since anti-unitary operators are anti-linear, we have

$$\langle \, \eta \mid (V \mid v \, \rangle) = \langle \, v \mid (V^\dagger \mid v \, \rangle) \tag{B.20}$$

which leads to the relation

$$\langle \, \hat{u} \mid \hat{B} \mid \hat{v} \, \rangle = \langle \, u \mid B \mid v \, \rangle^* \tag{B.21}$$

so that matrix elements are not invariant under V, but are transformed instead into their complex conjugates.

B.3   Time reversal

Equation (B.4) implies that an anti-unitary operator transforms a c-number into its complex conjugate:

$$AcA^\dagger = c^* AA^\dagger = c^* \tag{B.22}$$

so that, in particular, the basic commutators among dynamical operators change sign:

$$[\, p, q \,] = i\hbar \Longrightarrow [\, \hat{p}, \hat{q} \,] = -i\hbar \tag{B.23}$$

Let us call K_0 an operator which when applied on the coordinate and linear momentum operators yields

$$
\begin{aligned}
K_0 \mathbf{r} K_0^\dagger &= \mathbf{r} \\
K_0 \mathbf{p} K_0^\dagger &= -\mathbf{p}
\end{aligned}
\tag{B.24}
$$

This operator clearly changes the sign of the commutator, and its square is the identity. It is easy to see that it is anti-unitary. One can prove that $K_0^\dagger = K_0 = K_0^{-1}$. On the other hand we observe that its action on the coordinates and momenta, and accordingly on the angular momentum, coincides with the effect of reversing the sign of time, so we take K_0 as the time-reversal operator. In order to completely define K_0 we may demand that when it is applied on the kets of the basis set of eigenfunctions of H it leaves them invariant:

$$
K_0 \mid n \,\rangle = \mid n \,\rangle
\tag{B.25}
$$

which, when combined with its anti-linear properties implies that the coordinates in the basic set (that is, the wave function) of an arbitrary ket $\langle\, n \mid u \,\rangle$ is transformed by K_0 into its complex conjugate:

$$
\langle\, n \mid (K_0 \mid u \,\rangle) = \sum_m \langle\, n \mid (K_0 \{\mid m \,\rangle\langle\, m \mid u \,\rangle\}) =
$$
$$
\langle\, n \mid u \,\rangle^*
\tag{B.26}
$$

Also, the representation matrix elements of any linear operator are transformed by K_0 into their complex conjugates. In effect, we have seen that for an anti-unitary operator,

$$
\langle\, v \mid (A \mid u \,\rangle) = \{\, (\langle\, v \mid A) \mid u \,\rangle \,\}^*
$$

Then

$$
\langle\, \hat{m} \mid \hat{B} \mid \hat{n} \,\rangle = \langle\, m \mid K_0 B K_0 \mid n \,\rangle = \langle\, m \mid (K_0 B \mid n \,\rangle) =
$$
$$
[\, (\langle\, m \mid K_0) B \mid n \,\rangle \,]^* = \langle\, m \mid B \mid n \,\rangle^*
\tag{B.27}
$$

for any operator B.

If the Hamiltonian is invariant under time-reversal, we have

$$
[\, K_0, H \,] = 0 \ .
\tag{B.28}
$$

Consider now the time-dependent Schrödinger equation

$$
i\hbar \frac{\partial \mid \Psi(t)\rangle}{\partial t} = H \mid \Psi(t)\rangle
\tag{B.29}
$$

Applying K_0 on both sides we find that the state

$$
\mid \Psi_{rev}(t)\rangle = K_0 \mid \Psi(-t)\rangle
$$

satisfies the same equation B.29. In particular, consider an eigenstate $| \phi \rangle$ which
satisfies the stationary Schrödinger equation

$$H \, | \, \phi \rangle = E \, | \, \phi \rangle \tag{B.30}$$

with $E = real$, and apply K_0 on both sides. We find, taking into account that
$[H, K_0] = 0$ and that we have chosen the representation in which eigenkets are
real, that the wave function

$$\phi_m = \langle m \, | \, \phi \rangle \tag{B.31}$$

and its complex conjugate ϕ_m^* are degenerate. In the coordinate representation,
this implies that if $\Psi(r)$ is an eigenfunction of (B.30) then $\Psi^*(r)$ is another
solution with the same energy eigenvalue. Therefore, they are either proportional to each other or linearly independent; in the latter case they span a
two-dimensional degenerate subspace of that particular eigenvalue.

Index